Transition Metal Carbene Complexes

verlag
chemie

Transition Metal Carbene Complexes

With contributions by
Karl Heinz Dötz, Helmut Fischer,
Peter Hofmann, Fritz R. Kreissl,
Ulrich Schubert, Karin Weiss

verlag chemie

Weinheim
Deerfield Beach, Florida
Basel

CHEMISTRY

CIP-Kurztitelaufnahme der Deutschen Bibliothek

Transition metal carbene complexes : [this book is dedicated to
Prof. Dr. E. O. Fischer on the occasion of his 65th birthday / with contributions
by: Karl Heinz Dötz ... − Weinheim ; Deerfield Beach, Florida ; Basel :
Verlag Chemie, 1983.
    ISBN 3-527-26090-0 (Weinheim ...)
    ISBN 0-89573-073-1 (Deerfield Beach, Florida ...)
NE: Dötz, Karl Heinz [Mitverf.]; Fischer, Ernst O.: Festschrift

© Verlag Chemie GmbH, D-6940 Weinheim, 1983
All rights reserved (including those of translation into foreign languages). No part of
this book may be reproduced in any form − by photoprint, microfilm, or any other
means − nor transmitted or translated into a machine language without written per-
mission from the publishers.
Registered names, trademarks, etc. used in this book, even when not specifically
marked as such, are not to be considered unprotected by law.
Production Manager: Peter J. Biel
Composition: Mitterweger Werksatz GmbH, D-6831 Plankstadt
Printing: Zechnersche Buchdruckerei, D-6720 Speyer
Bookbinder: Georg Kränkl, D-6148 Heppenheim
Printed in the Federal Republic of Germany

This book is dedicated
to
Professor Dr. E. O. Fischer
on the occasion
of his 65th birthday.

Foreword

By means of this book, some of Professor E. O. Fischer's more recent co-workers take note of his 65th birthday on November 10, 1983 and express their sincere respect and heartfelt regard for their former teacher. Had this book been open to all of Professor Fischer's previous coworkers, writing on the broader topic of organometallic chemistry in general, a set of at least four volumes would have resulted. However, only one of the more recent areas of Professor Fischer's research was chosen for the subject of this book, that of transition metal carbene complexes. This is one of the several fields in which Professor Fischer was the pioneer who opened up the territory for others to follow. Such complexes had been prepared earlier by others, but their identity as carbene complexes, their significance and their potential had not been recognized. It was Professor Fischer's work which made transition metal carbene complexes a novel, important, dynamic and exciting area. After his first few papers reporting synthesis, spectroscopy, structure and chemistry, the "gold rush" was on. The subsequent investigation of the detailed chemistry of transition metal carbene complexes led to the discovery of the transition metal carbyne complexes, and it was no accident that this important discovery was made in Professor Fischer's laboratories.

Most chemists would consider themselves fortunate indeed if they made only one such major discovery during the course of their career, providing a new lead which literally hundreds of chemists throughout the world will follow. However, Professor Fischer has made major discoveries throughout his 35 year career in chemistry and it was his earlier pioneering research on metal-cyclopentadienyl and metal-arene complexes which provided the basis for the Nobel Prize which he shared with Geoffrey Wilkinson in 1973. This work on "sandwich complexes" was pace-setting to an even greater extent than that of transition metal carbene and carbyne complexes and it provided the impetus for the explosive growth of transition metal organic chemistry as a new and important area, with implications of vital importance to transition metal, lanthanide and actinide chemistry, to organic reactions and syntheses, to catalysis, to structure and bonding and even to biochemistry.

This book, meant as a special birthday present for Professor Fischer, honors a man recognized as an inspiring teacher and an exciting lecturer, as a research advisor par excellence, as a dedicated, brilliant and imaginative chemist and as a warm human being.

VIII

Together with the authors of this book, his many friends and admirers throughout the world honor Professor Fischer on this festive occasion and wish him many more happy, healthy and active years, with, perhaps, the opportunity to open up still another exciting area of chemistry.

Dietmar Seyferth
Massachusetts Institute of Technology

Authors' Preface

This book is dedicated to Professor E. O. Fischer on the occasion of his 65th birthday. The authors present a collection of "state of the art" articles on various aspects of carbene complex chemistry, which was founded by E. O. Fischer about twenty years ago. The articles are aimed to show principles. A comprehensive coverage of all papers dealing with carbene complexes was not intended.

The book is limited to complexes, which contain one or more terminally bonded CR_2 ligands, CR_2 being a neutral two-electron donor according to formal electron counting rules. Vinylidene, allenylidene and related complexes have been omitted, despite their similarity to carbene complexes. The terms "carbene" and "alkenylidene" occur as synonyms throughout this book.

The following abbreviations are used:

Me	methyl	Cp	η^5-cyclopentadienyl
Et	ethyl	cp*	η^5-C_5Me_5
Pr^n	n-propyl	MeCp	η^5-C_5H_4Me
Pr^i	iso-propyl	c-C_6H_{11}	cyclohexyl
Bu^n	n-butyl	TTP	5,10,15,20-tetraphenylporphinato
Bu^i	iso-butyl	THF	tetrahydrofurane
Bu^t	tert. butyl	DCCD	dicyclohexylcarbodiimide
Np	neopentyl ($-CH_2CMe_3$)	py	pyridine
Ph	phenyl	o	ortho
Bz	benzyl	m	meta
Tol	tolyl	p	para
Mes	mesityl	mer	meridional
Tos	tosyl	fac	facial

Contents

List of Contributors

Priv.-Doz. Dr. Karl Heinz Dötz
Anorganisch-chemisches Institut der Technischen Universität
Lichtenbergstrasse 4
D-8046 Garching
Federal Republic of Germany

Dr. Helmut Fischer
Anorganisch-chemisches Institut der Technischen Universität
Lichtenbergstrasse 4
D-8046 Garching
Federal Republic of Germany

Prof. Dr. Peter Hofmann
Anorganisch-chemisches Institut der Technischen Universtität
Lichtenbergstrasse 4
D-8046 Garching
Federal Republic of Germany

Priv.-Doz. Dr. Fritz Roland Kreissl
Anorganisch-chemisches Institut der Technischen Universität
Lichtenbergstrasse 4
D-8046 Garching
Federal Republic of Germany

Prof. Dr. Ulrich Schubert
Institut für Anorganische Chemie der Universität
Am Hubland
D-8700 Würzburg
Federal Republic of Germanny

Dr. Karin Weiss
Laboratorium für Anorganische Chemie der Universität
Universitätsstrasse 30
D-8580 Bayreuth
Federal Republic of Germany

The Synthesis
of Carbene Complexes

By Helmut Fischer

1 Introduction

Since the first planned synthesis and characterisation of a stable transition metal carbene complex by Fischer and Maasböl in 1964 (1), a whole series of new preparative routes have been developed. Several hundred carbene complexes have been isolated, characterised and studied. A number of reviews have appeared covering the area as a whole (2–7) or in part (8–16). This chapter will concentrate especially on (a) synthetic routes having a wide range of applicability and (b) preparative routes which have been elaborated only in recent years. Specifically only complexes of the type $L_nM = C(R^1)R^2$ are discussed. Although closely related to these complexes, vinylidene and allenylidene complexes are not considered because of limitations of space.

For the preparation of carbene (alkylidene) complexes, several strategies are possible (Scheme 1). New carbene complexes can be obtained from non-carbene complex precursors as well as by modification of pre-existing carbene complexes. All the routes summarised in Scheme 1 have been exploited, although with different degrees of success.

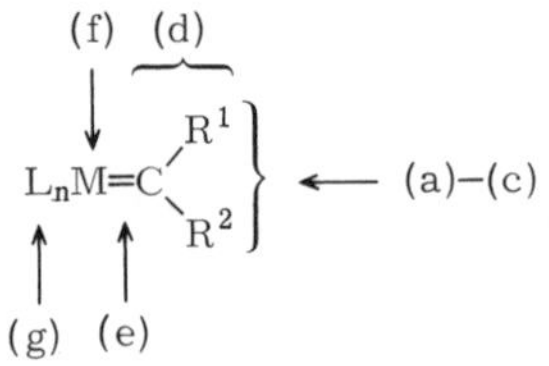

$$(f)\ (d)$$

$$L_nM{=}C\!\!\begin{array}{c} {}^{R^1} \\ {}_{R^2} \end{array}\!\!\Bigg\} \longleftarrow (a){-}(c)$$

$$(g)\ (e)$$

Scheme 1. Possible approaches for the synthesis of transition metal carbene complexes.

From non-carbene complex precursors:
(a) Transformation of a non-carbene ligand into a carbene ligand. (The carbene carbon atom is already attached to the metal in the complex precursor.)
(b) Addition of a carbene ligand precursor to a metal complex with concomitant transformation of the latter into a carbene ligand. (The carbene carbon atom is added to the metal in the course of the reaction.)

From pre-existing carbene complexes:
(c) Transfer of a carbene ligand from one metal centre to another metal.
(d) Modification of the carbene ligand.
(e) Insertion of an unsaturated organic molecule into the metal-carbene bond.

(f) Change of the oxidation state of the central metal.
(g) Modification of the metal-ligand framework.

2 Syntheses from Non-carbene Metal Complex Precursors by Transformation of a Non-carbene Ligand into a Carbene Ligand

2.1 Preparation from Metal Carbonyls

The most useful and general route for the direct synthesis of carbene complexes from non-carbene complex precursors involves addition of a molecule X − Y across the O = C bond in a metal carbonyl. This procedure may be carried out in two separate steps or in one single step. In either case, one of the two non-metal substituents on the carbene carbon (R^1 or R^2) in the resulting carbene complex will be an OR group.

The versatile two-step procedure was introduced by Fischer and Maasböl for the preparation of the first stable transition metal carbene complex recognized as such (1,17). It involves nucleophilic attack of the carbanion R^- in LiR (R = Me, Ph, etc.) on the carbon atom of a coordinated carbon monoxide molecule in a carbonyl metal complex. The product formed is a lithium acylmetallate (I), which in solution very likely forms an ion pair. This complex can be converted into the corresponding tetraalkylammonium salt and isolated in this more stable form. Acidification of I gives the labile hydroxycarbene complex II, which − at room temperature in solution − decomposes rapidly [forming aldehydes by a 1,2-hydrogen shift (17 − 19)], although it can be isolated at low temperatures (20). Alkylation of II (usually prepared only *in situ*) with diazomethane finally yields the thermally stable methoxycarbene complex III (Scheme 2).

By substituting [PhN$_2$]BF$_4$ for CH$_2$N$_2$ and Cr for W, the corresponding phenoxycarbene complex can be obtained in low yield (21). Since first being reported in 1964, this synthetic route has been modified several times and extended to a great number of different metal complexes, nucleophiles and alkylation agents. Better yields of III (> 80 %) than via the reaction sequence I→II→III are obtained by direct alkylation of I with trimethyloxonium tetrafluoroborate, [Me$_3$O]BF$_4$, (22) or MeOSO$_2$F (23). Furthermore, I and other related acylmetallates also react, for instance, with [Et$_3$O]BF$_4$ (24); Bu$_2^n$BCl (25); acyl halides (26 − 28); Me$_3$SiCl (29,30); Me$_3$SiOSO$_2$CF$_3$ (31); (Me$_3$Si)$_3$-SiBr (32); or Cp$_2$TiCl$_2$ (33,34) to give ethoxy-; di-n-butylboroxy-; acyloxy-; trimethylsiloxy-; tris(trimethylsilyl)siloxy-; and titanoxycarbene complexes.

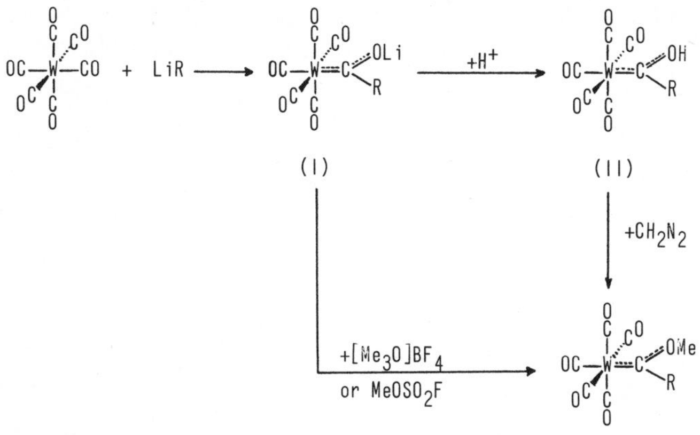

Scheme 2.   (III)

In the initial step (nucleophilic attack at a carbonyl carbon atom) a large number of nucleophiles may be used, *e.g.* the anions in alkyl- (17,35 – 42); cyclopropyl- (43,44); (+)-menthyl- (45,46); aryl- (17,47 – 53); alkenyl- (54 – 58); alkinyl- (28,59); furyl- (54,56,60); thienyl- (54,60); pyrrolyl- (54,60); ferrocenyl- (49,61 – 63); ruthenocenyl- (64); cymantrenyl- (63,65); benchrotrenyl- (66); triorganylsilyl- (67 – 73); dialkylamido- (52,74 – 78); and diphenylmethylenamidolithium (79) as well as dithianyllithium derivatives (80,81), potassium ethoxide (82), Grignard reagents (83,84) and trimethylphosphinemethylene (31,85). Among the binary or substituted metal carbonyls employed are $M(CO)_6$ [M = Cr, Mo, W] (17); $M_2(CO)_{10}$ [M = Mn, Tc, Re] (23,73,86 – 89); $Fe(CO)_5$ (52); $Ni(CO)_4$ (76); $(NO)Co(CO)_3$ (76); $(NO)_2Fe(CO)_2$ (71,76); $LM(CO)_5$ [M = Cr, Mo, W; L = PR_3 (83,90,91), $AsPh_3$, $SbPh_3$ (90), $CNBu^t$ (43), $C[N(Me)CH_2]_2$ (92 – 94); M = Mn, Re; L = $GePh_3$ (95,96), Cl, Br, I (97,98)]; $Ph_3XM(CO)_4$ [M = Fe; X = P (99); M = Co; X = Ge (39), Sn (100)]; $(\eta^6\text{-}C_6H_{6-n}Me_n)Cr(CO)_3$ [n = O, 1, 2, 3] (101); $(\eta^5\text{-}C_5H_4R)M(CO)_3$ [M = Mn; R = H (17,46,63), Me (78,85,102); M = Re; R = H (41)]; $Cp(Ph_3X)M(CO)_3$ [M = Mo, W; X = Ge, Sn (40)]; $Cp(NO)M(CO)_2$ [M = Cr, Mo, W (103)]; $Cp(GePh_3)Fe(CO)_2$ (40); $[(\eta^6\text{-}C_6Me_6)Mn(CO)_3]^+$ (104); $Os_3[1,2\text{-}\mu\text{-H}; 1,2\text{-}\mu\text{-O} = C(Me)](CO)_{10}$ (105); and $(PPh_3)_2Cl(N_2)Re(CO)_2$ (106).

From the experimental results the following generalisations can be made:

(a) With lithium carbanions as nucleophiles the yields of carbene complexes are essentially higher than with amido- or triorganylsilyllithium.

(b) When lithium reagents are employed, the reaction rates are very high. In most cases the addition as well as the alkylation reaction are almost instantaneous. Grignard compounds react much more slowly than organolithium reagents and produce lower yields.

(c) In complexes containing more than one carbonyl ligand, nucleophilic attack of the R^- in LiR always occurs at the CO group with the highest positive charge residing on the carbonyl carbon atom. In octahedral complexes of the type $(CO)_5ML$ [L being a ligand with a higher σ-donor/π-acceptor ratio than CO, *e.g.* PR_3, $AsPh_3$, CNR, $(CO)_5Mn$], this generally occurs at a *cis*-CO ligand. In complexes of the type $(CO)_4ML$ [L occupying an axial position of the trigonal bipyramid] it occurs at the *trans*-CO group. Therefore, the kinetically controlled reaction products obtained after alkylation are the *cis*-$L(CO)_4M[C(OR)R']$ (90,107) and *trans*-$L(CO)_3M[C(OR)R']$ complex (99) respectively. However, upon heating in solution ($>30\,°C$) *cis*-$(YR_3)(CO)_4$-$M[C(OR)R']$ [Y = P, As, Sb; M = Cr, W] isomerises to give a solution containing an equilibrium mixture of both *cis* and *trans* isomers (108,109) whose *cis/trans* equilibrium ratio varies depending on the steric requirements and electronic properties of the carbene and the YR_3 ligand, on the central metal and on the solvent used (110).

(d) The reaction rate of nucleophilic attack is governed by electronic factors and by the steric requirements of L, *e.g.* $[(PhO)_3P]W(CO)_5$ reacts with LiMe approximately three times faster than $[(c\text{-}C_6H_{11})_3P]W(CO)_5$ but 227 times slower than $W(CO)_6$ (111).

(e) The stability of the carbene complexes increases in the order [L = carbene ligand]: $LMo(CO)_5 < LCr(CO)_5 < LW(CO)_5$; $LRe_2(CO)_9 < LTc_2(CO)_9 < LMn_2(CO)_9$; and $LNi(CO)_3 < LFe(CO)_4 < LCr(CO)_5$.

In Fischer-type carbene complexes such as III (R = alkyl or aryl) the carbene carbon atom is strongly electrophilic. From MO calculations it follows that the lowest unoccupied molecular orbital (LUMO) is energetically isolated and spatially localised predominantly on the carbene carbon atom (112,113). This is in agreement with the results of ESR investigations on the complex radical anions $(CO)_5M[C(OMe)Ph]^{-\cdot}$ [M = Cr, Mo, W] (114). Accordingly, when attacking III, LiR will not add to the carbon atom of one of the carbonyl ligands but rather to the carbene ligand to give, for instance, pentacarbonyl[diphenyl(methoxy)methyl]tungstate(1-) [R = Ph] (60,115–117) and a carbene anion [R = Me] (118) (Scheme 3).

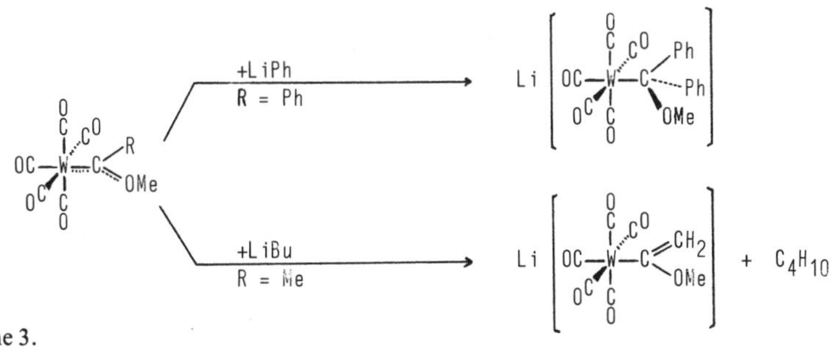

Scheme 3.

Thus, biscarbene complexes cannot usually be prepared by subsequent carbanion/carbonium ion addition to monocarbene complexes of type III. Both types of anions can, however, be used as precursors for the synthesis of new monocarbene complexes. In $(CO)_5ML$ compounds [M = Cr, Mo, W] containing the cyclic diaminocarbene ligand L = $=C[N(Me)CH_2]_2$, the electrophilicity of the carbene carbon atom is strongly reduced (due to the effective electron transfer from two amino substituents to the carbene carbon atom). Hence, the preparation of biscarbene complexes is possible by addition of LiMe and alkylation with $MeOSO_2F$ [yield: $40-60 \%$] (92–94). Currently, only relatively few other examples of the synthesis of biscarbene complexes from metal carbonyls are known:

(a) The reaction of $M(CO)_6$ [M = Cr, W] with $LiPMe_2/[Et_3O]BF_4$ gives *cis*-$(CO)_4M[C(OEt)PMe_2]_2$ [1.5 % yield] (119).

(b) The addition of *o*-dilithiobenzene to two *cis*-CO ligands in VIb hexacarbonyl complexes and subsequent alkylation of the resulting adduct with $[Et_3O]BF_4$ affords *cis*-1,4-chelated biscarbene complexes [$29-35 \%$] (120,121). When substituting 1,2-dilithio-1,2-diphenylethane for *o*-$Li_2C_6H_4$, a mixture [in the ratio $\approx 10:1$] of the 1,4-chelated mononuclear and binuclear biscarbene complexes (in which the carbene ligand bridges two $(CO)_5M$ fragments) is obtained (121).

(c) The reaction of $(CO)_5ReX$ [X = Cl, Br, I] with two equivalents of LiMe in THF at $-78\,°C$ yields the dianions *fac*-$(CO)_3XRe[C(=O)Me]_2^{2-}$, which upon protonation give the bis-hydroxycarbene complexes *fac*-$(CO)_3X$-$Re[C(OH)Me]_2$ (98).

The one-step transformation of a coordinated CO ligand into a carbene ligand can be performed employing metal amides (122) (Scheme 4).

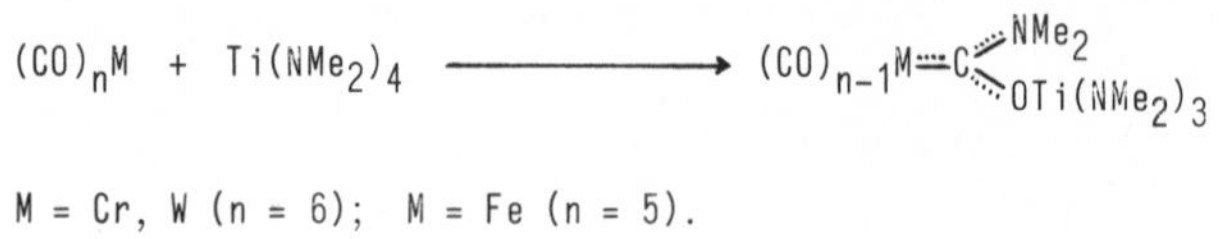

$$(CO)_nM \ + \ Ti(NMe_2)_4 \quad\longrightarrow\quad (CO)_{n-1}M\!=\!\!C\!\!\begin{array}{c} {}^{\text{NMe}_2} \\[-2pt] {}_{\text{OTi}(NMe_2)_3} \end{array}$$

M = Cr, W (n = 6); M = Fe (n = 5).

Scheme 4.

Other metal amides such as [Al(NMe₂)₃]₂ (123,124) or dimethylaminotin compounds (125,126) as well as Cp₂ZrH₂ (127) or silyl-substituted ylides [Me₃P = C(H)SiMe₃] (128) were likewise shown to add across the O = C bond of metal carbonyls giving respectively the amino-, secondary zirconoxy- and siloxycarbene complexes.

Complexes with cyclic amino(oxo)carbene or dioxocarbene ligands are obtained from the reaction of metal carbonyls − mainly cationic ones such as [CpFe(CO)₃]⁺, [Cp(NO)Mn(CO)₂]⁺ or [Cp(NO)LMn(CO)]⁺ (L = P(OPh)₃, CNC₆H₁₁) − with H₃NC₂H₄Br⁺/base or HOC₂H₄Br/NaH (129) or from [Cl(PPh₃)₂(CNMe)Os(CO)₂]⁺ and HOC₂H₄Cl/NEt₃ (130) (Scheme 5).

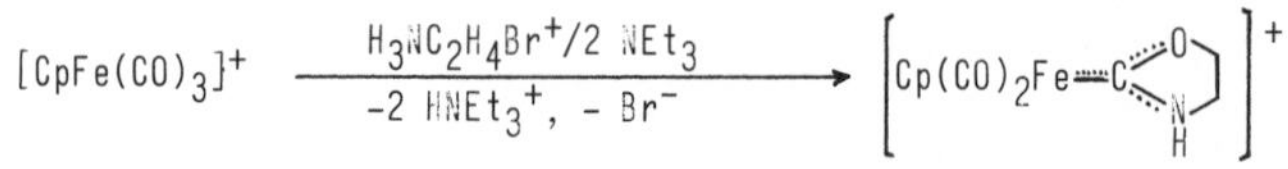

$$[CpFe(CO)_3]^+ \quad \xrightarrow[\text{-2 HNEt}_3{}^+,\ -\text{Br}^-]{H_3NC_2H_4Br^+/2\ NEt_3} \quad \left[Cp(CO)_2Fe\!=\!\!C\!\!\begin{array}{c}{}^{\text{O}}\\[-2pt]{}_{\text{N}}\\[-2pt]{}_{\text{H}}\end{array}\right]^+$$

Scheme 5.

When [Cp(NO)Mn(CO)₂]⁺ reacts with HOC₂H₄Br/NaH, a mixture of the mono- and the biscarbene complex results (129).

2.2 Preparation from Isonitrile Complexes

As long ago 1915, a "resonance stabilised" carbene complex, which was very likely the first one to be synthesized, was prepared by Tschugajeff and Skanawy-Grigorjewa from tetrakis(methylisonitrile)platinum and hydrazine (131,132). However, the resulting product − which can be protonated with HBF₄ (133 − 135) − was assigned the wrong structure. Only in 1970, *i.e.* well after Fischer and Maasböl's first successfully planned synthesis and characterisation of a stable tungsten carbene complex, were this Pt and other related Pt and Pd compounds recognised as cyclic carbene complexes (134,135) (Scheme 6).

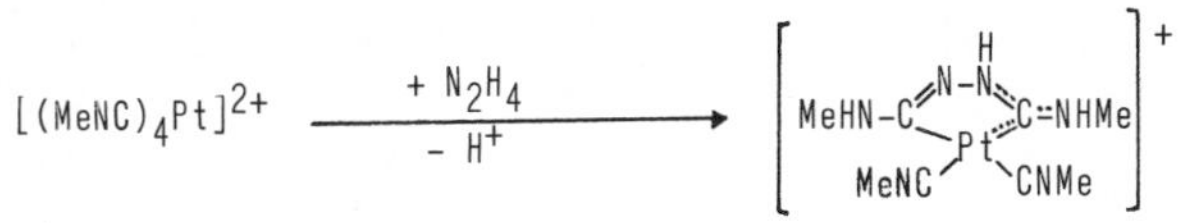

$$[(MeNC)_4Pt]^{2+} \xrightarrow[- H^+]{+ N_2H_4} \left[\begin{array}{c} H \\ MeHN-C \overset{N-N}{\underset{Pt}{\diagdown}} C=NHMe \\ MeNC \diagup \diagdown CNMe \end{array} \right]^+$$

Scheme 6.

The correct structure of Tschugajeff's salt was finally established by X-ray crystallographic analysis (136). Hydroxylamine reacts with $[(MeNC)_4Pt]^{2+}$ in a similar way (135). The one-step addition of a nucleophile across the $C \equiv N$ bond of a coordinated isonitrile can be performed with a number of isonitrile complexes (Scheme 7).

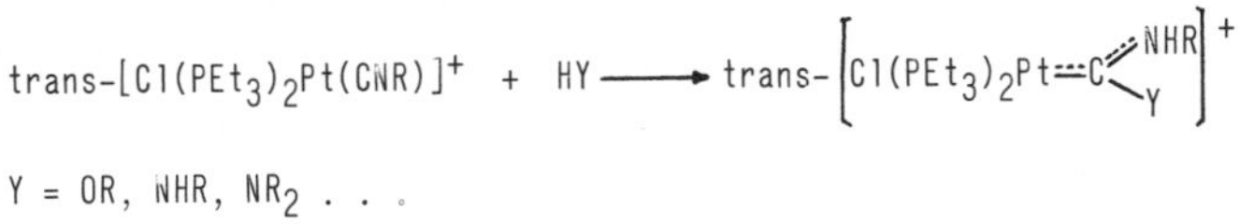

$$trans-[Cl(PEt_3)_2Pt(CNR)]^+ + HY \longrightarrow trans-\left[Cl(PEt_3)_2Pt = C \overset{NHR}{\underset{Y}{\diagdown}} \right]^+$$

$$Y = OR, NHR, NR_2 \ldots$$

Scheme 7.

These may be of Ni(II) (137,138), Pd(II) (137,139,140 − 145), Pt(II) (139,143,145 − 152), Fe(II) (153 − 162), Ru(II) (156,157), Os(II) (163), Au(I) (164 − 166), Au(III) (145), Rh(I) (167,168) or Rh(III) (158). Other nucleophiles, such as alcohols, primary or secondary amines and thiols can also be used.

It is characteristic of the resulting carbene complexes that:

(a) both of the non-metal substituents at the carbene carbon atom are bonded via a heteroatom ($R^1, R^2 = OR, NR_2, SR$) and

(b) one of the two substituents is always bonded via nitrogen.

The reactions of cis-$[Cl_2(PR_3)Pt(CNR)]$ with ethanol (150), of $[X_2Pd(CNR)_2]$ [X = Cl, Br, I] with different amines or alcohols (142) and of cis-$[X_2(L)Pd(CNR)]$ [X = Cl, Br; L = PPh_3, $AsPh_3$] with amines (169) were studied with particular care. It was found that electron-withdrawing groups on the nitrogen atom of the isocyanide ligand increase the reaction rate whereas on the nitrogen atom of the amine (or the oxygen atom of the alcohol) they decrease the rate. The ligands at the central metal have only a very small influence on the reaction rate.

These results are in agreement with a rate-determining nucleophilic attack of the amine (or alcohol) at the isocyanide carbon atom. However, when hexacoordinated Fe(II), Ru(II) and Os(II) complexes are employed steric factors become important. In the reaction of $[Cp(CO)Fe(CNMe)_2]^+$ with H_2NR the reaction rate drops significantly in the series R = Me, Et, Pr^i, Bu^t (160). When there is a choice between carbonyl and isonitrile ligands, nucleophilic attack of the amine usually occurs at the isonitrile group. From the reaction of H_2NMe with $[Cp(CO)(L)Fe(CNMe)]^+$, L = CN^-, CNR or PR_3, the corresponding diaminocarbene complexes are obtained. With L = CO, however, nucleophilic attack occurs at a carbonyl ligand to yield a carbamoyl complex (160,161). Compared to aryl- and alkyl(alkoxy)carbene complexes, the electrophilic character of the carbene carbon in carbene complexes derived from isonitrile compounds is strongly diminished owing to the electron donation from two heteroatoms. An additional nucleophile therefore does not attack at the carbene carbon atom but rather at a second isonitrile ligand. Thus, the synthesis of bis-, tris- and even tetrakiscarbene complexes becomes feasible. Back in 1967 the first non-chelated biscarbene complex was prepared from mercury acetate, methylisonitrile and secondary amines (170), presumably via a mercury-isonitrile intermediate. The reaction of $[(MeNC)_6M]^{2+}$ [M = Fe, Ru, Os] with amines gives mono-, chelated bis-, non-chelated bis- and meridional triscarbene complexes depending on the reaction conditions, the metal and the amine used (156,157,163). On warming solutions of IV, rearrangement takes place, probably via a chelated biscarbene intermediate (156) (Scheme 8).

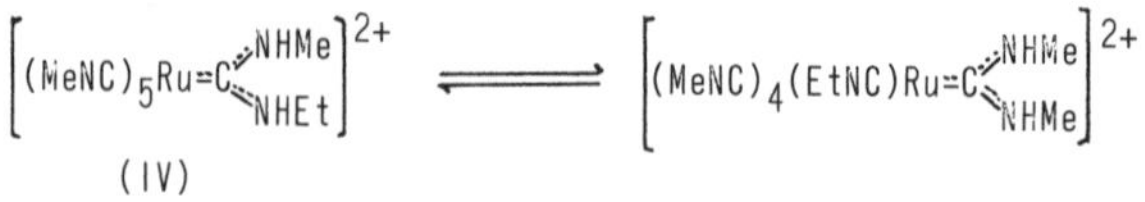

$$\left[(MeNC)_5Ru{=}C{\underset{NHEt}{\overset{NHMe}{<}}}\right]^{2+} \rightleftharpoons \left[(MeNC)_4(EtNC)Ru{=}C{\underset{NHMe}{\overset{NHMe}{<}}}\right]^{2+}$$

$$(IV)$$

Scheme 8.

Remarkably stable tetracarbene complexes were finally obtained by reacting $[(MeNC)_4M]^{2+}$ [M = Pd, Pt] with methylamine (139) (Scheme 9).

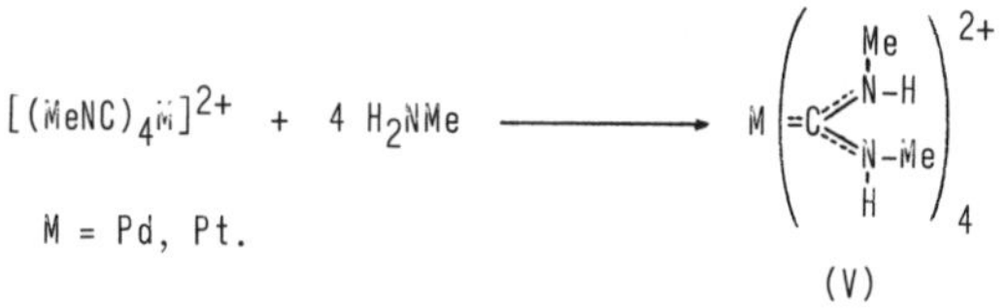

$$[(MeNC)_4M]^{2+} + 4\ H_2NMe \longrightarrow M\left(=C{\underset{\underset{H}{N-Me}}{\overset{\overset{Me}{N-H}}{<}}}\right)_4^{2+}$$

$$M = Pd, Pt.$$

$$(V)$$

Scheme 9.

X-ray crystallographic analysis of the platinum compound (V) (171) confirmed that the nitrogen substituents of the carbene ligand are in the *amphi*-configuration as shown for V [the preferred arrangement in all bis(amino)-

carbene complexes] and that the carbene planes form angles between 77° and 82° with the $Pt-(C_{Carb})_4$ plane. Addition of H_2NMe to *cis*-$[(CO)_4Mn(CNMe)_2]^+$ has been found to give a chelated biscarbene complex (172). Cyclic carbene compounds can be obtained using isonitriles with a β- or γ-hydroxy group in the reaction with Pd^{2+} or Pt^{2+} (145). Recently, even biscarbene complexes of molybdenum(0) formed by reaction of $[(NO)Mo-(CNR)_5]I$ with H_2NR' were prepared (173). A two-step route ($NaOR/HOR$ and then $HPF_6 \cdot Et_2O$) has been employed for the synthesis of two amino (alkoxy)carbene complexes of chromium from $[Cp(NO)_2Cr(CNMe)]^+$ (174).

2.3 Preparation from Thiocarbonyl Complexes

Compared to the great number of carbene complex syntheses based on metal carbonyls, only a few successful preparations of carbene complexes using thiocarbonyl complexes as precursors have been reported. Reaction of $(CO)_4Fe(CS)$ with $C(NMe_2)_4$ and subsequent alkylation with $MeOSO_2F$ in CH_2Cl_2 yields $(CO)_4Fe[C(SMe)NMe_2]$ (175). Series of dithiocarbene complexes were prepared from $(CO)_5W(CS)$ (176,177) (Scheme 10).

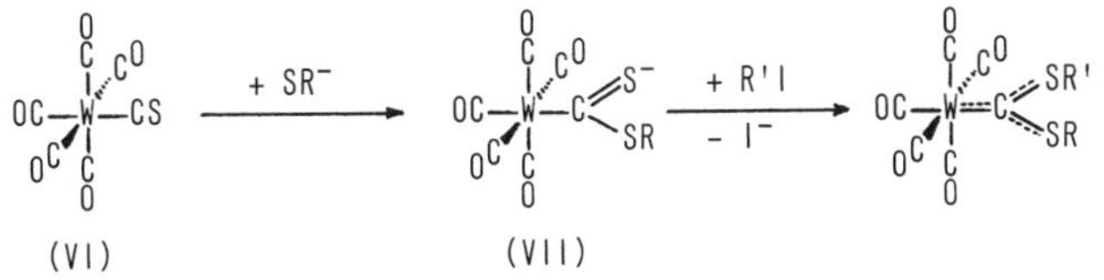

$$R = Me, Et, Pr^i, Bu^n, Bu^t; \quad R' = Me, Et, Pr^i, Bu^n.$$

Scheme 10.

Acidification of VII results in the reformation of VI. When employing other anions, such as $[PhS]^-$, $[PhSe]^-$ or $[MeO]^-$ instead of $[S\text{-alkyl}]^-$, no nucleophilic attack at the thiocarbonyl carbon atom was observed, whereas the reaction of VI with $[S-(CH_2)_n-SH]^-$ (n = 2, 3, 4) followed by alkylation with MeI gave cyclic dithiocarbene complexes (177). By substituting $[Cp(CO)_2Fe-(THF)]BF_4$ for R'I in the reaction with VII (R = Me) a bimetallic carbene complex can be prepared (177). *Cis*-phosphite-substituted $(CO)_4LW(CS)$ also reacts with SMe^-/MeI to give the corresponding *cis*-substituted dithiocarbene complexes. By contrast, no reaction was observed when using the *trans* isomers (177).

The reaction of $Cp(SnPh_3)(CO)Fe(CS)$ with $H_2NC_2H_4NH_2$ to give H_2S and the cyclic diaminocarbene complex $Cp(SnPh_3)(CO)Fe[C(NHCH_2)_2]$ can be envisioned as proceeding via nucleophilic attack of one amino nitrogen at the carbon atom of a thiocarbonyl ligand, followed by intramolecular cyclisation with extrusion of H_2S (178).

2.4 Preparation from Vinylidene Complexes

The addition of a nucleophile $H - Y$ across the $C = C$ double bond of a vinylidene complex was also established. $[Cp(CO)(PPh_3)Fe = C = CH_2]BF_4$ reacts with alcohols, thiols, secondary amines or hydrochloric acid to give the corresponding methylcarbene complexes (179). In THF solution VIII dimerises to form a 3:1 *syn/anti* ratio of IX. Upon substituting tricyclohexylphosphine for PPh_3 in VIII, the formation of only the *anti*-isomer analogue of IX could be detected (179) (Scheme 11).

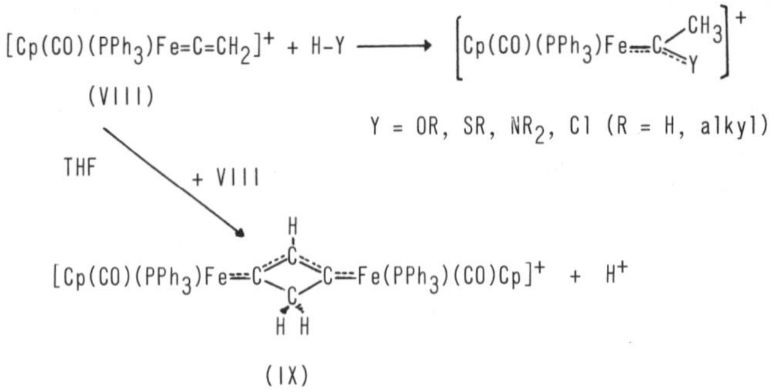

$$[Cp(CO)(PPh_3)Fe=C=CH_2]^+ + H-Y \longrightarrow \left[Cp(CO)(PPh_3)Fe\!\!=\!\!C\!\!\begin{smallmatrix} CH_3 \\ Y \end{smallmatrix} \right]^+$$

(VIII)

$$Y = OR, SR, NR_2, Cl \ (R = H, alkyl)$$

THF $\searrow$ + VIII

$$[Cp(CO)(PPh_3)Fe\!\!=\!\!C\!\!\begin{smallmatrix} H \\ | \\ C \\ / \ \backslash \\ H \ H \end{smallmatrix}\!\!C\!\!=\!\!Fe(PPh_3)(CO)Cp]^+ \ + \ H^+$$

(IX)

Scheme 11.

For the reaction of $[(CO)L_2(Cl)(Me)IrY]PF_6$ [L = PMe_2Ph, $PMePh_2$; Y = Me_2CO, MeOH] with terminal alkynes in methanol, vinylidene intermediates were also proposed. The replacement of the coordinated solvent by the alkyne is followed by rearrangement to a vinylidene complex which subsequently adds methanol to give the carbene complex (180). A similar route has been employed to synthesize alkoxy(alkyl)carbene complexes starting from halogeno-bridged binuclear platinum compounds (181) or from $[Cp(CO)_2-Fe(H_2C = CMe_2)]^+$ (182) and $HC \equiv CR/R'OH$.

The conversion of acetylide ligands into carbene ligands has also been formulated to proceed via vinylidene complex intermediates (see 2.9).

2.5 Preparation from Acyl Complexes

The second step in the classical Fischer two-step synthesis of carbene complexes involves addition of an electrophile to the acyl oxygen atom of the acyl-metallate. The same type of reaction can be carried out with neutral acyl complexes prepared by other means *e. g.* by reaction of carbonylmetal anions with acyl halides or by migratory insertion of CO into metal-alkyl bonds. These acyl complexes cannot in all cases be isolated but merely function as reaction intermediates.

Some of the earliest examples of carbene complex syntheses via alkylation or protonation of the acyl oxygen include the alkylation of diethylcarbamoyl-mercury with $[R_3O]BF_4$ (R = Me or Et) (183) and the reversible protonation (by HCl, HBr or HBF$_4$) and alkylation (by $[R_3O]BF_4$) of Cp(CO)(L)M-[C(=O)Me] (M = Fe, Ru; L = CO, PPh$_3$, P(C$_6$H$_{11}$)$_3$) to give respectively hydroxy- (184) and alkoxy(methyl)carbene complexes (185) (Scheme 12).

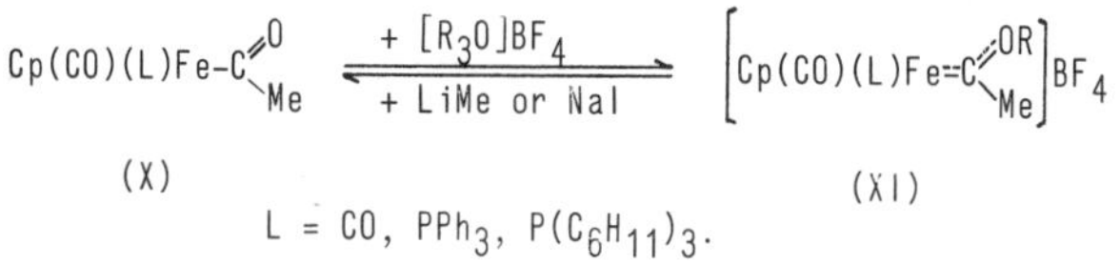

$$Cp(CO)(L)Fe-C{\nwarrow}^{O}_{Me} \underset{+\ LiMe\ or\ NaI}{\overset{+\ [R_3O]BF_4}{\rightleftharpoons}} \left[Cp(CO)(L)Fe=C{\nwarrow}^{OR}_{Me}\right]BF_4$$

(X)   (XI)

$$L = CO,\ PPh_3,\ P(C_6H_{11})_3.$$

Scheme 12.

Treatment of XI with LiMe (185) or NaI (186) reverses the alkylation and regenerates X. Subsequently, other protonating agents (*e. g.* [(MeO)$_2$HC]PF$_6$) (187) and alkylating reagents (MeOSO$_2$CF$_3$) (188) as well as organometallic Lewis bases (189) were employed successfully. Other iron (190 – 196) (Scheme 13), molybdenum (185,189,197), tungsten (189) or nickel compounds (198,199) were similarly used as precursors, as was a rhenium formyl complex (200).

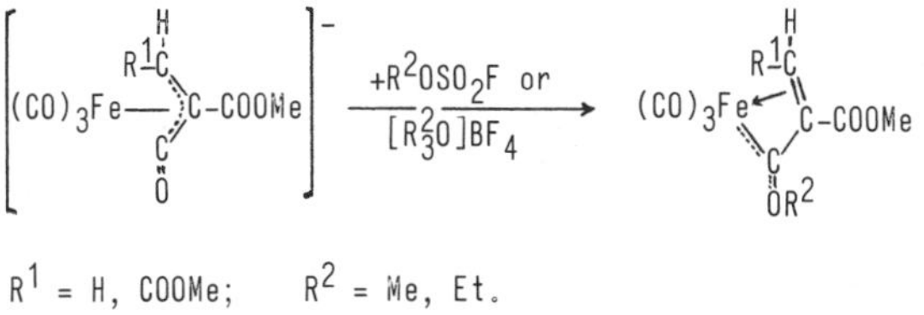

$$\left[(CO)_3Fe{-}{\overset{R^1\text{-}C{\nwarrow}^{H}}{\underset{\underset{O}{C}}{\vert}}}C\text{-}COOMe\right]^{-} \xrightarrow[{[R^2_3O]BF_4}]{+R^2OSO_2F\ or} (CO)_3Fe{\leftarrow}{\overset{R^1\text{-}C{\nwarrow}^{H}}{\underset{\underset{OR^2}{C}}{\Vert}}}C\text{-}COOMe$$

$$R^1 = H,\ COOMe;\quad R^2 = Me,\ Et.$$

Scheme 13.

In this way, non-cyclic dialkoxycarbene complexes which are difficult to prepare by nucleophilic attack at carbonyl ligands (82) become accessible in good yields (Scheme 14).

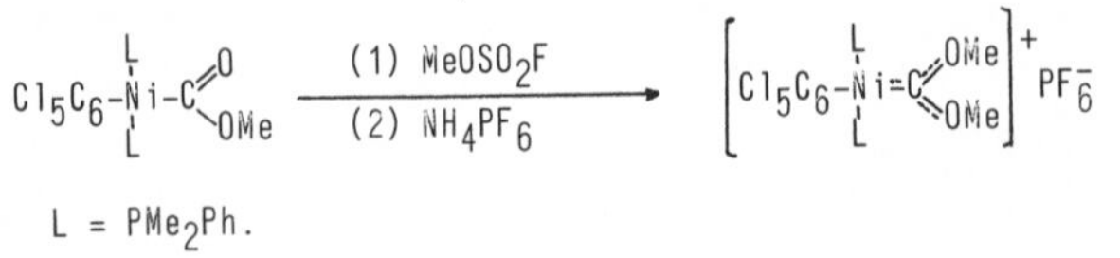

L = PMe₂Ph.

Scheme 14.

A series of *cis*-(CO)₄(L)Mn(alkoxycarbene) complexes were prepared start-ing from alkyl(pentacarbonyl)manganese via Lewis base promoted migratory insertion of CO into the Mn-C(alkyl) bond to give the corresponding acyl compounds, which subsequently can be protonated (201) or alkylated (197,202) (Scheme 15).

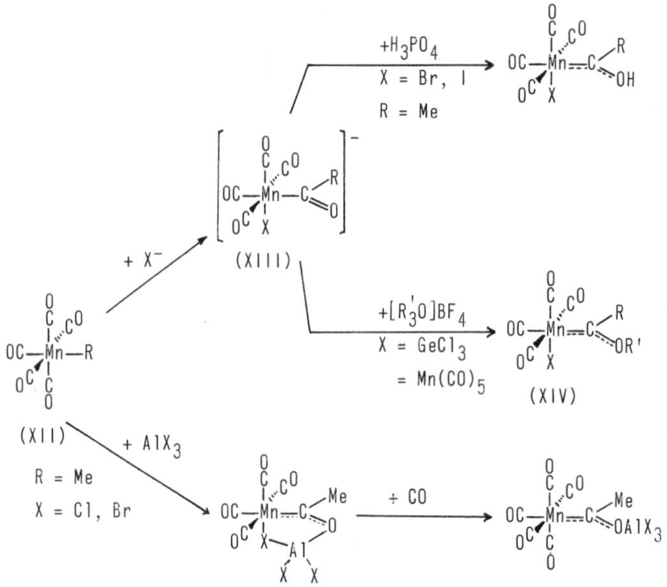

Scheme 15.

The reaction of XII with X⁻ = [(CO)₅Re]⁻, however, did not produce the corresponding manganese carbene complex XIV, but rather the rearranged rhenium compound *cis*-[(CO)₅Mn](CO)₄Re[C(OMe)Me] (89). Lewis acid induced migratory insertion in XII employing AlX₃ results in the formation of *cis*-substituted chelated halogenocarbene complexes (203 – 205); upon reaction with CO these again yield carbene(pentacarbonyl) complexes. An analogous reaction can also be carried out using aluminium oxide surfaces instead of AlX_3 (206) and $Cp(CO)_2Fe-Me$ or $Cp(CO)_3Mo-Me$ instead of XII (204).

In the case where $R = (CH_2)_nCl$ (n = 3, 4) in XII, the acyl complex (XIII) obtained after addition of $X = I$ spontaneously cyclises intramolecularly with extrusion of Cl^- to give a complex containing a cyclic (n + 2)-membered carbene ligand (207). A similar reaction was carried out in 1963 starting from $[(CO)_5Mn]^-$ and $Br(CH_2)_3Br$ (208). The resulting carbene complex, however, was not recognised as such. The correct structure was assigned only as late as 1970 (209). Besides compounds of type XII, several other transition metal alkyls – mainly molybdenum (197,210 – 212) and iron complexes (213,214) – were employed in the preparation of cyclic carbene complexes via this route (with X^- *e.g.* I^-, CN^-, PPh_3 ...). For the intramolecular cyclisation, a high nucleophilicity at the acyl oxygen atom is required (207) whereas for $R = O(CH_2)_2Cl$ in XIII $[X^- = CO]$ cyclisation has to be initiated by silver salts (215,216). For $R = (CH_2)_3Cl$ no reaction was observed, though XIII $[X^- = P(OMe)_3]$ and $Cp(CO)_2M[C(=O)R]$ (M = Fe, Ru) do cyclise for $R = (CH_2)_3Cl$ in the presence of silver salts (207). Furthermore, the size of the ring in the resulting carbene ligand is of considerable importance: cyclisation is easy when a five-membered ring is formed but difficult when a six-membered ring results (207). The preparation of an ethyl(hydroxy)carbene complex of osmium by reaction of an ethylene complex with excess HCl is also thought to proceed via a propionyl intermediate (217).

2.6 Preparation from Complexes $L_nM[C(=S)R]$

Closely related to alkylation of the acyl oxygen atom in acyl complexes is the S-alkylation of $L_nM[C(=S)R]$ compounds, M being usually Fe or Pt. Thus, *trans*-$[Cl(L)_2Pt-C(=S)Y]$ (L = PPh_3, $PMePh_2$; Y = OMe, SEt, NMe_2) can readily be converted into the corresponding carbene complexes with $MeOSO_2F$ or $[Et_3O]BF_4$ (218). Similarly, $Cp(CO)_2Fe-C(=S)Z$ (Z = OMe, OPh, SMe, SPh, SePh, $SRe(CO)_5$, $SFe(CO)_2Cp$) can be converted by reaction with $MeOSO_2F$ (219 – 221) or $MeOSO_2CF_3$ (222).

In most of these cationic compounds obtained by S-alkylation the carbene carbon atom is stabilised by two heteroatoms. One exception is a secondary carbene complex of osmium prepared by methylation of a thioformyl ligand (223) (Scheme 16).

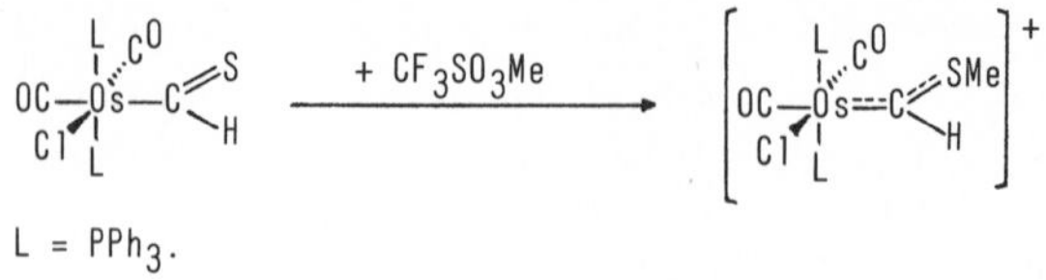

L = PPh$_3$.

Scheme 16.

The formation of $[(CO)_2I(PPh_3)Os[C(SMe)_2]]^+$ from $(CO)_2(PPh_3)_2Os(CS_2)$ and MeI (in excess) (224), of $Cp(PPh_3)(I)M[C(SMe)Me]^+$ (M = Rh, Ir) from $Cp(PPh_3)M(CS)$ and MeI (225), and of $(C_6F_5)(PPh_3)_2I_2Ir[C(SMe)Me]$ from $(C_6F_5)(PPh_3)_2Ir(CS)$ and MeI (226) were also suggested as proceeding via $[C(=S)R]$ intermediates (R = Me, SMe).

2.7 Preparation from Complexes $L_nM[C(=NR)R']$

The nitrogen atom of imidoyl ligands is also susceptible to electrophilic attack. Protonation (often reversible) as well as alkylation yields amino-carbene complexes. This has been demonstrated to occur preferentially with Ni (199,227), Pt (228,229), Fe (227,230), Ru (231,232), Ir (233) and Co compounds (227) as precursors. Closely related to this approach is protonation of the adducts formed by the reaction of carbodiimides with certain metal carbonyl anions $[(CO)_5Cr^{2-}, Cp(CO)_2Fe^-]$ (234,235) (Scheme 17).

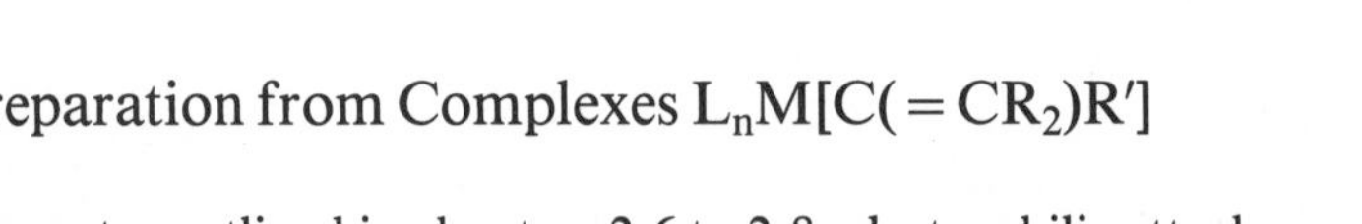

Scheme 17.

2.8 Preparation from Complexes $L_nM[C(=CR_2)R']$

In the routes outlined in chapters 2.6 to 2.8, electrophilic attack occurred at a heteroatom (O, S, N) connected via a double bond to a metal coordinated carbon atom. However, protonation at the β-carbon atom of a vinyl ligand or another ligand derived therefrom can also be used to generate carbene complexes. Thus, reaction of $Cp(CO)(L)Fe[C(=CH_2)H]$ with $HPF_6 \cdot Et_2O$ at $-80\,°C$ yields a cationic secondary carbene complex (187) (Scheme 18).

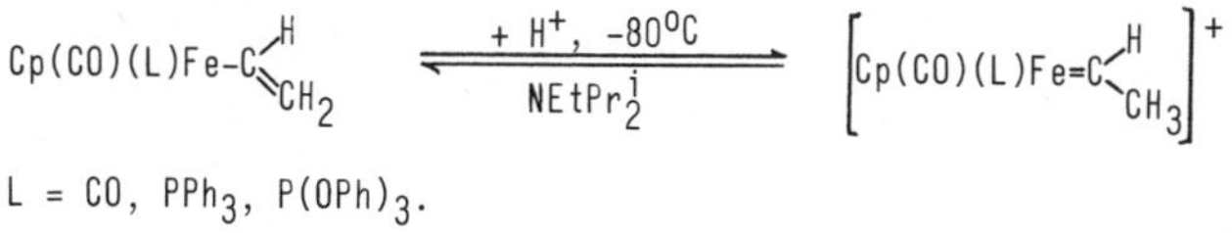

$$Cp(CO)(L)Fe-C\overset{H}{\underset{CH_2}{\Big\langle}} \quad \underset{NEtPr_2^i}{\overset{+ H^+, -80°C}{\rightleftharpoons}} \quad \left[Cp(CO)(L)Fe=C\overset{H}{\underset{CH_3}{\Big\langle}}\right]^+$$

L = CO, PPh$_3$, P(OPh)$_3$.

Scheme 18.

Amines reverse the H$^+$ addition. Cationic dimethylcarbene complexes were similarly prepared from Cp(CO)(L)Fe[C(=CH$_2$)Me] and HBF$_4$ (236) and other carbene complexes from substituted vinyl compounds of Ni(II), Pd(II) and Pt(II) using 60 % HClO$_4$ (237 – 240).

2.9 Preparation from Complexes L$_n$M(C≡CR)

Platinum(II) and nickel(II) acetylide complexes react with alcohol in the presence of acids to give alkoxycarbene complexes (241) (Scheme 19).

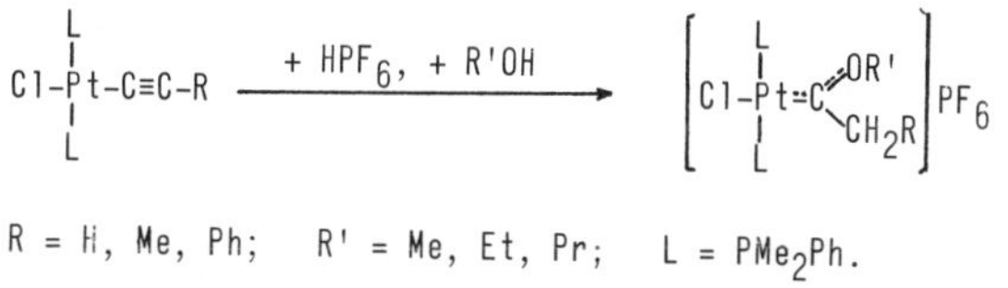

$$Cl-\overset{\overset{\displaystyle L}{|}}{\underset{\underset{\displaystyle L}{|}}{Pt}}-C≡C-R \quad \xrightarrow{+ HPF_6, + R'OH} \quad \left[Cl-\overset{\overset{\displaystyle L}{|}}{\underset{\underset{\displaystyle L}{|}}{Pt}}=C\overset{\nearrow OR'}{\underset{CH_2R}{\Big\backslash}}\right]PF_6$$

R = H, Me, Ph;   R' = Me, Et, Pr;   L = PMe$_2$Ph.

Scheme 19.

Various other Pt(II) (238, 242) and Ni(II) compounds (237,243) can be used as precursors. Related thereto is the reaction of Pt(II) halides with terminal alkynes and alcohols in the presence of silver salts (244 – 246). The use of non-coordinating anions is necessary to prevent conversion of the complex into a neutral acyl compound. Both types of reaction very likely proceed via a cationic vinylidene intermediate. Addition of the alcohol across the C=C double bond finally gives the carbene complex (246). This mechanistic scheme is confirmed by the finding that when HBF$_4$ · Me$_2$O (in excess) is added to Cp(CO)(PPh$_3$)Fe-C≡CPh (XV) in CH$_2$Cl$_2$ or toluene a cationic vinylidene complex can be detected spectroscopically which, upon addition of MeOH, gives the corresponding methoxycarbene compound with a 73 % overall yield (247) (Scheme 20).

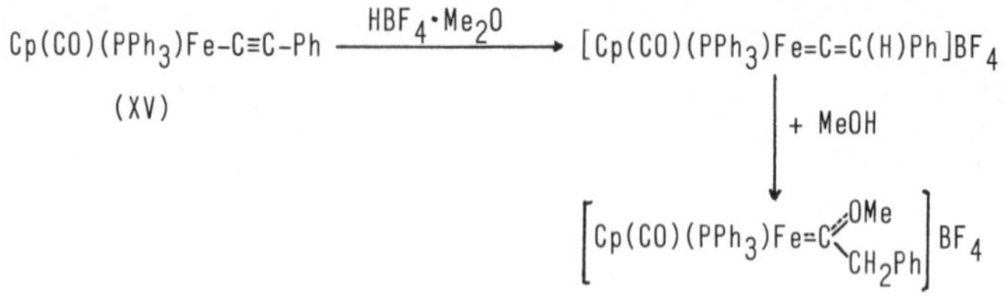

$$Cp(CO)(PPh_3)Fe-C{\equiv}C-Ph \xrightarrow{\text{HBF}_4 \cdot \text{Me}_2\text{O}} [Cp(CO)(PPh_3)Fe=C=C(H)Ph]BF_4$$

(XV)

+ MeOH ↓

$$\left[Cp(CO)(PPh_3)Fe=C\underset{CH_2Ph}{\overset{OMe}{\diagdown}} \right] BF_4$$

Scheme 20.

Reaction of XV with $HBF_4 \cdot Me_2O$ in methanol immediately gives the carbene complex with 82 % yield. The reaction of the dicarbonyl analogue of XV with $HBF_4 \cdot Me_2O$ in CH_2Cl_2 produces a dinuclear species related to IX.

2.10 Preparation from Complexes $L_nM{\equiv}CR$

The synthesis of the first carbyne complexes also opened up a new route to carbene complexes: addition of a nucleophile to the electrophilic carbyne carbon atom. Although the carbyne compounds initially must be prepared from carbene complexes, this method is especially useful since a series of carbene complexes can thereby be synthesised which are inaccessible via any other route. To date, mainly three types of carbyne compounds:

(a) $[A(CO)_2M{\equiv}CR]^+$ (M = Cr; A = η^6-$C_6H_{6-n}Me_n$ [n = 0, 1, 2, 3] and M = Mn, Re; A = η^5-C_5H_5, η^5-C_5H_4Me)

(b) $[(CO)_5Cr{\equiv}CNR_2]^+$ and

(c) *trans*-$[(CO)_5Re(CO)_4Re{\equiv}CSiPh_3]^+$ have been used.

Other cationic carbyne complexes — as well as some neutral ones — can also very likely function as precursors. This reaction approach, namely transformation of a carbyne (alkylidyne) ligand into a carbene (alkylidene) ligand, seems to promise considerable, though not yet fully explored, potential for the preparation of carbene and other complexes, as varying types of suitable carbyne complexes are being synthesised in increasing numbers.

In 1976, the first preparation of a series of carbene complexes using this approach was reported (101) (Scheme 21).

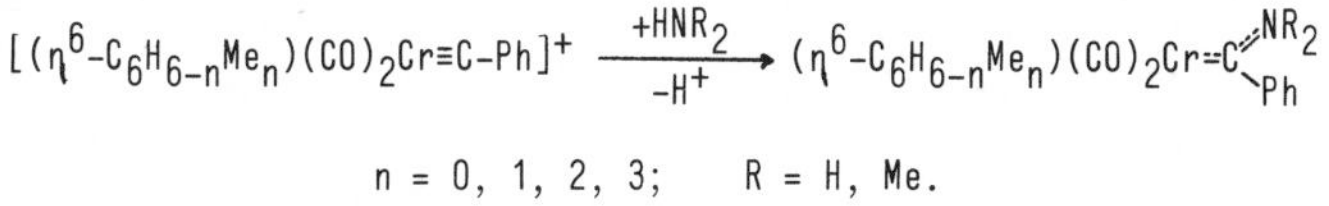

$$[(\eta^6\text{-}C_6H_{6-n}Me_n)(CO)_2Cr\equiv C\text{-}Ph]^+ \xrightarrow[\text{-}H^+]{\text{+HNR}_2} (\eta^6\text{-}C_6H_{6-n}Me_n)(CO)_2Cr\!=\!C\!\begin{smallmatrix}\nearrow NR_2\\ \searrow Ph\end{smallmatrix}$$

$$n = 0, 1, 2, 3; \qquad R = H, Me.$$

Scheme 21.

Similarly, the BF_4^-, BCl_4^- or $SbCl_6^-$ salts of $[Cp(CO)_2Mn\equiv C\text{-}Ph]^+$ (XVI) readily add anions such as CN^-, SCN^- (248), OCN^- (249), F^- (78), Me^- (250), alcoholates or phenolates (251) (Scheme 22).

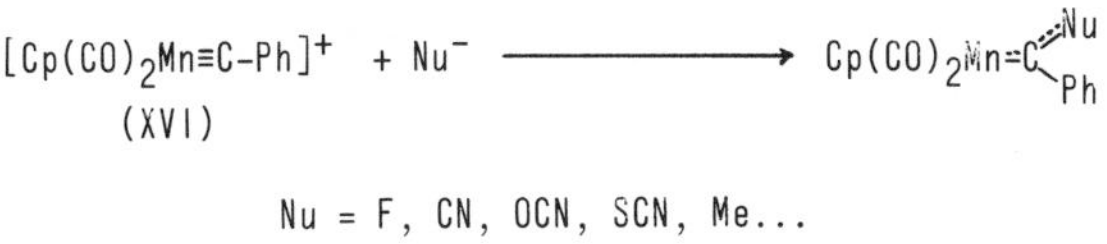

$$[Cp(CO)_2Mn\equiv C\text{-}Ph]^+ + Nu^- \longrightarrow Cp(CO)_2Mn\!=\!C\!\begin{smallmatrix}\nearrow Nu\\ \searrow Ph\end{smallmatrix}$$
$$(XVI)$$

$$Nu = F, CN, OCN, SCN, Me...$$

Scheme 22.

The reaction of XVI with $CN\text{-}Bu^t$ at low temperatures yields a keteniminyl complex which upon thermolysis at 25 °C forms $Cp(CO)_2Mn[C(CN)Ph]$ (252,253). The reaction of XVI or its p-tolyl analogue with the anions $[(CO)_5M]^-$ (M = Mn, Re) takes a different course. Instead of the carbene complexes, binuclear complexes are formed in which the Mn-M bond is bridged by the α-carbon atom of an arylketenyl group (254,255). However, when the acylmetallate $[Cp(CO)_2Mn\text{-}C(=O)Ph]^-$ is used as the nucleophile instead of $[(CO)_5M]^-$, a carbene anhydride complex is formed (256) (Scheme 23).

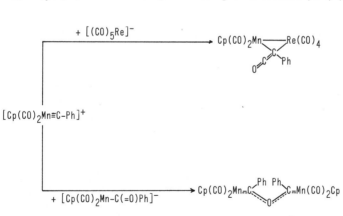

Scheme 23.

The C_5H_4Me analogue of XVI was found to add $C_5H_5^-$ (249) and $[(C_5H_4Me)(CO)_2Mn\equiv CNEt_2]^+$ was found to add H^- (from $LiAlH_4$) (257) to the carbyne carbon. Several nucleophiles react with $[Cp(CO)_2Re \equiv CPh]^+$ in a similar way (248, 256, 258). A stepwise reduction of the $Re \equiv CPh$ bond can be carried out using Et_2AlH (259) (Scheme 24).

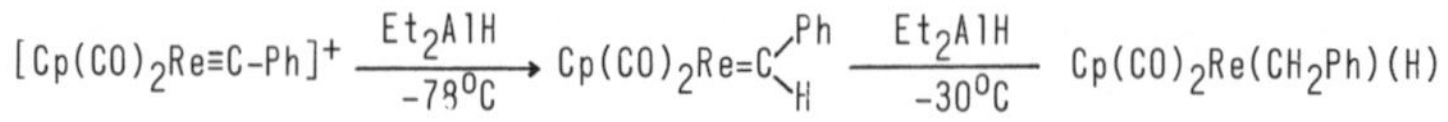

$$[Cp(CO)_2Re \equiv C-Ph]^+ \xrightarrow[-78°C]{Et_2AlH} Cp(CO)_2Re=C\overset{Ph}{\underset{H}{\diagdown}} \xrightarrow[-30°C]{Et_2AlH} Cp(CO)_2Re(CH_2Ph)(H)$$

Scheme 24.

The first dimethylcarbene complex, $Cp(CO)_2Mn[CMe_2]$, was also prepared by nucleophilic addition of Me^- to the carbyne carbon atom (250). When $[Cp(CO)_2Re \equiv C-SiPh_3]^+$ is reacted with EtOH or $HNMe_2$, a mixture of two different carbene complexes is obtained (260) (Scheme 25).

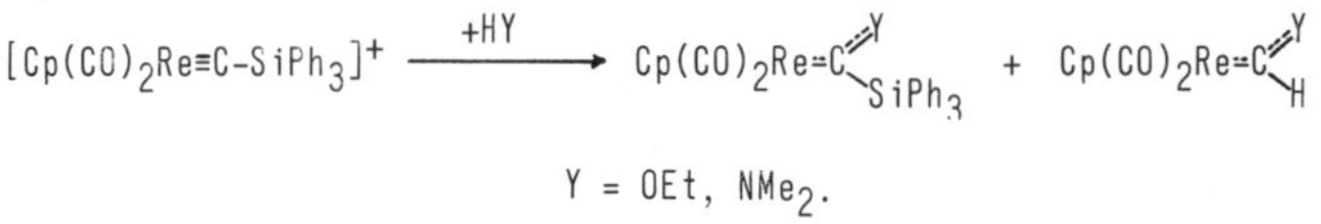

$$[Cp(CO)_2Re \equiv C-SiPh_3]^+ \xrightarrow{+HY} Cp(CO)_2Re=C\overset{Y}{\underset{SiPh_3}{\diagdown}} + Cp(CO)_2Re=C\overset{Y}{\underset{H}{\diagdown}}$$

$$Y = OEt, NMe_2.$$

Scheme 25.

MO calculations reveal that the LUMO in cationic carbyne complexes of the type $[(CO)_5Cr \equiv C-NR_2]^+$ is localised mainly on the carbyne ligand (261). These cations are therefore excellent precursors for the preparation of interesting and previously inaccessible carbene complexes. Even at low temperatures, they readily add a great variety of anionic nucleophiles, Nu^- [e.g. Nu = F (262), Cl (257), Br, I, NCO, NCS (263), CN, NCSe (264), SPh (265), SeR (R = aryl, 1-naphthyl) (266), TePh (267), $AsPh_2$ (268), $SnPh_3$ (269), $PbPh_3$ (270), $(CO)_5M[C(=O)R]$ (M = Cr, R = p-C_6H_4Me; M = W, R = Me) (271) (Scheme 26).

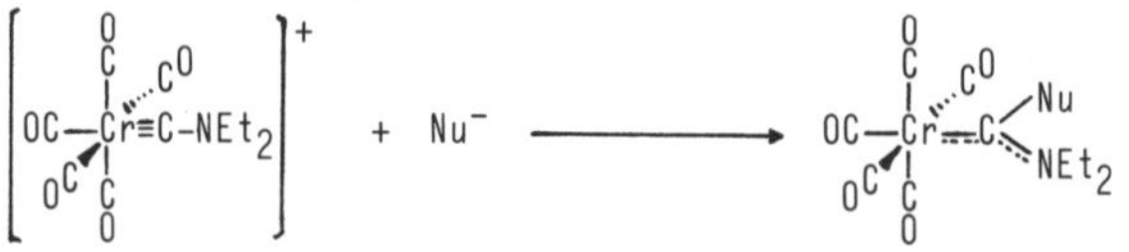

$$\left[\begin{array}{c} \overset{O}{\underset{C}{\|}} \\ OC-\underset{\underset{\overset{\|}{C}}{\underset{O}{}}}{\overset{CO}{\underset{|}{Cr}}} \equiv C-NEt_2 \end{array} \right]^+ + Nu^- \longrightarrow \begin{array}{c} \overset{O}{\underset{C}{\|}} \\ OC-\underset{\underset{\overset{\|}{C}}{\underset{O}{}}}{\overset{CO}{\underset{|}{Cr}}}\cdots C\overset{Nu}{\underset{NEt_2}{\diagdown}} \end{array}$$

(XVII)   (XVIII)

Scheme 26.

In some cases reaction conditions are important. The use of the BF_4^- salt of XVII is advisable since the BCl_4^- salt spontaneously gives XVIII (Nu = Cl) and BCl_3 even at temperatures above $-25\,°C$. In the compound XVIII, NCS and NCSe are bonded via nitrogen. However, when Nu = NCO is used for reaction with XVII, the product depends on the solvent: in MeOH predominantly the N-bonded complex XVIII (Nu = NCO) is formed whereas in CH$_2$Cl$_2$ the O-bonded complex is formed exclusively (263,272).

Similar addition reactions were carried out with (a) the NMe$_2$ analogue of XVII and CN$^-$, NMe$_2^-$ (273) or Cl$^-$ (257) and (b) the NC$_5$H$_{10}$ (257) or NPr$_2^i$ analogue of XVII (274) and Cl$^-$. Whereas attack of [AsPh$_2$]$^-$ on XVII yields (CO)$_5$Cr[C(AsPh$_2$)NEt$_2$], the reaction ov XVII with LiAsMe$_2$ leads to a dimeric carbene complex by reductive Cα-Cα coupling of the two fragments XVII (275) (Scheme 27).

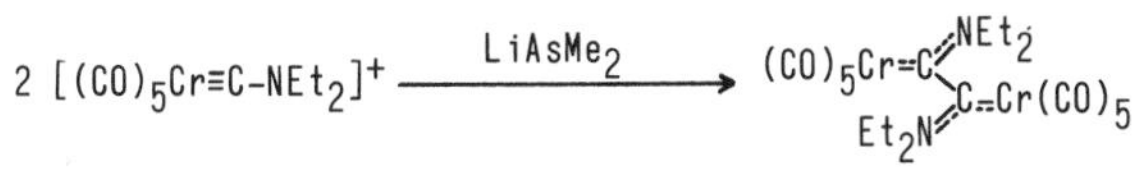

$$2 \; [(CO)_5Cr\equiv C-NEt_2]^+ \xrightarrow{\text{LiAsMe}_2} \begin{array}{c} \text{NEt}_2 \\ (CO)_5Cr=C \\ \text{Et}_2N \; C=Cr(CO)_5 \end{array}$$

Scheme 27.

Some (but not all) of the compounds XVIII (as well as the analogous NMe$_2$, NC$_5$H$_{10}$ or NPr$_2^i$ complexes with Nu = Cl) display an unusual feature. In solution, with the loss of one carbon monoxide, the Nu group migrates to the central metal to form neutral *trans*-tetracarbonyl(carbyne)(Nu) complexes (Scheme 28).

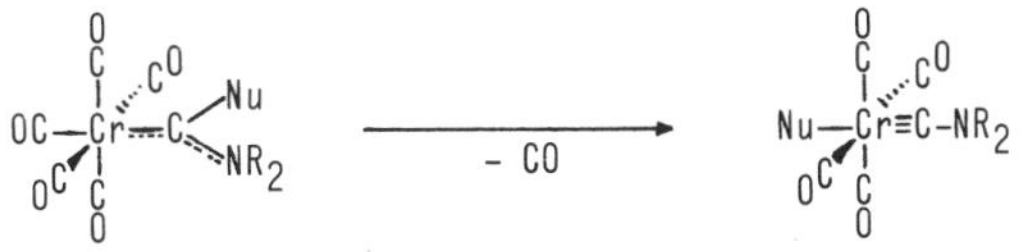

Scheme 28.

Based on the results of several kinetic investigations, an intramolecular mechanism has been deduced (267,270,276 – 279). This reaction principle (nucleophilic addition to cationic carbyne ligands) applied to chromium compounds of type XVII is not in general transferable to other metals. The reaction of the molybdenum and tungsten analogues of XVII with halides or selenides immediately yields the corresponding neutral carbyne complexes. Until now, only two tungsten complexes related to XVIII [*viz.* (CO)$_5$W[C(*p*-SeC$_6$H$_4$CF$_3$)NEt$_2$] (280) and (CO)$_5$W[C(AsPh$_2$)NEt$_2$] (281)] have been synthesised from [(CO)$_5$W$\equiv$C-NEt$_2$]$^+$ and a nucleophile Nu$^-$ (Nu = *p*-SeC$_6$H$_4$CF$_3$ and AsPh$_2$, respectively).

Another prerequisite for this particular route is the availability of suitable and relatively stable cationic carbyne precursors. The stabilisation of cationic carbyne complexes can be achieved by

(a) blocking the *trans* position to the carbyne ligand by groups with a higher σ-donor/π-acceptor ratio than CO (the carbyne ligand is a better acceptor than CO) or

(b) having a π-donor substituent at the carbyne carbon atom (e.g. NR_2).

Both effects have to be balanced carefully. Whereas the reaction of *trans*-$[(PMe_3)(CO)_4Cr \equiv C\text{-Me}]^+$ with anhydrous ethanol gives *trans*-(PMe_3)-$(CO)_4Cr[C(OEt)Me]$ (282), the reaction of *trans*-PPh_3-substituted XVII with $Nu^- = Cl^-$, Br^- or I^- does not yield a carbene complex but rather a carbyne complex, *mer*-$(PPh_3)(Nu)(CO)_3Cr \equiv C\text{-}NEt_2$, PPh_3 being now *cis* to $CNEt_2$ (283). Surprisingly, when *trans*-$[(CO)_5Re(CO)_4Re \equiv C\text{-}SiPh_3]^+$ is reacted in CH_2Cl_2 with alcohols or amines in excess, a mixture of *cis* and *trans* carbene complexes is obtained (260,284). When halides are employed as nucleophiles, only the *trans* compounds are formed (260).

In some cases the conversion of neutral carbyne into carbene complexes was also achieved by:

(a) Reduction of a metal-carbon triple bond with molecular hydrogen (30 psi, in CH_2Cl_2) (285) (Scheme 29).

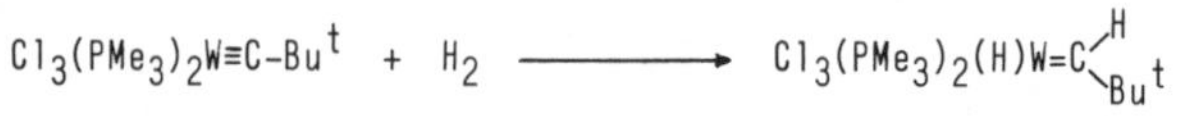

$$Cl_3(PMe_3)_2W \equiv C\text{-}Bu^t \ + \ H_2 \ \longrightarrow \ Cl_3(PMe_3)_2(H)W{=}C\underset{Bu^t}{\overset{H}{\diagdown}}$$

Scheme 29.

(b) Addition of a proton to the carbyne (alkylidene) carbon atom to yield a cationic carbene complex (286) (Scheme 30).

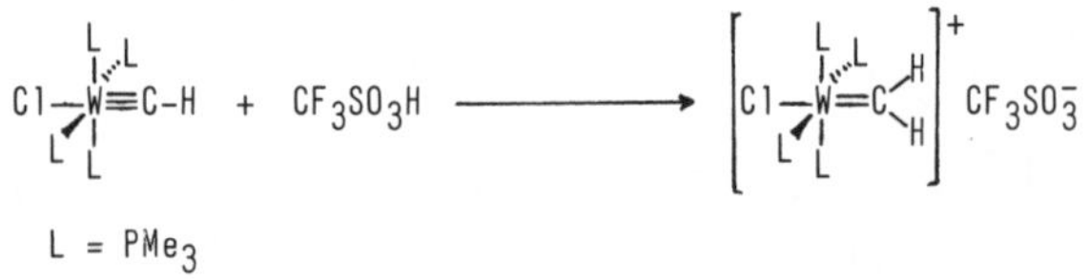

$$Cl\underset{L}{\overset{L}{\underset{|}{-}}}\overset{L}{\underset{|}{W}}{\equiv}C\text{-}H \ + \ CF_3SO_3H \ \longrightarrow \ \left[Cl\underset{L}{\overset{L}{\underset{|}{-}}}\overset{L}{\underset{|}{W}}{=}C\overset{H}{\underset{H}{\diagdown}} \right]^+ CF_3SO_3^-$$

$$L \ = \ PMe_3$$

Scheme 30.

HCl, however, reacts with $Cl(PMe_3)_4W \equiv CH$ by substitution of a PMe_3 ligand to give the methylidyne hydride complex $Cl_2(PMe_3)_3(H)W \equiv CH$ (286).

(c) Addition of $HClO_4$ across the $Os \equiv C$ triple bond of $(CO)Cl(PPh_3)_2$-$Os \equiv C\text{-}(p\text{-}C_6H_4Me)$ to form $(CO)Cl(PPh_3)_2(ClO_4)Os[C(H)p\text{-}C_6H_4Me]$. Metal halides, such as CuI, AgCl or ClAu(PPh$_3$), can also add across the Os$\equiv$C bond giving dimetallacyclopropane derivatives (287).

2.11 Preparation from Complexes L$_n$M(CS$_2$)

Apart from routes already described in previous chapters, we come to a method which utilises the cycloaddition of alkynes to a ligand having 1,3-dipolar character. Cyclic dithiocarbene complexes are obtained by this method from, for instance, L$_2$(CO)$_2$Fe(CS$_2$) and acetylenes (288 – 290) (Scheme 31).

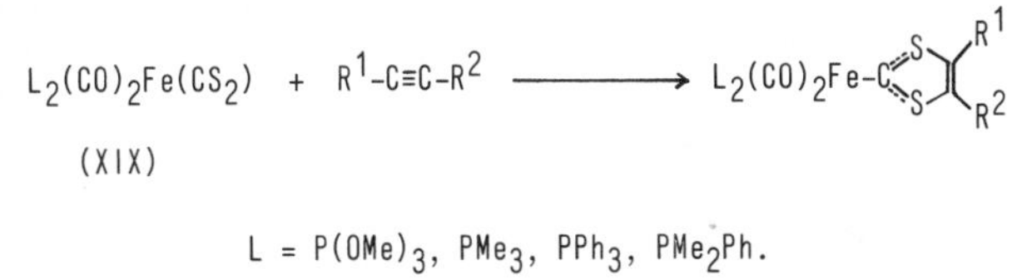

$$L_2(CO)_2Fe(CS_2) \; + \; R^1-C\equiv C-R^2 \; \longrightarrow \; L_2(CO)_2Fe-C\underset{S}{\overset{S}{<}}\Big\rangle\overset{R^1}{\underset{R^2}{<}}$$

(XIX)

$$L = P(OMe)_3, \; PMe_3, \; PPh_3, \; PMe_2Ph.$$

Scheme 31.

The addition takes place only if at least one of the two alkyne substituents is an electron-withdrawing group. For R^1 = R^2 = COOMe and L = PMe$_3$, PMe$_2$Ph the alkyne addition is reversible. In such cases the kinetically controlled product XX slowly isomerizes, leading to an equilibrium mixture of a heterometallacycle XXI and the carbene complex XX. Note that XXI:XX = 100:0 for L = PMe$_3$ and 18:82 for PMe$_2$Ph; in C$_6$D$_6$ (290) (Scheme 32). For L = P(OMe)$_3$ or PPh$_3$, XX is both the kinetically and thermodynamically controlled product (290). A mechanism for this transformation was proposed which involves retro-cycloaddition (XX → XIX + MeOOC-C$\equiv$C-COOMe) followed by 1,3-dipolar addition of the alkyne to Fe$^+$-C-S$^-$. Similar cycloadditions of alkynes to the CS$_2$ ligand were employed for the preparation of other cyclic dithiocarbene complexes from e.g. Cp(CO)(PPh$_3$)Mn(CS$_2$) (291) or ring-substituted (η^6-C$_6$H$_{6-n}$R$_n$)(CO)$_2$Cr(CS$_2$) (292). No isomerisation was observed in the case of these compounds.

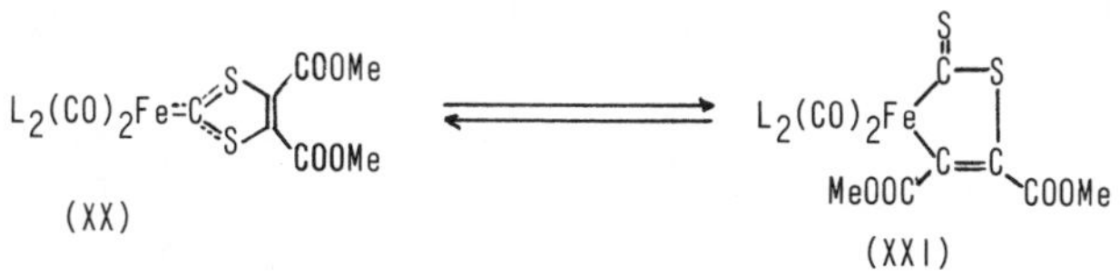

$$L_2(CO)_2Fe=C\underset{S}{\overset{S}{<}}\Big\rangle\overset{COOMe}{\underset{COOMe}{<}} \;\; \rightleftharpoons \;\; L_2(CO)_2Fe\underset{MeOOC}{\overset{\overset{\overset{S}{\|}}{C-S}}{\underset{\underset{COOMe}{C=C}}{|}}}$$

(XX)   (XXI)

Scheme 32.

Other dithiocarbene complexes were obtained via oxidative addition of MeI or 1,2-dibromethane to osmium compounds (224,293) or of MeI to $(PPh_3)_2Pt(CS_2)$ (294).

2.12 Preparation from Complexes $L_nM(CR^1R^2R^3)$

The abstraction of an atom or a group bonded to the α-carbon atom of a transition metal alkyl or its derivatives has proved to be another versatile method for the synthesis of carbene complexes. Non-heteroatom stabilised iron carbene complexes can be prepared via methoxide abstraction from $Cp(CO)(L)Fe[CH(OMe)R]$ employing $[Ph_3C]PF_6$ (295), HPF_6 (187), $Me_3SiO-SO_2CF_3$ (188), CF_3COOH, or other acids (236,295) (Scheme 33).

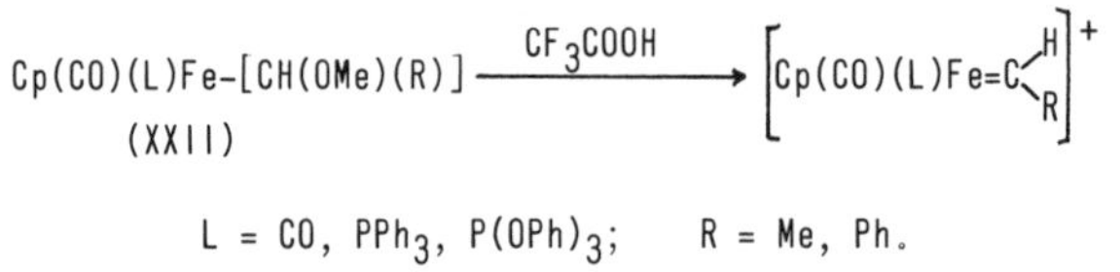

$$Cp(CO)(L)Fe-[CH(OMe)(R)] \xrightarrow{CF_3COOH} \left[Cp(CO)(L)Fe=C\begin{array}{c}H\\R\end{array} \right]^+$$

(XXII)

$$L = CO, PPh_3, P(OPh)_3; \quad R = Me, Ph.$$

Scheme 33.

The corresponding cationic methylene complex $[Cp(CO)_2Fe=CH_2]^+$ has been postulated several times as a reaction intermediate. It is said to be formed on acid treatment of the methoxymethyl iron compound, but has never been isolated nor detected because of its instability (295 – 297). The comparable cation $[Cp(Ph_2PC_2H_4PPh_2)Fe=CH_2]^+$, however, can be generated by the same method from the ethoxymethyl compound and characterised spectroscopically (298,299).

With R = H in XXII (L = CO, PPh_3), hydride abstraction was found in the reaction with $[Ph_3C]^+$ (300). The same observation was made with other alkyl iron compounds (301,302) or $Cp(PPh_3)(NO)Re-L$ [L = Me, CH_2OMe] (200,303) as precursors. The reaction of the analogous rhenium compound (L = CH_2OMe) with 0.5 equivalent $MeOSO_2F$ yields a 1.1:1 mixture of two complexes, the cationic methoxycarbene and the neutral methyl complex (304).

Selective alkoxide or hydride abstraction could be achieved with some molybdenum and tungsten complexes by altering the reaction conditions (305) (Scheme 34).

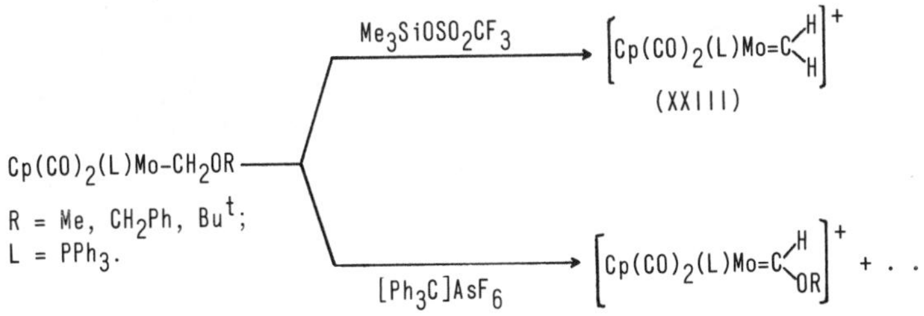

Scheme 34.

The cation XXIII decomposes at temperatures above $-70\,°C$. Thermal abstraction of a zirconoxy group was used to prepare a stable phenylcarbene tungsten complex (306) (Scheme 35).

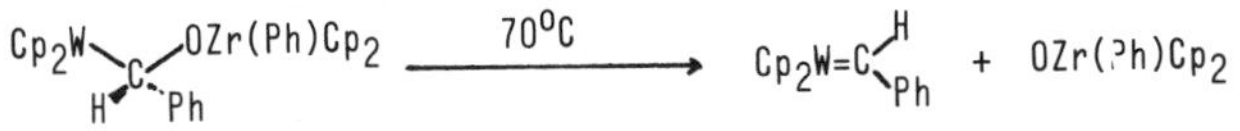

Scheme 35.

Dithiocarbene complexes were synthesised by abstraction of SR^- from $Cp(CO)_2Fe[C(SR)_3]$ using acids (307) and abstraction of PPh_3 from $(CO)_5Cr[C(SPh)_2PPh_3]$ employing S_6 (308). Fluoride abstraction from $Cp(CO)_2LMoCF_3$ (L = CO, PPh_3) by SbF_5 leads to difluorocarbene complexes, which, however, could not be isolated but could be observed by NMR spectroscopy (309). A stable difluorocarbene complex, $(CO)Cl_2(PPh_3)_2$-$Ru=CF_2$, was synthesised from $(MeCN)(CO)Cl(PPh_3)_2Ru-CF_3$ and dry HCl in benzene (310). Hydride abstraction by $[Ph_3C]^+$, although not from the α-carbon atom, was likewise employed in the synthesis of some cyclic iron and carbene tungsten complexes (311 – 313) (Scheme 36).

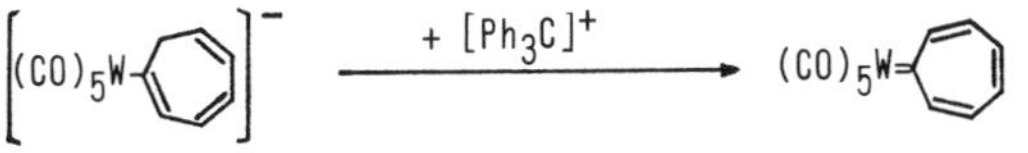

Scheme 36.

Stable non-heteroatom stabilised secondary carbene complexes – alkylidene complexes of the Schrock type – can be obtained via α-hydrogen abstraction from an alkyl ligand. In this important class of compounds the carbene (alkylidene) carbon atom is nucleophilic in contrast to the electrophilic carbene carbon atom in Fischer-type compounds. The first example of a nucleophilic carbene complex was prepared by Schrock in 1974 by reaction of Cl_2Np_3Ta ($Np = CH_2CMe_3$) with two equivalents of LiNp (314,315) (Scheme 37).

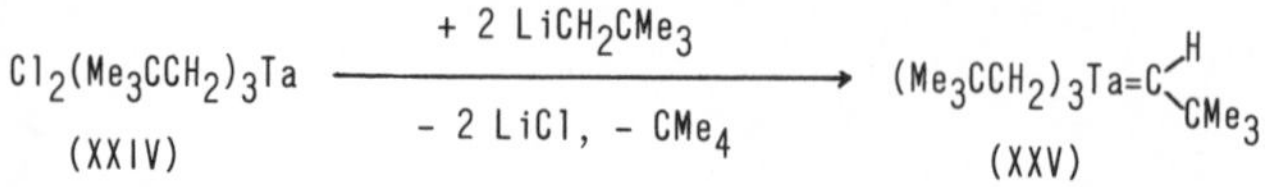

$$Cl_2(Me_3CCH_2)_3Ta \xrightarrow[- 2\ LiCl,\ -\ CMe_4]{+\ 2\ LiCH_2CMe_3} (Me_3CCH_2)_3Ta=C\underset{CMe_3}{\overset{H}{\diagup}}$$

(XXIV)   (XXV)

Scheme 37.

The rate determining step is believed to be the formation of $ClNp_4Ta$ (XXVI) from XXIV and one equivalent of LiNp. Compound XXVI, which can also be synthesised from XXV and HCl at $-78\,°C$, then reacts rapidly (compared to the rate of its formation from XXIV) with LiNp giving XXV, LiCl and CMe_4. In this fast hydrogen abstraction step a relatively nucleophilic axial alkyl α-carbon atom of the trigonal bipyramidal intermediate removes a relatively acidic proton from an equatorial alkyl α-carbon atom. Steric crowding around the alkyl metal precursor is believed to be an important factor in determining when the α-hydrogen abstraction occurs (315). The analogous, thermally unstable niobium compound has also been prepared (315,316). In general, alkylidene niobium complexes appear to be less stable than their tantalum analogues. Another tantalum complex was prepared by reacting $Cl_3[Me_3Si)_2N]_2Ta$ with three equivalents of $LiCH_2SiMe_3$ (317). Bisalkylidene complexes can be obtained by reaction of $Cl_3(PMe_3)_2Ta[=C(H)CMe_3]$ with $MgNp_2$ dioxane (318) or by reaction of $Cl(\eta^5\text{-}C_5Me_5)(PMe_3)_2Ta\equiv C\text{-}CMe_3$ with LiNp (319). Compound XXV can also be prepared directly from $TaCl_5$ and five equivalents of LiNp, though less conveniently and with lower yield (315). The similar reaction of $MoCl_5$ with five equivalents of $LiCH_2SiMe_3$ yields a mixture of $(Me_3SiCH_2)_6Mo_2$, $(Me_3SiCH_2)_3Mo\equiv C\text{-}SiMe_3$ and $(Me_3SiCH_2)_5Mo[=C(H)SiMe_3]$ (320). Hydrogen abstraction can also be induced by other means such as:

(a) Substitution of halide ligands with cyclopentadienyl ligands in neopentyl or benzyl complexes (318,321 – 325) (Scheme 38).

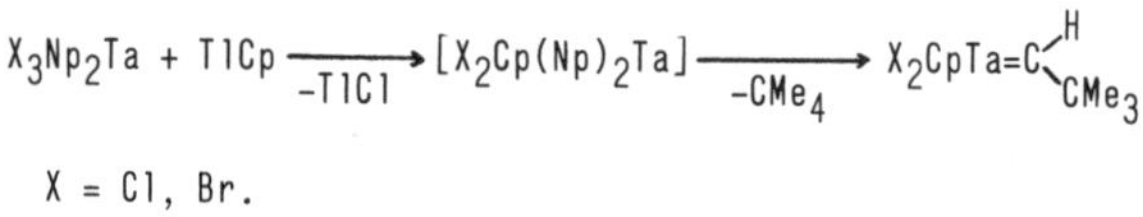

$$X_3Np_2Ta + TlCp \xrightarrow[-TlCl]{} [X_2Cp(Np)_2Ta] \xrightarrow[-CMe_4]{} X_2CpTa=C\underset{CMe_3}{\overset{H}{\diagup}}$$

X = Cl, Br.

Scheme 38.

A detailed investigation of this reaction concluded that the α-hydrogen abstraction in these compounds is intramolecular. The reaction rate is strongly dependent on the nature of X (Cl or Br), the solvent and on whether TlCp or TlC$_5$Me$_5$ is used. A deuterium isotope effect with $k_H/k_D \approx 6$ was observed.

(b) Addition of PMe$_3$ to neopentyl or benzyl complexes (318,319,326,327) (Scheme 39).

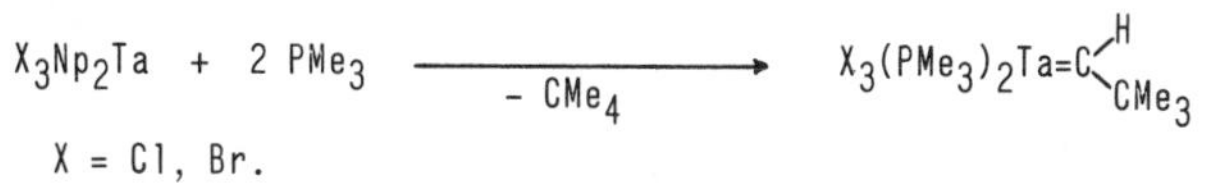

$$X_3Np_2Ta \ + \ 2\ PMe_3 \ \xrightarrow[-\ CMe_4]{} \ X_3(PMe_3)_2Ta{=}C{\overset{H}{\underset{CMe_3}{\diagup}}}$$

$$X = Cl,\ Br.$$

Scheme 39.

X$_3$(PMe$_3$)$_2$Ta[C(H)CMe$_3$] slowly reacts further with X$_3$Np$_2$Ta to yield dimeric [X$_2$(PMe$_3$)Ta($=$CHCMe$_3$)]$_2$(μ-X)$_2$.

(c) Addition of THF or pyridine to neopentyl complexes (326).

An unusual alkyl-alkylidene-alkylidyne complex is obtained from the reaction of Np$_3$W$\equiv$C-CMe$_3$ with PMe$_3$ (328) (Scheme 40).

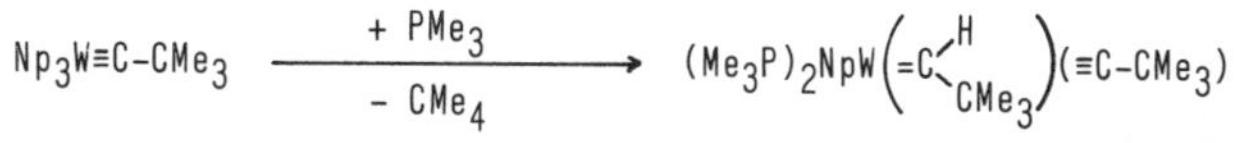

$$Np_3W{\equiv}C{-}CMe_3 \ \xrightarrow[-\ CMe_4]{+\ PMe_3} \ (Me_3P)_2NpW\left(=C{\overset{H}{\underset{CMe_3}{\diagup}}}\right)({\equiv}C{-}CMe_3)$$

Scheme 40.

In general, the preparation of alkylidene complexes containing β-hydrogen atoms necessitates the use of other techniques since β-hydrogen atoms are lost more readily than α-hydrogen atoms. However, the reaction of Cl$_2$(Et)Np$_2$Ta with PMe$_3$ was found to yield a mixture of two tautomers in equilibrium with each other (329) (Scheme 41).

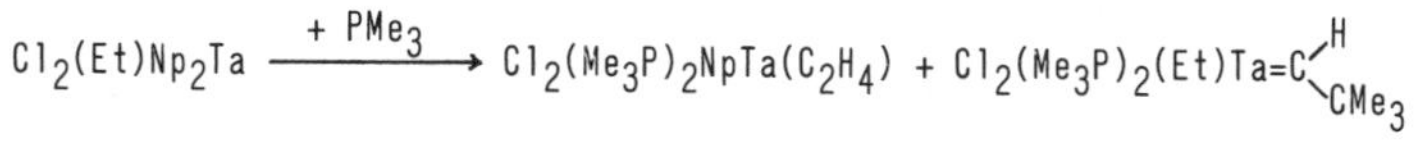

$$Cl_2(Et)Np_2Ta \ \xrightarrow{+\ PMe_3} \ Cl_2(Me_3P)_2NpTa(C_2H_4) \ + \ Cl_2(Me_3P)_2(Et)Ta{=}C{\overset{H}{\underset{CMe_3}{\diagup}}}$$

Scheme 41.

Surprisingly, the compound X$_3$(Mes)MeTa (X = Cl or Br) reacts with PMe$_3$ by γ-hydrogen abstraction to form a benzylidene complex, X$_3$(Me$_3$P)$_2$-Ta[= C(H)C$_6$H$_3$Me$_2$] (330).

The abstraction of a proton from sterically crowded cationic or neutral alkyl compounds can also be carried out using external bases such as

Me_3PCH_2, $LiN(SiMe_3)_2$, NaOMe or alkyl Li. The first transition metal methylene complex was obtained by this method (321,331) (Scheme 42).

$$[Cp_2TaMe_2]BF_4 + Me_3PCH_2 \longrightarrow Cp_2MeTa=CH_2 + [Me_4P]BF_4$$

Scheme 42.

In addition to methylene complexes, other related alkylidene [*e.g.* benzylidene (325) or trimethylsilylmethylene (331,332)] complexes can be prepared by a similar procedure.

3 Syntheses by Addition of a Carbene Ligand Precursor to a Metal Complex

All the synthetic strategies described in chapter 2 take advantage of pre-existing non-carbene ligands which are subsequently transformed by various means into carbene ligands. A different approach is to use a suitable carbene ligand precursor (*e.g.* dimers, halides or diazo compounds) which will be attached to a metal complex during synthesis and simultaneously modified. In some cases this procedure is associated with a change of the formal oxidation number of the central metal. Whereas in the former methods the metal-carbon bond of the carbene complex already exists in the precursor — at least in principle — the metal-carbon bond is now built up in the course of the synthesis.

3.1 Preparation from Precursors "X_2CR_2"

The reaction of anionic metal complexes with certain organic salts or neutral compounds having highly ionic bonds has been used several times for the preparation of carbene complexes. The first report of a successful carbene complex synthesis by this method was that of Öfele as long ago as 1968 (333) (Scheme 43).

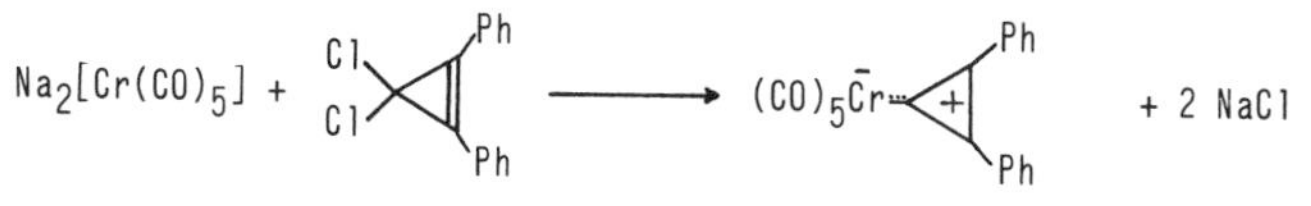

$$Na_2[Cr(CO)_5] + \underset{Cl}{\overset{Cl}{\diagdown}}\!\!\diagup\!\!\overset{Ph}{\underset{Ph}{\diagup}} \longrightarrow (CO)_5\bar{Cr}\!\cdots\!\overset{Ph}{\underset{Ph}{\diagup}} \;+\; 2\ NaCl$$

Scheme 43.

Carbene complexes of iron have been obtained by similar means (334). Dichloro-2,3-diphenylcyclopropene also reacts with metallic palladium to give a dimeric chlorine-bridged species, which on heating in pyridine forms a monomeric carbene complex (335,336). These compounds are thermally very stable, probably due to the aromaticity of the diphenylcyclopropenium cation.

A series of tetraphenylporphinato(carbene)iron complexes were synthesized from 5,10,15,20-tetraphenylporphinatoiron(II), (TPP)Fe(II), and $Cl_2CR^1R^2$ in the presence of an excess of reducing agent (Fe or $Na_2S_2O_4$) (337 – 343) (Scheme 44).

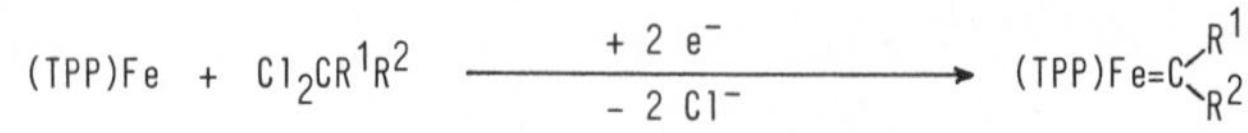

$$(TPP)Fe \; + \; Cl_2CR^1R^2 \xrightarrow[\;-\;2\;Cl^-\;]{+\;2\;e^-} (TPP)Fe{=}C\underset{R^2}{\overset{R^1}{\diagup}}$$

$R^1 = Cl; \quad R^2 = Cl, \; COOEt, \; CN, \; Me, \; CH_2OH, \; CF_3, \; SPh, \; SCH_2Ph, \; . . .$

$R^1 = Br; \quad R^2 = CH_2OH; \qquad R^1 = H; \quad R^2 = SPh.$

Scheme 44.

It was suggested that the reaction proceeds via a radical pathway involving (TPP)Fe(III)Cl and ClCR^{1}R^2 radicals which then react further with (TPP)Fe(II) to form (TPP)Fe(III)[CClR^{1}R^2]. Dissociation of the Cl$^-$ finally affords the carbene complexes. From the reaction of (TPP)Fe(II) with Cl$_3$C[C(C$_6$H$_4$Cl)$_2$H] the carbene complex was not isolated (it is probably formed as an intermediate) but rather a vinylidene complex (343). Similarly, a carbene complex is a very likely intermediate in the reactions of (TPP)Fe(II) with Cl$_3$CSiMe$_3$ and Fe (344) or with CI$_4$/Fe (or Na$_2$S$_2$O$_4$) (345). In both cases a carbide complex is the isolated product (Scheme 45).

$$(TPP)Fe \; + \; CI_4 \xrightarrow{\;+\;Fe\;} (TPP)Fe{=}C{=}Fe(TPP)$$

Scheme 45.

A large number of aminocarbene complexes can be prepared from neutral and anionic metal complexes employing immonium salts of the type [Me$_2$N-=C(R)Cl]Cl (R =H, Cl, NMe$_2$) as carbene ligand precursors. By this means secondary carbene complexes (R = H) of Cr, Fe, Mo, W, Mn, Re, Co, Ru, Ir, Pt, Rh and even V can be conveniently obtained, although sometimes in relatively low yields (346 – 349) (Scheme 46).

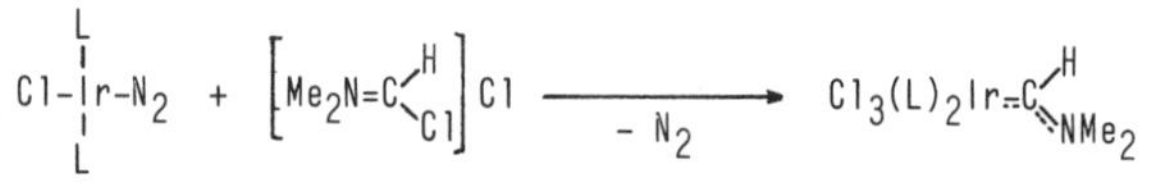

$$Cl{-}\underset{L}{\overset{L}{Ir}}{-}N_2 \; + \; \left[Me_2N{=}C\overset{H}{\underset{Cl}{\diagdown}}\right]Cl \xrightarrow{\;-\;N_2\;} Cl_3(L)_2Ir{=}C\overset{H}{\underset{NMe_2}{\diagdown}}$$

$L = PPh_3.$

Scheme 46.

Amino(chloro)carbene complexes of chromium, manganese or rhodium were synthesised by a similar method (348) (Scheme 47).

$$[(CO)_5Cr]^{2-} \; + \; \left[Me_2N{=}C\overset{Cl}{\underset{Cl}{\diagdown}}\right]Cl \xrightarrow{\hspace{2cm}} (CO)_5Cr{=}C\overset{Cl}{\underset{NMe_2}{\diagdown}} \; + \; 2\;Cl^-$$

Scheme 47.

A diaminocarbene complex was prepared from $Fe_2(CO)_9$ and $[(Me_2N)_2C-Cl]Cl$ (350).

Oxidative addition of 2-chloroimidazolium, -pyrazolium, -triazolium, -tetrazolium, or -thiazolium tetrafluoroborate to metal complexes or metal salts was also used for the preparation of cyclic diheteroatom-stabilised carbene complexes of various metals (233,351 – 353). The reaction of $Hg(CCl_3)_2$ with $(CO)Cl(PPh_3)_3M(H)$ [M = Ru, Os] (354,355) or $Cl_2(PPh_3)_3Ir(H)$ (356) finally yields dichlorocarbene complexes (Scheme 48).

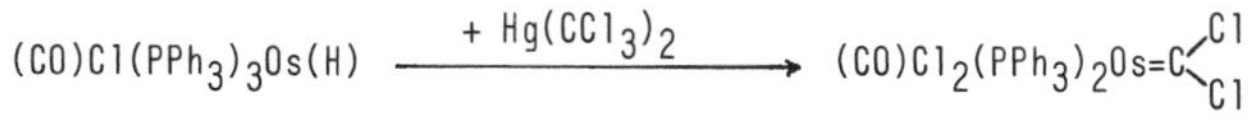

$$(CO)Cl(PPh_3)_3Os(H) \xrightarrow{+ Hg(CCl_3)_2} (CO)Cl_2(PPh_3)_2Os=C\begin{smallmatrix}Cl\\Cl\end{smallmatrix}$$

Scheme 48.

3.2 Preparation from Precursors $Cl(R)C = NR'$

Imidoyl chlorides can also be employed as precursors. Thus, from the reaction of $[(PPh_3)(CO)ClRh]_2$ with $Cl(Ph)C = NMe$ and HCl an aminocarbene complex, $(PPh_3)(CO)Cl_3Rh[C(NHMe)Ph]$, is obtained (357). However, in the complete absence of HCl, the reaction of the dicarbonyl compound $[(CO)_2ClRh]_2$ with two equivalents of $Cl(Ph)C = NR'$ affords a chelated rhodium(III) carbene complex, the chelate bridge of which can be cleaved with tertiary phosphines (358). Recently, carbonyl(cyclopentadienyl)aminocarbene complexes of iron and molybdenum were prepared by a similar route from carbonyl(cyclopentadienyl)metallates and $Cl(Ph)C = NR$ [R = Me, Ph, CH_2Ph, $CHMe_2$] (359 – 361) (Scheme 49).

$$[Cp(CO)_2Fe]^- + Cl(Ph)C=NR \xrightarrow[\text{(2) } NH_4PF_6]{\text{(1) } HCl/THF} \left[Cp(CO)_2Fe=C\begin{smallmatrix}Ph\\NHR\end{smallmatrix}\right]PF_6$$

Scheme 49.

3.3 Preparation from Precursors $[R^1R^2C = Y]^+$
   ## $(Y = NR_2$ or $OR)$

Carbonylmetallates react with $[H_2C = NMe_2]^+$ to yield secondary aminocarbene complexes (349,362). An analogous iron compound (among others) can be isolated as a product of the reaction of $[MeS(H)C = NMe_2]^+$ with

$[(CO)_4Fe]^{2-}$ (362). However, when $[(CO)_5M]^{2-}$ (M = Cr, Mo, W) reacts with $[Ph_2C=NMe_2]^+$ and subsequently with CF_3COOH, diphenylcarbene complexes are obtained (363) (Scheme 50).

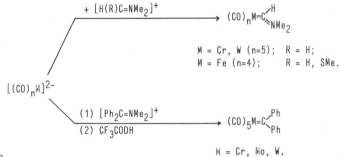

Scheme 50.

Remarkably stable cyclic diaminocarbene complexes can be prepared from imidazolium salts and carbonyl(hydrido)metal anions with the elimination of molecular hydrogen (364,365) (Scheme 51).

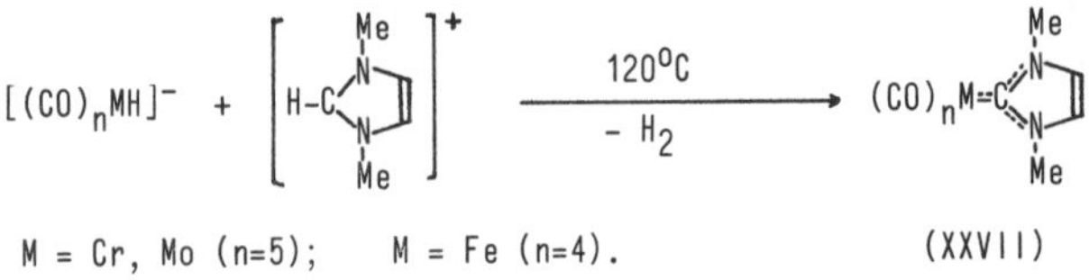

$$[(CO)_nMH]^- \;+\; \left[\begin{array}{c} Me \\ N \\ H-C \\ N \\ Me \end{array} \right]^+ \xrightarrow[-\,H_2]{120^\circ C} (CO)_nM=C\begin{array}{c} Me \\ N \\ N \\ Me \end{array}$$

M = Cr, Mo (n=5);　　M = Fe (n=4).　　　　　　　(XXVII)

Scheme 51.

On heating, XXVII [M = Cr, Mo] disproportionates to yield a mixture of the hexacarbonyl and *cis*-tetracarbonyl(biscarbene) complexes (366). Irradiation converts the latter into the corresponding *trans* isomers which isomerise thermally back to the *cis* isomers (353). Imidazolium salts can also be made to react with mercury salts affording mono- or biscarbene complexes (367,368). Related to XXVII are carbene complexes of ruthenium obtained by acid-catalysed rearrangement of the N-bonded imidazole or xanthine compounds (369,370). Apart from the reaction with immonium salts, carbonylmetallates were also shown to react with oxonium salts (371).

3.4 Preparation from Precursors $R_2C = Y$ (Y = S, NR′, N_2, PR_3)

Starting from $R_2C = Y$-type precursors, several carbene iron complexes have been prepared, *e.g.* by reaction of $Fe_2(CO)_9$ with $(NH_2)_2C = S$ (372) and 2,3-diphenylcyclopropenylthioketone (373) and reaction of $Fe(CO)_5$ with cyclic esters of thiocarbonic acid (374) or its complexes (375). The reaction of carbodiimides with (a) carbonylmetallates and subsequently with HCl (162,234) or (b) certain organometallic nickel compounds (376) also leads to carbene complexes (see 2.7). A number of non-heteroatom-stabilised carbene complexes can be synthesised from $Cp(CO)_2Mn(THF)$ or its methylcyclopentadienyl analogue and diazo compounds (377 – 380) (Scheme 52).

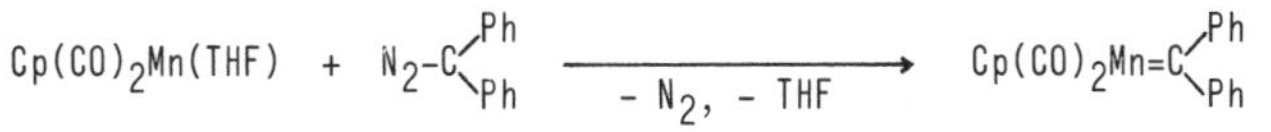

$$Cp(CO)_2Mn(THF) \;+\; N_2\text{-}C\overset{Ph}{\underset{Ph}{}} \quad \xrightarrow[-N_2, \; -THF]{} \quad Cp(CO)_2Mn\text{=}C\overset{Ph}{\underset{Ph}{}}$$

Scheme 52.

Oligomeric biscarbene complexes result if diazocyclopentadiene is employed in the reaction with $Cp(CO)_2Mn(THF)$ (380). Electrophilic ethylidene complexes cannot, in general, be prepared by α-hydrogen abstraction since β-hydrogen atoms are lost more readily from ethyl ligands than α-hydrogen atoms, but can be obtained from $R_3P = C(H)Me$ and tantalum compounds (381) (Scheme 53).

$$Cp_2(PMe_3)Ta(Me) \;+\; Et_3P\text{=}C\overset{H}{\underset{Me}{}} \quad \xrightarrow[-PMe_3, \; -PEt_3]{} \quad Cp_2(Me)Ta\text{=}C\overset{H}{\underset{Me}{}}$$

Scheme 53.

3.5 Preparation from Precursors $[R_2C\text{-}Y]^-$

Only a few examples are known for the preparation of carbene complexes using anionic organic precursors:

(a) Reaction of $(CO)_5M(NCMe)$ (M = Cr, W) with $Na[TosNNC(SEt)_2]$ giving $(CO)_5M[C(SEt)_2]$ (382) and

(b) Reaction of $(CO)_5M(THF)$ (M = Cr, W) with $Li[C(SPh)_3]$ giving $(CO)_5M[C(SPh)_2]$ (383).

In both cases dithiocarbene complexes are formed and for both types of reaction the formation of bis(organylthio)carbene intermediates has been proposed.

3.6 Preparation from Electron-rich Olefins

Cyclic diamino- or amino(thio)carbene complexes can be prepared via bridge-cleaving reactions of binuclear and/or ligand substitution reactions of mononuclear metal compounds with electron-rich olefins such as XXVIII – XXX (13,384) (Scheme 54).

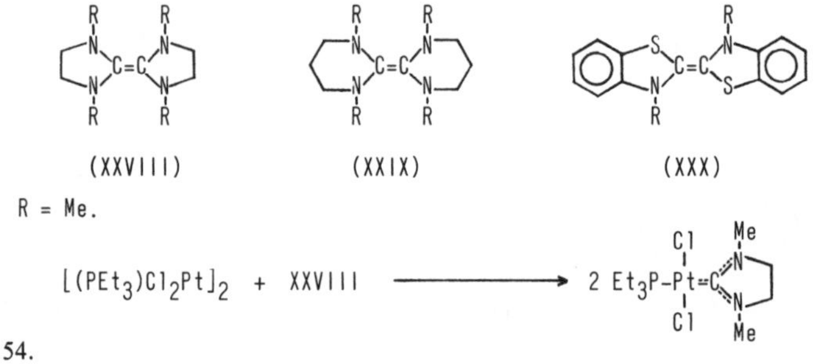

(XXVIII)   (XXIX)   (XXX)

R = Me.

$$[(PEt_3)Cl_2Pt]_2 \; + \; XXVIII \longrightarrow 2 \; Et_3P\text{-}Pt{=}C$$

Scheme 54.

Similar carbene rhodium complexes were found to be isolatable intermediates in the metathesis of two different electron-rich olefins (385). Many carbene complexes were synthesised via replacement of CO, PR$_3$, CNR, norbornadiene or halides by cyclic carbene ligands arising from electron-rich olefins (13) (Scheme 55).

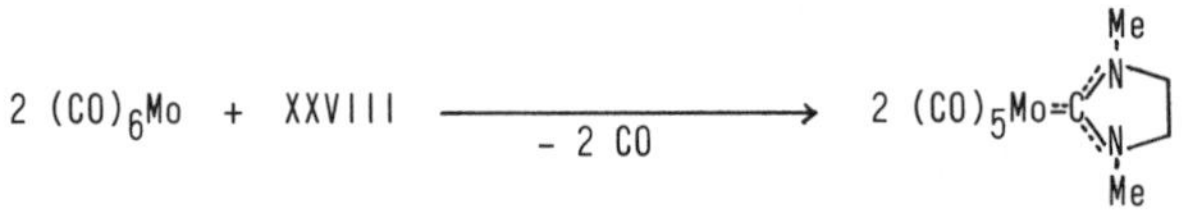

$$2 \; (CO)_6Mo \; + \; XXVIII \xrightarrow{\; - \; 2 \; CO \;} 2 \; (CO)_5Mo{=}C$$

Scheme 55.

A mechanistic scheme was proposed for this reaction involving replacement of one CO ligand by XXVIII. The olefin, initially N-bonded, rearranges in a fast second step to a C-bonded species which may then fragment to form the carbene metal complex with expulsion of a resonance-stabilised carbene fragment.

Above 100 °C the monocarbene molybdenum complex disproportionates to form $(CO)_6Mo$ and the *cis*-biscarbene(tetracarbonyl) compound, which upon irradiation gives the corresponding *trans* isomer. At 25 °C this isomerisation can be reversed (93).

A great number of mono-, bis-, tris- and even tetrakiscarbene complexes of various transition metals, *e.g.* Cr (92), Mo (93), W (94), Mn, Fe, Co (386), Ni (386–388), Ru (386,387,389), Rh (79,388), Ir (388), Os (387), Au (390), Pd (384,391) and Pt (384,388,391), were synthesised from electron-rich olefins. In addition, the N-Et, N-p-Tol and N-CH$_2$Ph analogues of XXVIII and even a macrocyclic derivative (392) were employed. There are, nevertheless, significant differences in reactivity for the various olefins: the reactivity decreases in the order XXVIII > XXX > XXIX.

If the N-Ph analogue of XVIII is used in the reaction with $(PPh_3)_3Cl_2Ru$, a pentacoordinated carbene complex is formed containing an *ortho*-metallated N-arylcarbene ligand (393). Similar *ortho*-metallated products are obtained with some palladium (394) and platinum compounds (395).

4 Syntheses Using Pre-existing Carbene Complexes

4.1 Carbene Ligand Transfer from One Metal to Another

In some cases carbene complexes can be obtained by transfer of a carbene ligand from one element to another. This was shown for the first time when Cp(NO)(CO)Mo(carbene) complexes were irradiated in the presence of Fe(CO)$_5$ (52,103) (Scheme 56).

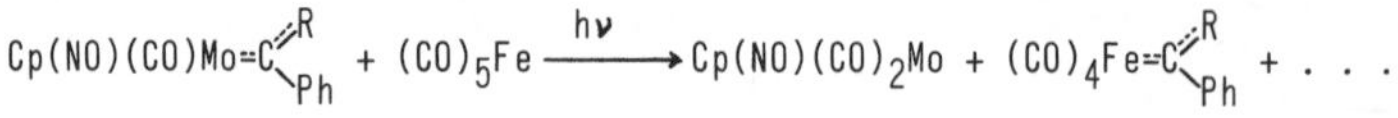

$$Cp(NO)(CO)Mo=C\langle^R_{Ph} + (CO)_5Fe \xrightarrow{h\nu} Cp(NO)(CO)_2Mo + (CO)_4Fe=C\langle^R_{Ph} + \ldots$$

R = OMe, OEt, NMe$_2$.

Scheme 56.

Several carbene iron complexes hitherto not available could thus be conveniently prepared by this method. Recently, the carbene ligand could also be transferred thermally from carbene tungsten compounds to Cp(CO)$_2$Mn(THF) (396) and to HAuCl$_4$ (397) (Scheme 57).

$$(CO)_5W=C\langle^X_{Ph} + HAuCl_4 \longrightarrow Cl-Au=C\langle^X_{Ph} + (CO)_4WCl_2 + \ldots$$

X = OMe, NH$_2$, NMe$_2$ . .

Scheme 57.

In the case of the system pentacarbonyl(2-oxacyclopentylidene)chromium/(CO)$_6$W the thermally induced transfer of the carbene ligand from chromium to tungsten was found to be reversible and the equilibrium constant measured at 140 °C (4) (Scheme 58).

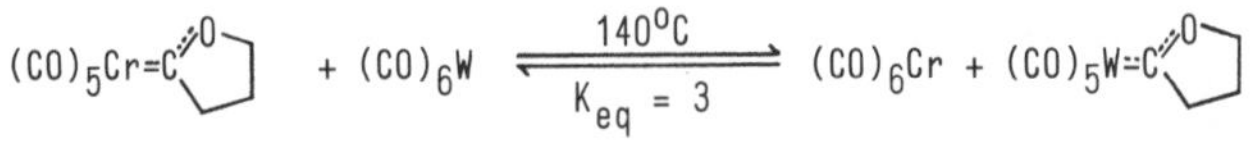

$$(CO)_5Cr=C\overbrace{\qquad}^{O} + (CO)_6W \underset{K_{eq} = 3}{\overset{140°C}{\rightleftharpoons}} (CO)_6Cr + (CO)_5W=C\overbrace{\qquad}^{O}$$

Scheme 58.

Carbene ligand transfer must also be involved in the thermal disproportionation of some pentacarbonyl complexes containing a cyclic diaminocarbene ligand to form (CO)$_6$M and *cis*-(CO)$_4$M(carbene)$_2$ (93,353,366,398) [see also 3.3 and 3.6] as well as in the thermolysis of (CO)$_5$M(carbene) [as deduced from the organic reaction products and a kinetic investigation (399)].

Similarly, in the reaction of a cyclic carbene complex of manganese with $(PMeBu_2^t)Pt(C_2H_4)_2$, transfer of the carbene ligand from Mn to Pt must occur since in the resulting dimetallacyclopropane derivative the carbene ligand is bonded to platinum (400).

The transfer of a neopentylidene ligand from tantalum to tungsten was also accomplished (401 – 403). However, up to the present carbene ligand transfer has not been generally applied in the synthesis of carbene complexes but has rather been confined to special cases.

4.2 Modification of the Carbene Ligand

In contrast to the synthetic approaches outlined in the previous sections, there are those which use pre-existing carbene ligands by modifying them in such a way that new carbene complexes are formed. This can be accomplished, for instance, by exchange of one or both of the non-metal substituents on the carbene carbon atom (R^1 or R^2 or $R^1 = R^2$ in Scheme 1) by other groups. Ammonia, primary and secondary amines react with alkoxy(organyl)carbene complexes to yield amino(organyl)carbene complexes and the corresponding alcohol (in most cases methanol) ["aminolysis"] (Scheme 59).

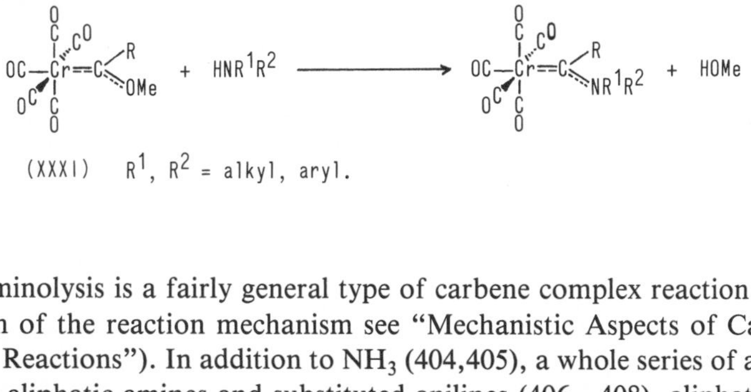

(XXXI) R^1, R^2 = alkyl, aryl.

Scheme 59.

This aminolysis is a fairly general type of carbene complex reaction (for a discussion of the reaction mechanism see "Mechanistic Aspects of Carbene Complex Reactions"). In addition to NH_3 (404,405), a whole series of amines including aliphatic amines and substituted anilines (406 – 408), aliphatic and aromatic diamines (409), ω-enamines (410), optically active amines (53,186) and amino acid esters (411,412) were employed. With secondary amines, however, steric factors become important. The reaction of $(CO)_5Cr[C(Me)OMe]$ with the sterically hindered diisopropylamine did not yield the expected diisopropylamino(methyl)carbene but instead, via propylene elimination, the methyl(monoisopropylamino)carbene complex (406). Other alkoxycarbene complexes of Cr, Mo, W (40,91), Re (72) and $(CO)_5M[C(SiR_3)OR']$ (M = Cr,

Mo, W; R = Ph, Me; R' = Me, Et) (67 – 69), as well as a cyclic carbene complex of Pt (413), react with amines in the same way. An interesting competition takes place in the reaction of $HNMe_2$ with $(CO)_5M[C(OEt)C≡CPh]$ (M = Cr, W). Whereas at very low temperatures (Cr: $-115\,°C$; W: $-60\,°C$) aminolysis occurs, at $-20\,°C$ conjugate addition of one molecule of amine to the triple bond is observed. At room temperature, the aminolysis product (but not the product of the conjugate addition) can be converted with another molecule of amine into $(CO)_5M[C(NMe_2)CH = C(Ph)NMe_2]$ (414).

Other N-nucleophiles have also been used. Reaction of XXXI (R = Me) with benzophenoneimine gave $(CO)_5Cr[C(Me)N = CPh_2]$ (415). Similar imino-substituted carbene complexes were isolated from XXXI (R = Me) and benzaldoxime (415) and from XXXI (R = Ph) and 1-aminoethanol (416). However, in the reaction of XXXI (R = Me) with substituted hydrazines the hoped for hydrazinocarbene complex could not be isolated. Instead a nitrile complex was obtained as the product of rearrangement and nitrogen-nitrogen bond cleavage (417).

The reaction principle of aminolysis is not transferable to PH_3 or primary and secondary phosphines. Reaction of phosphines with XXXI, for instance, results either in simultaneous substitution of the carbene and one CO ligand by PH_3 (418) or addition of HPR_2 to the carbene carbon atom with subsequent rearrangement to give phosphine complexes (419). Closely related to aminolysis is the reaction of XXXI or the tungsten analogue (R = Me) with thiols (420) or – in a two-step procedure – with thiolates and subsequently with acids (421) to form organyl(organylthio)carbene complexes ("thiolysis").

Replacement of the MeO group in XXXI (R = Me) or its tungsten analogue by a MeSe substituent ("selenolysis") can be achieved by reaction with MeSeH (422). Unlike methylselenol, the more acidic phenylselenol yields a rearranged product in which the ligand is bonded via selenium to the metal (423).

Exchange of OR with OR' in alkoxycarbene complexes can also be carried out, *e.g.* $(CO)_5Cr[C(OEt)Me]$ reacts with methanol in the presence of catalytic amounts of NaOMe to give $(CO)_5Cr[(OMe)Me]$ (424). Similarly, $(CO)_5W[C(OMe)p-C_6H_4Me]$ and 3-buten-1-ol in the presence of base and a molecular sieve (to remove the resulting methanol from the solution) yield $(CO)_5W[C(OCH_2CH_2CH = CH_2)p-C_6H_4Me]$ (425). Whereas some Pt compounds do react with alcohols in the same way (241), $(CO)_5Cr$-$[C(OEt)CH_2SiMe_3]$ and HOR (R = H, Me) produce $(CO)_5Cr[C(OEt)Me]$ (42). Two-step reaction of alkoxycarbene complexes with enolate anions and subsequently with acids was also used to generate new carbene complexes. Thus, $(CO)_5W[C(OMe)Ph]$ reacts with the enolate of isobutyrophenone/H^+ with replacement of MeO by $Me_2C = C(Ph)O$ (84).

On the other hand, if methoxy(vinyl)carbene complex derivatives [XXXI: R = CH=CMe$_2$, CH=C(H)Ph] are employed in the reaction with the enolates of cyclopentanone, isobutyrophenone or dimethylmalonate, and subsequently with H$^+$, (instead of MeO substitution) addition of the enolate to the vinylic double bond is observed (84) (Scheme 60).

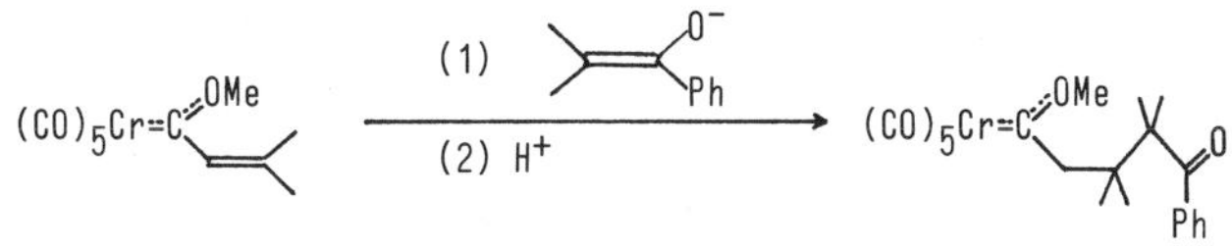

Scheme 60.

The reaction with the lithium enolate of acetone takes yet another course; organic products result arising from addition of the enolate to the carbene carbon atom (84). Since the carbene carbon-oxygen bond in acetoxycarbene complexes is more reactive than that in alkoxycarbene complexes, aminolysis and thiolysis reactions can easily be carried out with acetoxycarbene complexes. Through reaction with NaOPh, organyl(phenoxy)carbene complexes of chromium or tungsten, which are difficult to prepare by other routes, become accessible in relatively high yields (up to 92 %) (26,426).

In the reaction of a trimethylsiloxycarbene complex with methanol, the Me$_3$SiO group is not replaced by MeO but rather the Me$_3$Si by H to produce a hydroxycarbene complex (30). (CO)$_5$Cr[C(OSiMe$_3$)Me], however, was found to react with H$_2$NMe by aminolysis to yield (CO)$_5$Cr[C(NHMe)Me] (427).

A very important type of reaction involving alkoxycarbene complexes is that with carbanions. The actual product depends on whether alkoxy(alkyl)- or alkoxy(aryl)carbene complexes are used as starting compounds (see also 2.1). Nucleophilic attack of phenyllithium on (CO)$_5$W[C(OMe)Ph] at − 78 °C, for instance, gives Li[(CO)$_5$W-CPh$_2$OMe] (115) which, in solution, decomposes rapidly at room temperature, though it may be isolated at low temperatures (117). Treatment of this adduct at − 78 °C with acids such as HCl (115,116) or silicagel (60) induces loss of methanol and a diphenylcarbene complex is formed (Scheme 61).

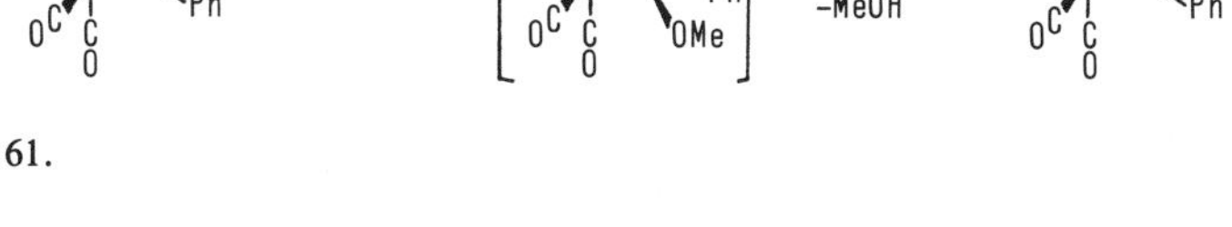

Scheme 61.

This reaction sequence can be carried out with substituted phenyl-, furyl-, thienyl- and pyrrolyl(methoxy)carbene complexes of chromium and tungsten as precursors as well as with other organolithium compounds, including substituted phenyl-, furyl-, thienyl- and pyrrolyllithium (60). Attempts to synthesise a phenyl(styryl)carbene complex from $(CO)_5Cr[C(OMe)-CH=C(H)Ph]$ and LiPh/H$^+$ failed. Phenyllithium adds to the C=C double bond to give the conjugate addition product $(CO)_5Cr[C(OMe)CH_2CHPh_2]$ (428). If the carbanion in alkyllithium is used as the nucleophile, the resulting alkyl(phenyl)carbene complex obtained after protonation of the adduct is unstable and rearranges immediately with 1,2-hydrogen migration to form an olefin complex (429) (Scheme 62).

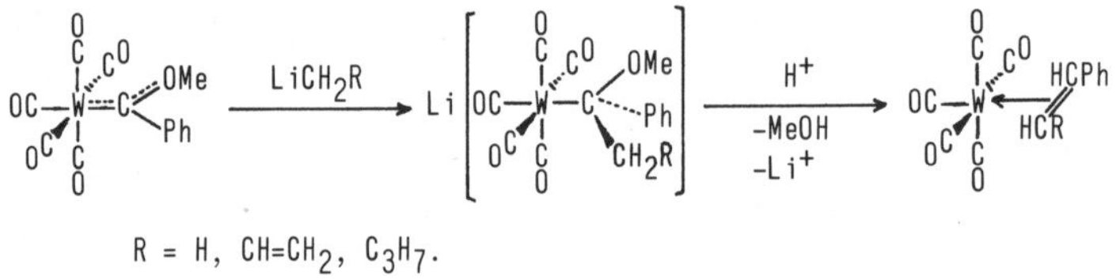

R = H, $CH=CH_2$, C_3H_7.

Scheme 62.

Alkoxy(alkyl)carbene complexes are remarkably acidic. In alkaline deuteriomethanol, the hydrogen atoms attached to the α-carbon atom undergo rapid H/D exchange (424). This is explained by the intermediary existence of a carbene anion, which can be generated stoichiometrically from carbene complexes by reaction with NaOMe or LiBu (118) (see also 2.1) and isolated as the bis(triphenylphosphine)imminium salt (430). The moderate reactivity of these carbene anions towards nucleophiles [including epoxides (431), ethylene sulphide (432), enol ethers (433), aldehydes (118,434), α-bromoesters, α,β-unsaturated carbonyl compounds (432), acetylchloride (432,435), MeOSO$_2$F (118) and even PhCH$_2$I (4) or Br$_2$ (432)] can be used to prepare a series of carbene complexes inaccessible by other synthetic routes (Scheme 63).

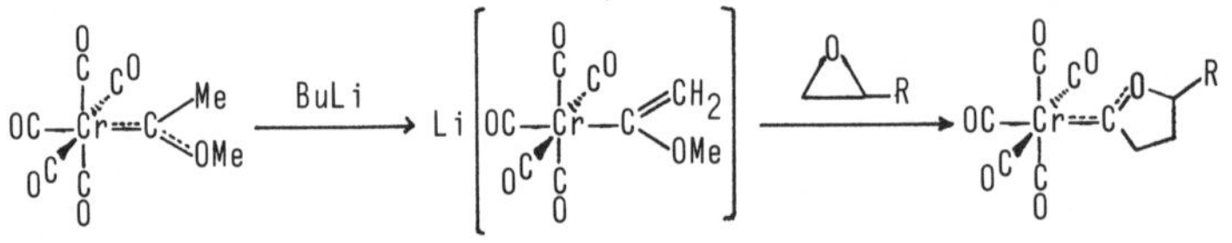

(XXXII)

Scheme 63.

The nucleophilic attack by the carbene anion occurs at the least hindered carbon of propylene oxide (R = Me). Acetyl chloride also reacts with metal carbene anions. If the initial product contains an enolisable hydrogen an enol ester is isolated (118) (Scheme 64).

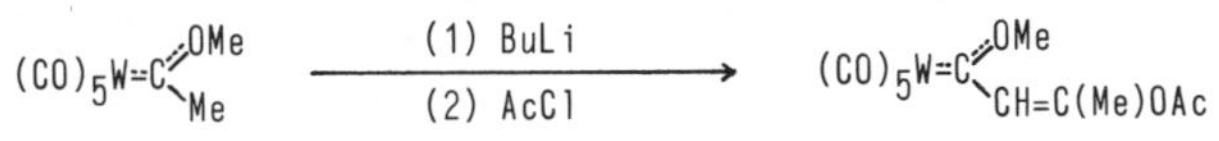

$$(CO)_5W=C\diagdown^{OMe}_{Me} \quad \xrightarrow[\text{(2) AcCl}]{\text{(1) BuLi}} \quad (CO)_5W=C\diagdown^{OMe}_{CH=C(Me)OAc}$$

Scheme 64.

Reaction of the carbene anion from XXXII with acetaldehyde yields a vinylcarbene complex and with formaldehyde a dimeric species (434). But sometimes problems arise from dialkylation as a side reaction in the alkylation of carbene anions, *e.g.* with α-bromoesters or chloromethoxymethane. Carbene anions from neutral alkoxy(alkyl)carbene complexes of manganese (118,435) or cobalt (39) have also been generated and reacted with $MeOSO_2F$, MeI and $[Et_3O]BF_4$. On the other hand, $(CO)_5W[C(OMe)Me]$ reacts with LiMe at $-40\,°C$ in diethyl ether and subsequently with CF_3COOH to give the binuclear, bridged complex $[(CO)_5W]_2[C(H)\text{-}CH=CMe_2]$ (436).

In addition to carbanions, hydrides have been used as nucleophiles. Thus, from the reaction of $(CO)_5W[C(OMe)Ph]$ with the anion in $K[HB(OPr^i)_3]$ a tungstate complex can be isolated (437) which on reaction with CF_3COOH at $-78\,°C$ gave $(CO)_5W[C(H)Ph]$ (438,439). Since this compound decomposes at $-56\,°C$ with a half-life of 24 min, it could not be obtained in an analytically pure form but could be trapped with PBu_3^n (439). Earlier, from the related reaction of $(CO)_5Cr[C(OMe)Me]$ with $Li[HAl(OBu^t)_3]$ at room temperature in THF, a vinylcarbene complex, $(CO)_5Cr[C(OMe)CH=CH\text{-}CH=C(Me)OMe]$ had been obtained, but only with very low yield ($<0.1\,\%$) (440).

Some unusual compounds result from the reaction of hydroxycarbene complexes with dicyclohexylcarbodiimide (DCCD). The product formed depends on the central metal and on whether a hydroxy(methyl)- or a hydroxy(phenyl)-carbene complex is used as the precursor (441,442) (Scheme 65).

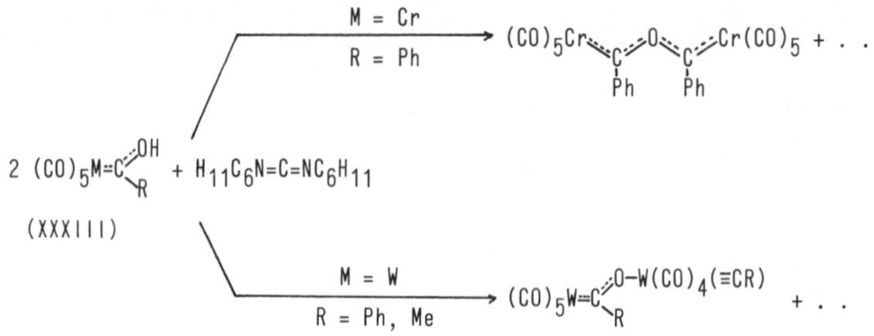

$$\text{2 (CO)}_5\text{M=C}\overset{\text{OH}}{\underset{\text{R}}{\diagup}} + \text{H}_{11}\text{C}_6\text{N=C=NC}_6\text{H}_{11}$$

(XXXIII)

$$\xrightarrow[\text{R = Ph}]{\text{M = Cr}} \quad \text{(CO)}_5\text{Cr} \cdots \underset{\text{Ph}}{\overset{}{\text{C}}} \cdots \text{O} \cdots \underset{\text{Ph}}{\overset{}{\text{C}}} \cdots \text{Cr(CO)}_5 + . .$$

$$\xrightarrow[\text{R = Ph, Me}]{\text{M = W}} \quad \text{(CO)}_5\text{W=C}\overset{\text{O--W(CO)}_4(\equiv\text{CR})}{\underset{\text{R}}{\diagup}} \quad + . .$$

Scheme 65.

Mixtures of hydroxy(*p*-tolyl)carbene chromium and hydroxy(phenyl)-carbene tungsten yielded with DCCD the carbenecarbynechromiumtungsten compound. When employing XXXIII (M = Cr; R = Me) in the reaction with DCCD, again other products are isolated (443) (Scheme 66).

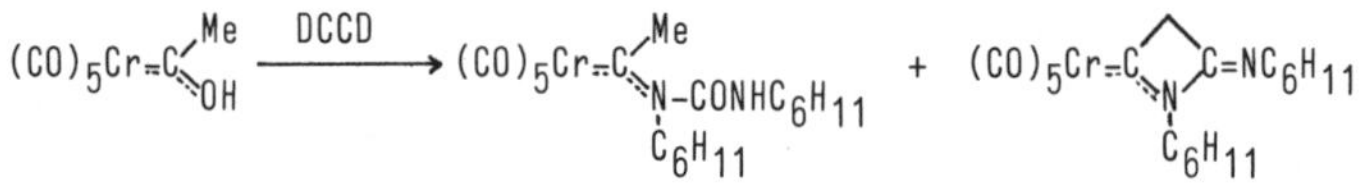

$$\text{(CO)}_5\text{Cr=C}\overset{\text{Me}}{\underset{\text{OH}}{\diagup}} \xrightarrow{\text{DCCD}} \text{(CO)}_5\text{Cr=C}\overset{\text{Me}}{\underset{\underset{\text{C}_6\text{H}_{11}}{\text{N--CONHC}_6\text{H}_{11}}}{\diagup}} \quad + \quad \text{(CO)}_5\text{Cr=C}\overset{}{\underset{\underset{\text{C}_6\text{H}_{11}}{\text{N}}}{\diagup}}\text{C=NC}_6\text{H}_{11}$$

Scheme 66.

It was proposed that the cyclic carbene complex is formed by (2 + 2) cyc-loaddition of DCCD to a vinylidene complex intermediate.

In the reaction of *cis*-Br(CO)$_4$Mn[C(OH)Me] with BBr$_3$ an unusual metal-lacyclic carbene complex [related to those obtained from (CO)$_5$MnMe and AlX$_3$; see 2.5] is formed (12).

In comparison with the carbene ligand in alkoxy-, hydroxy- and acetoxy(or-ganyl)carbene complexes, that in amino(organyl)carbene complexes is far less reactive. Replacement of the amino group by OR, SR or SeR has not been observed. The treatment of the *E* isomer of (CO)$_5$Cr[C(NHMe)Me] with KOH-MeOH, NaOMe/MeOH or NaOCMe$_3$/HOBu^t results in isomerisation giving a mixture of the *E* and *Z* isomers (427). The compounds (CO)$_5$Cr[C(NH$_2$)R] can be converted with

(a) LiMe and subsequently with [Me$_3$O]BF$_4$ stepwise into the monomethyl-amino- and dimethylaminocarbene complexes (for R = Me) (427) and

(b) Me-C≡C-NEt$_2$ into the alkylideneaminocarbene complexes (CO)$_5$-Cr{C[N=C(Et)NEt$_2$]R} (R = Me, Ph) (444).

The carbene ligand in cyclic and non-cyclic diaminocarbene complexes under comparable conditions is chemically inert. Yet, for compounds containing a carbene ligand of the type [C(SMe)YR] (YR = OMe, OPh, SMe, SPh, SePh), several replacement reactions of Y'R' for SMe or YR have been reported. The dithiocarbene complex $(CO)_5W[C(SMe)_2]$ reacts with HNR^1R^2. The product depends on the nature of the amine: primary amines yield iso-nitrile complexes but secondary amines form amino(methylthio)carbene complexes (177). If the carbene compound $Cp(CO)_2Fe[C(SMe)YR]^+$ (XXXIV) is used as the precursor, similar observations are made for YR = SMe, SPh, SePh (219,445). In the case of competition between SMe and YR = OMe, OPh or SePh as leaving groups in the reaction with secondary amines, for YR = OMe the SMe group and for YR = SePh the SePh group will be exchanged. For YR = OPh a mixture of the SMe and OPh substituted products is isolated (445). In contrast to simple primary amines, the reaction of XXXIV (YR = SMe) with primary diamines such as $H_2N(CH_2)_3NH_2$ or $o\text{-}C_6H_4(NH_2)_2$ yields cyclic diaminocarbene complexes (219). Similarly, XXXIV (YR = SMe) reacts with (a) $HS(CH_2)_nSH$ (n = 2, 3) to give cyclic dithiocarbene complexes, (b) HSPh to give a mixture of the mono- and disubstituted product and (c) MeOH to give $[Cp(CO)_3Fe]^+$ (220). If the reaction conditions are altered, XXXIV (YR = SMe) and HSPh produce a neutral complex resulting from addition of PhS^- to the carbene carbon atom (307). In the case where YR = OPh or OMe the product of the reaction with MeOH is $Cp(CO)_2\text{-}Fe[C(OMe)_2]^+$ (178,220).

The reaction characteristics of dithiocarbene complexes towards amines are paralleled by those of dihalocarbene complexes. Whereas dichlorocarbene compounds of Fe (446), Ru (354), Os (355) and Ir (356) react with primary amines to give isonitrile complexes, the reaction of Ru or Ir dichlorocarbene complexes and $(CO)Cl_2(PPh_3)_2Ru[CF_2]$ with $HNMe_2$ lead to amino(halo-geno)carbene compounds (310,354,356). A variety of products results if $(CO)Cl_2(PPh_3)_2Ru[CF_2]$ is reacted with HOR, depending on the nature of R (310) (Scheme 67).

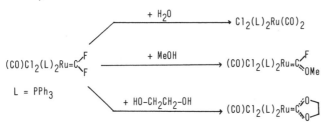

Scheme 67.

The analogous CCl_2 compound also reacts with (a) HSR and HSC_2H_4SH forming respectively non-cyclic and cyclic dithiocarbene complexes or (b) H_2X (X = O, S, Se) respectively converting the CCl_2 ligand into a CO, CS or CSe group (354). Similar reactions were observed with a dichlorocarbene iridium complex (356). The chloro substituent at the carbene carbon atom of $(CO)_5Cr[C(Cl)NR_2]$ also turned out to be exchangeable: the replacement of Cl by CN [R = Me] (273) and of Cl by NCO, NCS, NCSe or F (R = Et) has been demonstrated (272). The preparation of $(CO)_5Cr[C(Cl)NEt_2]$ from $(CO)_5Cr[C(OEt)NEt_2]$ and BCl_3 probably proceeds via a cationic carbyne complex (447). In some amino(ethoxy)carbene complexes of palladium, the exchange of OEt by substituted hydrazines or benzalhydrazone could be brought about (448), a reaction which was not feasible with alkoxy(organyl)-carbene complexes (417).

4.3 Insertions into the Metal-Carbene Bond

The insertion of a suitable molecule into the metal-carbene bond was demonstrated to be another route for the modification of pre-existing carbene complexes. Thus, aminoacetylenes react with pentacarbonyl(carbene) complexes of chromium, molybdenum and tungsten by insertion of the triple bond into the metal-carbene bond and redistribution of the π-electrons to give aminocarbene complexes (449,450) (Scheme 68).

$$(CO)_5M{=}C\underset{R'}{\overset{OMe}{\big<}} \quad + \quad R{-}C{\equiv}C{-}NEt_2 \longrightarrow (CO)_5M{=}C\underset{\underset{R}{\big\backslash}C{=}C\underset{OMe}{\overset{R'}{\big/}}}{\overset{NEt_2}{\big<}}$$

Scheme 68.

In this stereoselective reaction, it is predominantly the insertion product with the *E* configuration which is obtained. Other ynamines as well as other carbene compounds can also be employed (58,451 – 453); similarly, ethoxy-acetylene may be used for reaction with certain carbene complexes of tungsten, namely with $(CO)_5W[C(Ph)R]$ (R = H, Ph, OMe) (439). Bis(di-ethylamino)acetylene (R = NEt_2), however, reacts with $(CO)_5M[C(OMe)Ph]$ (M = Cr, W) not only by insertion but additionally with substitution of one *cis*-CO ligand by NEt_2 forming a chelated carbene complex (454).

These insertions are not confined to molecules containing carbon-carbon triple bonds. Carbon-nitrogen triple bonds can also be inserted into the metal-carbene bond giving methyleneaminocarbene complexes (455 – 458) (Scheme 69).

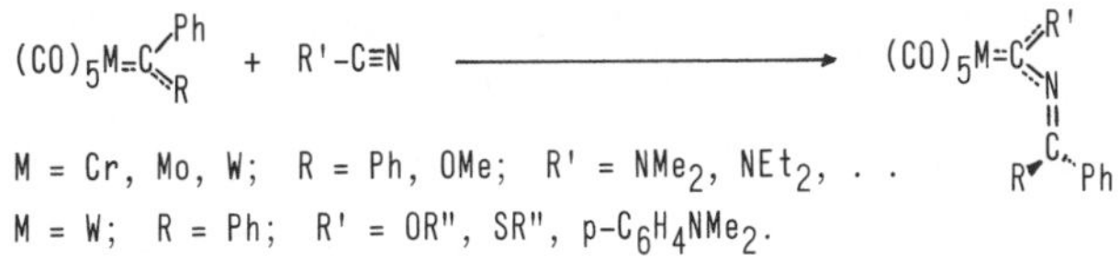

$$(CO)_5M=C\overset{Ph}{\underset{R}{\diagdown}} \quad + \quad R'-C\equiv N \quad\xrightarrow{\hspace{3cm}}\quad (CO)_5M=C\overset{R'}{\underset{\underset{R\overset{\diagdown}{C}\diagdown Ph}{N}}{\diagup}}$$

M = Cr, Mo, W; R = Ph, OMe; R' = NMe$_2$, NEt$_2$, . .

M = W; R = Ph; R' = OR", SR", p-C$_6$H$_4$NMe$_2$.

Scheme 69.

Prerequisites for $N\equiv C$-insertion into the M-C(carbene) bond are a strongly polarised triple bond and high basicity of the nitrogen atom. From the results of kinetic investigations of ynamine and cyanamide insertions, an associative mechanism has been deduced, nucleophilic addition of the triple bond systems to the carbene carbon atom being the rate determining step (274,453,459).

$(CO)_5Cr[C(OMe)Me]$ also adds isocyanides probably by insertion to give keteneimine complexes which — at least in the case of $(CO)_5Cr[H_{11}C_6-N=C=C(Me)OMe]$ — react further with HCl, MeOH or $(PhCO)_2O_2$ to yield new aminocarbene complexes (460,461). Although diphenylacetylene and less polar alkynes do not react, for instance, with $(CO)_5Cr[C(OMe)Ph]$ to give new carbene complexes but rather by substitution of a CO ligand and subsequent cyclisation to form substituted tricarbonyl(naphthol)chromium compounds [see *e.g.* (462)], Ph-C$\equiv$C-Ph does react with an electrophilic neopentylidene complex by insertion of the C$\equiv$C bond into the Ta$=$C bond of $CpCl_2Ta[=C(H)CMe_3]$ to yield a new carbene complex. Only one isomer is formed (324). Owing to the reversed polarisation of the metal-alkylidene carbon bond in this compound in relation to Fischer-type carbene complexes, the product of the reaction with nitriles is not a carbene but rather an imido complex (324).

This reaction principle (insertion of triple bonds into metal-carbene bonds) is not generally transferable to double bond systems such as enamines or enol ethers. From the reaction of these types of olefins (and some non-activated ones) with electrophilic carbene complexes, cyclopropane derivatives and/or metathesis-like scission products are obtained. Nevertheless, some strained cyclic enol ethers, such as 1-ethoxycyclopentene (463) or 2-ethoxynorbornene (464), and 1-pyrrolidino-1-cyclopentene (465) were found to react with $(CO)_5W[CPh_2]$ by insertion forming respectively new alkoxy- and aminocarbene complexes (Scheme 70).

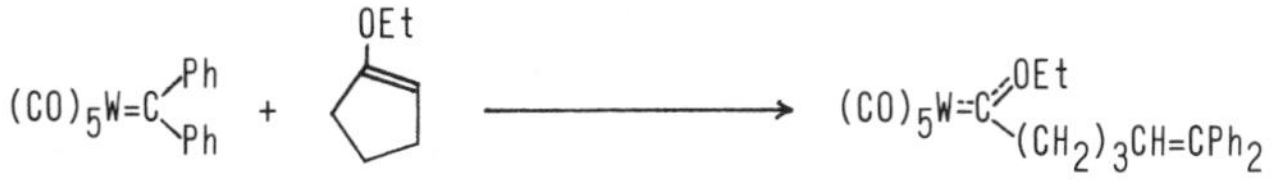

$$(CO)_5W=C\overset{Ph}{\underset{Ph}{\diagdown}} \quad + \quad \overset{OEt}{\bigcirc} \quad\xrightarrow{\hspace{3cm}}\quad (CO)_5W=C\overset{OEt}{\underset{(CH_2)_3CH=CPh_2}{\diagdown}}$$

Scheme 70.

Cyclic enol ethers with six- and eight-membered rings failed to give similar products (463).

A formal carbonyl insertion was observed in the reaction of $(CO)_5Cr[C(O-Me)Ph]$ with NaOMe. On subsequent alkylation with oxonium salts, a neutral carbene complex was obtained (466,467). The reaction of $(CO)_6M$ [M = Cr, W] with $LiCH(SR^1)R^2$ solutions does not yield simple carbene complexes on alkylation but rather, upon further carbonyl insertion, heterometallacyclic chelates (81).

4.4 Oxidation and Reduction of Carbene Complexes

One-electron oxidation of carbene complexes by electrochemical means or by use of silver salts has been employed in a few cases to generate paramagnetic carbene complexes. Some of the resulting iron(I) and chromium(I) salts containing cyclic diaminocarbene ligands are thermally stable (38,468). ESR measurements on these Fe(I) compounds indicate that the unpaired electron is centred mainly on the iron atom whereas the carbene ligand contributes to stability by its strong Fe-C bond and delocalisation of the positive charge (468). By reduction of $(CO)_5M[C(R)R']$ (M = Cr, Mo, W; R = Ph, OMe; R' = aryl) with K-Na alloy in THF at low temperature unstable carbenemetal(1-) complexes were obtained and studied by ESR spectroscopy (114).

The reduction of a alkylidene tantalum complex was also investigated. Although in the reaction of $(PMe_3)_2Cl_3Ta[C(H)CMe_3]$ with Na/Hg in the presence of PMe_3 under argon the formation of $(PMe_3)_4ClTa[C(H)CMe_3]$ is observed (469), under nitrogen a dinuclear dinitrogen-bridged bisalkylidene complex is obtained (470).

4.5 Modification of the Metal-Ligand Framework

In addition to the synthetic routes already mentioned, several methods concerning modification of the metal-non-carbene ligand framework are available, including ligand substitution, isomerisation or oxidative addition to the metal complex. By thermally or photochemically induced ligand substitution, the displacement of CO by PR_3, AsR_3, SbR_3 or $P(OR)_3$ can be carried out for many carbene complexes. In some cases the stereochemistry of the product depends on the reaction conditions and the steric and electronic requirements of the carbene complex and of the incoming group. The reverse process, the exchange of PR_3 for carbon monoxide, has also been observed for some rhodium complexes (471) as well as the replacement of two CO ligands by a chelating bisphosphine molecule [see *e.g.* (102)]. A multitude of other ligand substitution reactions have been reported, especially for carbene complexes containing cyclic diaminocarbene ligands. These include displacement of halides by other halides, carbanions, hydrides [see *e.g.* (472)], CO, PF_3, $P(OR)_3$ [see *e.g.* (389)] or even by organometallic bases (473). Intramolecular substitution reactions have also been observed [see *e.g.* (144)]. For the preparation of certain carbene complexes, thermally or photochemically induced isomerisation was employed (93,108,109,353). Rapid isomerisation in some neutral bis(diaminocarbene) complexes of molybdenum and tungsten containing cyclic carbene ligands could be induced by electrochemical oxidation. After isomerisation, the resulting cation was immediately reduced again to the neutral carbene compound (474,475).

Oxidative addition to carbene complexes has also been used for the modification of carbene complexes. By this method, gold(III) complexes can be synthesised from gold(I) compounds [*e.g.* (164)] or platinum(IV) from platinum(II) complexes (476). Among others, MeI (413), I_2 (94), H_2 or HCl (351) can be added to the metal complexes.

5 References

(1) E. O. Fischer, A. Maasböl, Angew. Chem. **76** (1964) 645; Angew. Chem. Int. Ed. Engl. **3** (1964) 580.
(2) D. J. Cardin, B. Cetinkaya, M. F. Lappert, Chem. Rev. **72** (1972) 545 – 574.
(3) F. A. Cotton, C. M. Lukehart, Progr. Inorg. Chem. **16** (1972) 487 – 613.
(4) C. P. Casey, in H. Alper (Ed.): Transition Metal Organometallics in Organic Synthesis. Academic Press, New York 1976, vol. 1, pp. 189 – 233.
(5) F. J. Brown, Progr. Inorg. Chem. **27** (1980) 1 – 122.
(6) H. Fischer, in F. R. Hartley, S. Patai (Ed.): The Chemistry of the Metal-Carbon Bond. John Wiley and Sons, Chichester 1982, pp. 181 – 231.
(7) K. H. Dötz, in P. S. Braterman (Ed.): Reactions of Coordinated Ligands. Plenum Press, New York, in the press.
(8) E. O. Fischer, Pure Appl. Chem. **24** (1970) 407 – 423.
(9) E. O. Fischer, Pure Appl. Chem. **30** (1972) 353 – 372.
(10) C. G. Kreiter, E. O. Fischer, In XXIIIrd International Congress of Pure and Applied Chemistry. Butterworths, London 1971, pp. 151 – 168.
(11) E. O. Fischer, U. Schubert, H. Fischer, Pure Appl. Chem. **50** (1978) 857 – 870.
(12) E. O. Fischer, Angew. Chem. **86** (1974) 651 – 663; Adv. Organomet. Chem. **14** (1976) 1 – 32.
(13) M. F. Lappert, J. Organomet. Chem. **100** (1975) 139 – 159.
(14) C. P. Casey, in M. Jones, Jr., R. A. Moss (Ed.): Reactive Intermediates. John Wiley and Sons, New York 1981, vol. 2, pp. 135 – 174.
(15) E. O. Fischer, H. Fischer, in J. J. Zuckerman (Ed.): Inorganic Reactions and Methods. Verlag Chemie, Weinheim, in the press.
(16) R. R. Schrock, Acc. Chem. Res. **12** (1979) 98 – 104.
(17) E. O. Fischer, A. Maasböl, Chem. Ber. **100** (1967) 2445 – 2456.
(18) M. Ryang, I. Rhee, S. Tsutsumi, Bull. Chem. Soc. Jap. **37** (1964) 341 – 343.
(19) E. O. Fischer, V. Kiener, J. Organomet. Chem. **23** (1970) 215 – 223.
(20) E. O. Fischer, G. Kreis, F. R. Kreissl, J. Organomet. Chem. **56** (1973) C37 – C40.
(21) E. O. Fischer, W. Kalbfus, J. Organomet. Chem. **46** (1972) C15 – C18.
(22) R. Aumann, E. O. Fischer, Angew. Chem. **79** (1967) 900 – 901; Angew. Chem. Int. Ed. Engl. **6** (1967) 879 – 880.
(23) C. P. Casey, C. R. Cyr, R. A. Boggs, Synth. Inorg. Met.-Org. Chem. **3** (1973) 249 – 254.

(24) E. O. Fischer, A. Maasböl, J. Organomet. Chem. **12** (1968) P15 – P17.

(25) E. O. Fischer, S. G. Gibbins, W. Kellerer, J. Organomet. Chem. **218** (1981) C51 – C53.

(26) J. A. Connor, E. M. Jones, J. Chem. Soc. A (1971) 3368 – 3372.

(27) E. O. Fischer, T. Selmayr, F. R. Kreissl, Chem. Ber. **110** (1977) 2947 – 2955.

(28) H. Berke, P. Härter, G. Huttner, J. v. Seyerl, J. Organomet. Chem. **219** (1981) 317 – 327.

(29) E. Moser, E. O. Fischer, J. Organomet. Chem. **12** (1968) P1 – P2.

(30) E. O. Fischer, T. Selmayr, F. R. Kreissl, U. Schubert, Chem. Ber. **110** (1977) 2574 – 2583.

(31) S. Voran, H. Blau, W. Malisch, U. Schubert, J. Organomet. Chem. **232** (1982) C33 – C40.

(32) U. Schubert, M. Wiener, F. H. Köhler, Chem. Ber. **112** (1979) 708 – 716.

(33) E. O. Fischer, S. Fontana, J. Organomet. Chem. **40** (1972) 159 – 162.

(34) H. G. Raubenheimer, E. O. Fischer, J. Organomet. Chem. **91** (1975) C23 – C26.

(35) E. O. Fischer, H. Fischer, Chem. Ber. **107** (1974) 657 – 672.

(36) E. O. Fischer, D. Plabst, Chem. Ber. **107** (1974) 3326 – 3331.

(37) M. Y. Darensbourg, D. J. Darensbourg, Inorg. Chem. **9** (1970) 32 – 39.

(38) M. F. Lappert, R. W. McCabe, J. J. MacQuitty, P. L. Pye, P. I. Riley, J. Chem. Soc., Dalton Trans. (1980) 90 – 98.

(39) F. Carre, G. Cerveau, E. Colomer, R. J. P. Corriu, J. C. Young, L. Ricard, R. Weiss, J. Organomet. Chem. **179** (1979) 215 – 226.

(40) K. W. Dean, W. A. G. Graham, Inorg. Chem. **16** (1977) 1061 – 1067.

(41) E. O. Fischer, A. Riedel, Chem. Ber. **101** (1968) 156 – 161.

(42) J. A. Connor, E. M. Jones, J. Organomet. Chem. **60** (1973) 77 – 86.

(43) J. A. Connor, E. M. Jones, J. Chem. Soc., Dalton Trans. (1973) 2119 – 2124.

(44) E. O. Fischer, N. H. Tran-Huy, D. Neugebauer, J. Organomet. Chem. **229** (1982) 169 – 177.

(45) S. Fontana, O. Orama, E. O. Fischer, U. Schubert, F. R. Kreissl, J. Organomet. Chem. **149** (1978) C57 – C62.

(46) S. Fontana, U. Schubert, E. O. Fischer, J. Organomet. Chem. **146** (1978) 39 – 44.

(47) E. O. Fischer, H. J. Kollmeier, C. G. Kreiter, J. Müller, R. D. Fischer, J. Organomet. Chem. **22** (1970) C39 – C42.

(48) E. O. Fischer, C. G. Kreiter, H. J. Kollmeier, J. Müller, R. D. Fischer, J. Organomet. Chem. **28** (1971) 237 – 258.

(49) G. A. Moser, E. O. Fischer, M. D. Rausch, J. Organomet. Chem. **27** (1971) 379 – 382.

(50) G. M. Bodner, S. B. Kahl, K. Bork, B. N. Storhoff, J. E. Wuller, L. J. Todd, Inorg. Chem. **12** (1973) 1071 – 1074.

(51) E. O. Fischer, A. Schwanzer, H. Fischer, D. Neugebauer, G. Huttner, Chem. Ber. **110** (1977) 53 – 66.

(52) E. O. Fischer, H.-J. Beck, C. G. Kreiter, J. Lynch, J. Müller, E. Winkler, Chem. Ber. **105** (1972) 162 – 172.

(53) H. Brunner, J. Doppelberger, E. O. Fischer, M. Lappus, J. Organomet. Chem. **112** (1976) 65 – 78.

(54) J. A. Connor, E. M. Jones, J. Chem. Soc. A (1971) 1974 – 1979.

(55) J. W. Wilson, E. O. Fischer, J. Organomet. Chem. **57** (1973) C63 – C66.

(56) K. H. Dötz, R. Dietz, Chem. Ber. **111** (1978) 2517 – 2526.

(57) E. O. Fischer, W. R. Wagner, F. R. Kreissl, D. Neugebauer, Chem. Ber. **112** (1979) 1320 – 1328.

(58) K. H. Dötz, B. Fügen-Köster, D. Neugebauer, J. Organomet. Chem. **182** (1979) 489 – 498.

(59) E. O. Fischer, F. R. Kreissl, J. Organomet. Chem. **35** (1972) C47 – C51.

(60) E. O. Fischer, W. Held, F. R. Kreissl, A. Frank, G. Huttner, Chem. Ber. **110** (1977) 656 – 666.

(61) J. A. Connor, J. P. Lloyd, J. Chem. Soc., Dalton Trans. (1972) 1470 – 1476.

(62) E. O. Fischer, M. Schluge, J. O. Besenhard, P. Friedrich, G. Huttner, F. R. Kreissl, Chem. Ber. **111** (1978) 3530 – 3541.

(63) E. O. Fischer, V. N. Postnov, F. R. Kreissl, J. Organomet. Chem. **231** (1982) C73 – C77.

(64) E. O. Fischer, F. J. Gammel, J. O. Besenhard, A. Frank, D. Neugebauer, J. Organomet. Chem. **191** (1980) 261 – 282.

(65) E. O. Fischer, V. N. Postnov, F. R. Kreissl, J. Organomet. Chem. **127** (1977) C19 – C21.

(66) E. O. Fischer, F. J. Gammel, D. Neugebauer, Chem. Ber. **113** (1980) 1010 – 1019.

(67) E. O. Fischer, H. Hollfelder, F. R. Kreissl, W. Uedelhofen, J. Organomet. Chem. **113** (1976) C31 – C34.

(68) E. O. Fischer, H. Hollfelder, P. Friedrich, F. R. Kreissl, G. Huttner, Chem. Ber. **110** (1977) 3467 – 3480.

(69) E. O. Fischer, H. Hollfelder, F. R. Kreissl, Chem. Ber. **112** (1979) 2177 – 2189.

(70) E. O. Fischer, P. Rustemeyer, D. Neugebauer, Z. Naturforsch. **35b** (1980) 1083 – 1087.

(71) E. O. Fischer, P. Stadler, Z. Naturforsch. **36b** (1981) 781 – 782.

(72) E. O. Fischer, P. Rustemeyer, J. Organomet. Chem. **225** (1982) 265 – 277.

(73) E. O. Fischer, P. Rustemeyer, Z. Naturforsch. **37b** (1982) 627 – 630.

(74) E. O. Fischer, H.-J. Kollmeier, Angew. Chem. **82** (1970) 325; Angew. Chem. Int. Ed. Engl. **9** (1970) 309 – 310.

(75) E. O. Fischer, E. Winkler, C. G. Kreiter, G. Huttner, B. Krieg, Angew. Chem. **83** (1971) 1021 – 1022; Angew. Chem. Int. Ed. Engl. **10** (1971) 922 – 924.

(76) E. O. Fischer, F. R. Kreissl, E. Winkler, C. G. Kreiter, Chem. Ber. **105** (1972) 588 – 598.

(77) E. O. Fischer, W. Kleine, G. Kreis, F. R. Kreissl, Chem. Ber. **111** (1978) 3542 – 3551.

(78) E. O. Fischer, W. Kleine, W. Schambeck, U. Schubert, Z. Naturforsch. **36b** (1981) 1575 – 1579.

(79) M. J. Doyle, M. F. Lappert, G. M. McLaughlin, J. McMeeking, J. Chem. Soc., Dalton Trans. (1974) 1494 – 1501.

(80) H. G. Raubenheimer, S. Lotz, J. Coetzer, J. Chem. Soc., Chem. Commun. (1976) 732 – 734.

(81) H. G. Raubenheimer, S. Lotz, H. W. Viljoen, A. A. Chalmers, J. Organomet. Chem. **152** (1978) 73 – 84.

(82) E. O. Fischer, K. Scherzer, F. R. Kreissl, J. Organomet. Chem. **118** (1976) C33 – C34.

(83) D. J. Darensbourg, M. Y. Darensbourg, Inorg. Chim. Acta **5** (1971) 247 – 253.

(84) C. P. Casey, W. R. Brunsvold, Inorg. Chem. **16** (1977) 391 – 396.

(85) W. Malisch, H. Blau, U. Schubert, Angew. Chem. **92** (1980) 1065 – 1066; Angew. Chem. Int. Ed. Engl. **19** (1980) 1020 – 1021.

(86) E. O. Fischer, E. Offhaus, Chem. Ber. **102** (1969) 2449 – 2455.

(87) E. O. Fischer, E. Offhaus, J. Müller, D. Nöthe, Chem. Ber. **105** (1972) 3027 – 3035.

(88) E. O. Fischer, T. L. Lindner, H. Fischer, G. Huttner, P. Friedrich, F. R. Kreissl, Z. Naturforsch. **32b** (1977) 648 – 652.

(89) C. P. Casey, C. R. Cyr, R. L. Anderson, D. F. Marten, J. Am. Chem. Soc. **97** (1975) 3053 – 3059.

(90) E. O. Fischer, R. Aumann, Chem. Ber. **102** (1969) 1495 – 1503.

(91) C. L. Hyde, D. J. Darensbourg, Inorg. Chim. Acta **7** (1973) 145 – 149.

(92) P. B. Hitchcock, M. F. Lappert, P. L. Pye, J. Chem. Soc., Dalton Trans. (1977) 2160 – 2172.

(93) M. F. Lappert, P. L. Pye, G. M. McLaughlin, J. Chem. Soc., Dalton Trans. (1977) 1272 – 1282.

(94) M. F. Lappert, P. L. Pye, J. Chem. Soc., Dalton Trans. (1977) 1283 – 1291.

(95) M. J. Webb, R. P. Stewart, Jr., W. A. G. Graham, J. Organomet. Chem. **59** (1973) C21 – C23.

(96) M. J. Webb, M. J. Bennett, L. Y. Y. Chan, W. A. G. Graham, J. Am. Chem. Soc. **96** (1973) 5931 – 5932.

(97) K. P. Darst, C. M. Lukehart, J. Organomet. Chem. **171** (1979) 65 – 71.

(98) K. P. Darst, P. G. Lenhert, C. M. Lukehart, L. T. Warfield, J. Organomet. Chem. **195** (1980) 317 – 324.

(99) H. L. Conder, M. Y. Darensbourg, Inorg. Chem. **13** (1974) 506 – 511.

(100) D. J. Darensbourg, M. Y. Darensbourg, Inorg. Chem. **9** (1970) 1691 – 1694.

(101) E. O. Fischer, P. Stückler, H.-J. Beck, F. R. Kreissl, Chem. Ber. **109** (1976) 3089 – 3098.

(102) E. O. Fischer, G. Besl, J. Organomet. Chem. **157** (1978) C33 – C34.

(103) E. O. Fischer, H.-J. Beck, Chem. Ber. **104** (1971) 3101 – 3107.

(104) G. A. M. Munro, P. L. Pauson, Israel J. Chem. **15** (1976/1977) 258 – 261.

(105) C. M. Jensen, T. J. Lynch, C. B. Knobler, H. D. Kaesz, J. Am. Chem. Soc. **104** (1982) 4679 – 4680.

(106) J. Chatt, G. J. Leigh, C. J. Pickett, D. R. Stanley, J. Organomet. Chem. **184** (1980) C64 – C66.

(107) M. Y. Darensbourg, H. L. Conder, D. J. Darensbourg, C. Hasday, J. Am. Chem. Soc. **95** (1973) 5919 – 5924.

(108) E. O. Fischer, H. Fischer, H. Werner, Angew. Chem. **84** (1972) 682 – 683; Angew. Chem. Int. Ed. Engl. **11** (1972) 644 – 645.

(109) E. O. Fischer, K. Richter, Chem. Ber. **109** (1976) 1140 – 1157.

(110) H. Fischer, E. O. Fischer, Chem. Ber. **107** (1974) 673 – 679.

(111) G. R. Dobson, J. R. Paxson, J. Am. Chem. Soc. **95** (1973) 5925 – 5930.

(112) T. F. Block, R. F. Fenske, C. P. Casey, J. Am. Chem. Soc. **98** (1976) 441 – 443.

(113) T. F. Block, R. F. Fenske, J. Organomet. Chem. **139** (1977) 235 – 269.

(114) P. J. Krusic, U. Klabunde, C. P. Casey, T. F. Block, J. Am. Chem. Soc. **98** (1976) 2015 – 2018.

(115) C. P. Casey, T. J. Burkhardt, J. Am. Chem. Soc. **95** (1973) 5833 – 5834.

(116) C. P. Casey, T. J. Burkhardt, C. A. Bunnell, J. C. Calabrese, J. Am. Chem. Soc. **99** (1977) 2127 – 2134.

(117) E. O. Fischer, W. Held, F. R. Kreissl, Chem. Ber. **110** (1977) 3842 – 3848.

(118) C. P. Casey, R. A. Boggs, R. L. Anderson, J. Am. Chem. Soc. **94** (1972) 8947 – 8949.

(119) E. O. Fischer, F. R. Kreissl, C. G. Kreiter, E. W. Meineke, Chem. Ber. **105** (1972) 2558 – 2564.

(120) E. O. Fischer, W. Röll, U. Schubert, K. Ackermann, Angew. Chem. **93** (1981) 582 – 583; Angew. Chem. Int. Ed. Engl. **20** (1981) 611 – 612.

(121) E. O. Fischer, W. Röll, N. Hoa Tran Huy, K. Ackermann, Chem. Ber. **115** (1982) 2951 – 2964.

(122) W. Petz, J. Organomet. Chem. **72** (1974) 369 – 375.

(123) W. Petz, G. Schmid, Angew. Chem. **84** (1972) 997 – 998; Angew. Chem. Int. Ed. Engl. **11** (1972) 934 – 935.

(124) W. Petz, J. Organomet. Chem. **55** (1973) C42 – C44.

(125) W. Petz, A. Jonas, J. Organomet. Chem. **120** (1976) 423 – 432.

(126) W. Petz, J. Organomet. Chem. **165** (1979) 199 – 207.

(127) P. T. Wolczanski, R. S. Threlkel, J. E. Bercaw, J. Am. Chem. Soc. **101** (1979) 218 – 220.

(128) W. Malisch, H. Blau, S. Voran, Angew. Chem. **90** (1978) 827 – 828; Angew. Chem. Int. Ed. Engl. **17** (1978) 780.

(129) H. Motschi, R. J. Angelici, Organometallics **1** (1982) 343 – 349.

(130) K. R. Grundy, W. R. Roper, J. Organomet. Chem. **113** (1976) C45 – C48.

(131) L. Tschugajeff, M. Skanawy-Grigorjewa, J. Russ. Chem. Soc. **47** (1915) 776.

(132) L. Tschugajeff, M. Skanawy-Grigorjewa, A. Posnjak, Z. Anorg. Chem. **148** (1925) 37 – 42.

(133) G. Rouschias, B. L. Shaw, J. Chem. Soc., Chem. Commun. (1970) 183.

(134) G. Rouschias, B. L. Shaw, J. Chem. Soc. A (1971) 2097 – 2104.

(135) A. Burke, A. L. Balch, J. H. Enemark, J. Am. Chem. Soc. **92** (1970) 2555 – 2557.

(136) W. M. Butler, J. H. Enemark, J. Parks, A. L. Balch, Inorg. Chem. **12** (1973) 451 – 457.

(137) C. H. Davies, C. H. Game, M. Green, F. G. A. Stone, J. Chem. Soc., Dalton Trans. (1974) 357 – 363.

(138) M. Wada, S.-I. Kanai, R. Maeda, M. Kinoshita, K. Oguro, Inorg. Chem. **18** (1979) 417 – 421.

(139) J. S. Miller, A. L. Balch, Inorg. Chem. **11** (1972) 2069 – 2074.

(140) B. Crociani, T. Boschi, U. Belluco, Inorg. Chem. **9** (1970) 2021 – 2025.

(141) F. Bonati, G. Minghetti, T. Boschi, B. Crociani, J. Organomet. Chem. **25** (1970) 255 – 260.

(142) G. A. Larkin, R. P. Scott, M. G. H. Wallbridge, J. Organomet. Chem. **37** (1972) C21 – C23.

(143) L. Busetto, A. Palazzi, B. Crociani, U. Belluco, E. M. Badley, B. J. L. Kilby, R. L. Richards, J. Chem. Soc., Dalton Trans. (1972) 1800 – 1805.

(144) R. Zanella, T. Boschi, B. Crociani, U. Belluco, J. Organomet. Chem. **71** (1974) 135 – 143.

(145) K. Bartel, W. P. Fehlhammer, Angew. Chem. **86** (1974) 588 – 589; Angew. Chem. Int. Ed. Engl. **13** (1974) 600 – 601.

(146) E. M. Badley, J. Chatt, R. L. Richards, G. A. Sim, J. Chem. Soc., Chem. Commun. (1969) 1322 – 1323.

(147) E. M. Badley, J. Chatt, R. L. Richards, J. Chem. Soc. A (1971) 21 – 25

(148) E. M. Badley, B. J. L. Kilby, R. L. Richards, J. Organomet. Chem. **27** (1971) C37 – C38.

(149 F. Bonati, G. Minghetti, J. Organomet. Chem. **24** (1970) 251 – 256.

(150) J. Chatt, R. J. Richards, G. H. D. Royston, Inorg. Chim. Acta **6** (1972) 669 – 670.

(151) H. C. Clark, L. E. Manzer, Inorg. Chem. **11** (1972) 503 – 510.

(152) C. J. Cardin, D. J. Cardin, M. F. Lappert, K. W. Muir, J. Organomet. Chem. **60** (1973) C70 – C73.

(153) R. J. Angelici, L. M. Charley, J. Organomet. Chem. **24** (1970) 205 – 209.

(154) J. Miller, A. L. Balch, J. H. Enemark, J. Am. Chem. Soc. **93** (1971) 4613 – 4614.

(155) A. L. Balch, J. Miller, J. Am. Chem. Soc. **94** (1972) 417 – 420.

(156) D. J. Doonan, A. L. Balch, J. Am. Chem. Soc. **95** (1973) 4769 – 4771.

(157) D. J. Doonan, A. L. Balch, Inorg. Chem. **13** (1974) 921 – 927.

(158) P. R. Branson, R. A. Cable, M. Green, M. K. Lloyd, J. Chem. Soc., Dalton Trans. (1976) 12 – 17.

(159) B. V. Johnson, J. E. Shade, J. Organomet. Chem. **179** (1979) 357 – 366.

(160) B. V. Johnson, D. P. Sturtzel, J. E. Shade, J. Organomet. Chem. **154** (1978) 89 – 111.

(161) R. J. Angelici, P. A. Christian, B. D. Dombeck, G. A. Pfeffer, J. Organomet. Chem. **67** (1974) 287 – 294.

(162) W. P. Fehlhammer, A. Mayr, G. Christian, J. Organomet. Chem. **209** (1981) 57 – 67.

(163) J. Chatt, R. L. Richards, G. H. D. Royston, J. Chem. Soc., Dalton Trans. (1973) 1433 – 1439.

(164) G. Minghetti, F. Bonati, J. Organomet. Chem. **54** (1973) C62 – C63.

(165) F. Bonati, G. Minghetti, J. Organomet. Chem. **59** (1973) 403 – 410.

(166) R. Usón, A. Laguna, J. Vicente, J. Garcia, B. Bergareche, J. Organomet. Chem. **173** (1979) 349 – 355.

(167) P. R. Branson, M. Green, J. Chem. Soc., Dalton Trans. (1972) 1303 – 1310.

(168) P. R. Branson, R. A. Cable, M. Green, M. K. Lloyd, J. Chem. Soc., Chem. Commun. (1974) 364 – 365.

(169) B. Crociani, T. Boschi, M. Nicolini, U. Belluco, Inorg. Chem. **11** (1972) 1292 – 1296.

(170) U. Schöllkopf, F. Gerhart, Angew. Chem. **79** (1967) 990; Angew. Chem. Int. Ed. Engl. **6** (1967) 970.

(171) S. Z. Goldberg, R. Eisenberg, J. S. Miller, Inorg. Chem. **16** (1977) 1502 – 1507.

(172) T. Sawai, R. J. Angelici, J. Organomet. Chem. **80** (1974) 91 – 102.

(173) J. A. McCleverty, J. Williams, Transition Met. Chem. **1** (1976) 288 – 294.

(174) F. J. Regina, A. Wojcicki, Inorg. Chem. **19** (1980) 3803 – 3807.

(175) W. Petz, J. Organomet. Chem. **205** (1981) 203 – 210.

(176) R. A. Pickering, R. A. Jacobson, R. J. Angelici, J. Am. Chem. Soc. **103** (1981) 817 – 821.

(177) R. A. Pickering, R. J. Angelici, Inorg. Chem. **20** (1981) 2977 – 2983.

(178) M. H. Quick, R. J. Angelici, J. Organomet. Chem. **160** (1978) 231 – 239.

(179) B. E. Boland-Lussier, R. P. Hughes, Organometallics **1** (1982) 635 – 639.

(180) H. C. Clark, L. E. Manzer, J. Organomet. Chem. **47** (1973) C17 – C20.

(181) G. K. Anderson, R. J. Cross, L. Manojlović-Muir, K. W. Muir, R. A. Wales, J. Chem. Soc., Dalton Trans. (1979) 684 – 689.

(182) D. J. Bates, M. Rosenblum, S. B. Samuels, J. Organomet. Chem. **209** (1981) C55 – C59.

(183) U. Schöllkopf, F. Gerhart, Angew. Chem. **79** (1967) 578; Angew. Chem. Int. Ed. Engl. **6** (1967) 560 – 561.

(184) M. L. H. Green, C. R. Hurley, J. Organomet. Chem. **10** (1967) 188 – 190.

(185) M. L. H. Green, L. C. Mitchard, M. G. Swanwick, J. Chem. Soc. A (1971) 794 – 797.

(186) A. Davison, D. L. Reger, J. Am. Chem. Soc. **94** (1972) 9237 – 9238.

(187) T. Bodnar, A. R. Cutler, J. Organomet. Chem. **213** (1981) C31 – C36.

(188) M. Brookhart, J. R. Tucker, G. R. Husk, J. Am. Chem. Soc. **103** (1981) 979 – 981.

(189) S. J. LaCroce, A. R. Cutler, J. Am. Chem. Soc. **104** (1982) 2312 – 2314.

(190) R. E. Stimson, D. F. Shriver, Inorg. Chem. **19** (1980) 1141 – 1145.

(191) H. Felkin, B. Meunier, C. Pascard, T. Prange, J. Organomet. Chem. **135** (1977) 361 – 372.

(192) T. Bodnar, G. Coman, S. LaCroce, C. Lambert, K. Menard, A. Cutler, J. Am. Chem. Soc. **103** (1981) 2471 – 2472.

(193) A. N. Nesmeyanov, T. N. Sal'nikova, Yu. T. Struchkov, V. G. Andrianov, A. A. Pogrebnyak, L. V. Rybin, M. I. Rybinskaya, J. Organomet. Chem. **117** (1976) C16 – C20

(194) T.-A. Mitsudo, H. Nakanishi, T. Inubushi, I. Morishima, Y. Watanabe, Y. Takegami, J. Chem. Soc., Chem. Commun. (1976) 416 – 417.

(195) T.-A. Mitsudo, Y. Watanabe, H. Nakanishi, I. Morishima, T. Inubushi, Y. Takegami, J. Chem. Soc., Dalton Trans. (1978) 1298 – 1304.

(196) W. P. Fehlhammer, P. Hirschmann, H. Stolzenberg, J. Organomet. Chem. **224** (1982) 165 – 180.

(197) W. K. Dean, W. A. G. Graham, J. Organomet. Chem. **120** (1976) 73 – 86.

(198) M. Wada, N. Asada, K. Oguro, Inorg. Chem. **17** (1978) 2353 – 2357.

(199) M. Wada, S. Kanai, R. Maeda, M. Kinoshita, K. Oguro, Inorg. Chem. **18** (1979) 417 – 421.

(200) W. Tam, G.-Y. Lin, W.-K. Wong, W. A. Kiel, V. K. Wong, J. A. Gladysz, J. Am. Chem. Soc. **104** (1982) 141 – 152.

(201) J. R. Moss, M. Green, F. G. A. Stone, J. Chem. Soc., Dalton Trans. (1973) 975 – 977.

(202) C. P. Casey, R. L. Anderson, J. Am. Chem. Soc. **93** (1971) 3554 – 3555.

(203) S. B. Butts, E. M. Holt, S. H. Strauss, N. W. Alcock, R. E. Stimson, D. F. Shriver, J. Am. Chem. Soc. **101** (1979) 5864 – 5866.

(204) S. B. Butts, S. H. Strauss, E. M. Holt, R. E. Stimson, N. W. Alcock, D. F. Shriver, J. Am. Chem. Soc. **102** (1980) 5093 – 5100.

(205) T. G. Richmond, F. Basolo, D. F. Shriver, Inorg. Chem. **21** (1982) 1272 – 1273.

(206) F. Correa, R. Nakamura, R. E. Stimson, R. L. Burwell, Jr., D. F. Shriver, J. Am. Chem. Soc. **102** (1980) 5112 – 5114.

(207) C. H. Game, M. Green, J. R. Moss, F. G. A. Stone, J. Chem. Soc., Dalton Trans. (1974) 351 – 357.

(208) R. B. King, J. Am. Chem. Soc. **85** (1963) 1922 – 1926.

(209) C. P. Casey, J. Chem. Soc., Chem. Commun. (1970) 1220 – 1221.

(210) F. A. Cotton, C. M. Lukehart, J. Am. Chem. Soc. **93** (1971) 2672 – 2676.

(211) T. Kruck, L. Liebig, Chem. Ber. **106** (1973) 1055 – 1061.

(212) N. A. Bailey, P. L. Chell, A. Mukhopadhyay, H. E. Tabbron, M. J. Winter, J. Chem. Soc., Chem. Commun. (1982) 215 – 217.

(213) J. R. Moss, J. Organomet. Chem. **231** (1982) 229 – 235.

(214) J. P. Collman, R. K. Rothrock, R. G. Finke, E. J. Moore, F. Rose-Munch, Inorg. Chem. **21** (1982) 146 – 156.

(215) M. Green, J. R. Moss, I. W. Nowell, F. G. A. Stone, J. Chem. Soc., Chem. Commun. (1972) 1339 – 1340.

(216) D. H. Bowen, M. Green, D. M. Grove, J. R. Moss, F. G. A. Stone, J. Chem. Soc., Dalton Trans. (1974) 1189 – 1194.

(217) K. R. Grundy, W. R. Roper, J. Organomet. Chem. **216** (1981) 255 – 262.

(218) E. D. Dobrzynski, R. J. Angelici, Inorg. Chem. **14** (1975) 1513 – 1518.

(219) F. B. McCormick, R. J. Angelici, Inorg. Chem. **18** (1979) 1231 – 1235.

(220) F. B. McCormick, R. J. Angelici, Inorg. Chem. **20** (1981) 1111 – 1117.

(221) L. Busetto, A. Palazzi, M. Monari, J. Chem. Soc., Dalton Trans. (1982) 1631 – 1634.

(222) L. Busetto, A. Palazzi, M. Monari, J. Organomet. Chem. **228** (1982) C19 – C20.

(223) T. J. Collins, W. R. Roper, J. Chem. Soc., Chem. Commun. (1976) 1044 – 1045.

(224) K. R. Grundy, R. O. Harris, W. R. Roper, J. Organomet. Chem. **90** (1975) C34 – C36.

(225) F. Faraone, G. Tresoldi, G. A. Loprete, J. Chem. Soc., Dalton Trans. (1979) 933 – 937.

(226) G. Tresoldi, F. Faraone, P. Piraino, J. Chem. Soc., Dalton Trans. (1979) 1053 – 1056.

(227) Y. Yamamoto, H. Yamazaki, Bull. Chem. Soc. Jap. **48** (1975) 3691 – 3697.

(228) P. M. Treichel, W. J. Knebel, Inorg. Chem. **11** (1972) 1285 – 1288.

(229) D. F. Christian, H. C. Clark, R. F. Stepaniak, J. Organomet. Chem. **112** (1976) 227 – 241.

(230) W. P. Fehlhammer, P. Hirschmann, A. Mayr, J. Organomet. Chem. **224** (1982) 153 – 164.

(231) D. F. Christian, G. R. Clark, W. R. Roper, J. M. Waters, K. R. Whittle, J. Chem. Soc., Chem. Commun. (1972) 458 – 459.

(232) D. F. Christian, G. R. Clark, W. R. Roper, J. Organomet. Chem. **81** (1974) C7 – C8.

(233) P. J. Fraser, W. R. Roper, F. G. A. Stone, J. Chem. Soc., Dalton Trans. (1974) 102 – 105.

(234) W. P. Fehlhammer, A. Mayr, M. Ritter, Angew. Chem. **89** (1977) 660 – 665; Angew. Chem. Int. Ed. Engl. **16** (1977) 641 – 642.

(235) W. P. Fehlhammer, G. Christian, A. Mayr, J. Organomet. Chem. **199** (1980) 87 – 98.

(236) C. P. Casey, W. H. Miles, H. Tukada, J. M. O'Connor, J. Am. Chem. Soc. **104** (1982) 3761 – 3762.

(237) K. Oguro, M. Wada, R. Okawara, J. Organomet. Chem. **159** (1978) 417 – 429.

(238) M. Wada, Y. Koyama, J. Organomet. Chem. **201** (1980) 477 – 491.

(239) M. Wada, Y. Koyama, K. Sameshima, J. Organomet. Chem. **209** (1981) 115 – 121.

(240) M. Wada, K. Sameshima, K. Nishiwaki, Y. Kawasaki, J. Chem. Soc., Dalton Trans. (1982) 793 – 797.

(241) R. A. Bell, M. H. Chisholm, D. A. Couch, L. A. Rankel, Inorg. Chem. **16** (1977) 677 – 686.

(242) M. H. Chisholm, D. A. Couch, J. Chem. Soc., Chem. Commun. (1974) 42 – 43.

(243) M. Wada, K. Oguro, Y. Kawasaki, J. Organomet. Chem. **178** (1979) 261 – 271.

(244) M. H. Chisholm, H. C. Clark, Inorg. Chem. **10** (1971) 1711 – 1716.

(245) T. G. Attig, H. C. Clark, Can. J. Chem. **53** (1975) 3466 – 3470.

(246) M. H. Chisholm, H. C. Clark, J. Am. Chem. Soc. **94** (1972) 1532 – 1539.

(247) A. Davison, J. P. Solar, J. Organomet. Chem. **155** (1978) C8 – C12.

(248) E. O. Fischer, P. Stückler, F. R. Kreissl, J. Organomet. Chem. **129** (1977) 197 – 202.

(249) E. O. Fischer, G. Besl, Z. Naturforsch. **34b** (1979) 1186 – 1189.

(250) E. O. Fischer, R. L. Clough, G. Besl, F. R. Kreissl, Angew. Chem. **88** (1976) 584 – 585; Angew. Chem. Int. Ed. Engl. **15** (1976) 543 – 544.

(251) E. O. Fischer, E. W. Meineke, F. R. Kreissl, Chem. Ber. **110** (1977) 1140 – 1147.

(252) E. O. Fischer, W. Schambeck, F. R. Kreissl, J. Organomet. Chem. **169** (1979) C27 – C30.

(253) E. O. Fischer, W. Schambeck, J. Organomet. Chem. **201** (1980) 311 – 318.

(254) O. Orama, U. Schubert, F. R. Kreissl, E. O. Fischer, Z. Naturforsch. **35b** (1980) 82 – 85.

(255) J. Martin-Gil, J. A. K. Howard, R. Navarro, F. G. A. Stone, J. Chem. Soc., Chem. Commun. (1979) 1168 – 1169.

(256) E. O. Fischer, J. Chen, U. Schubert, Z. Naturforsch. **37b** (1982) 1284 – 1288.

(257) A. Motsch, Thesis, Technische Universität München, 1980.

(258) E. O. Fischer, R. L. Clough, P. Stückler, J. Organomet. Chem. **120** (1976) C6 – C8.

(259) E. O. Fischer, A. Frank, Chem. Ber. **111** (1978) 3740 – 3744.

(260) E. O. Fischer, P. Rustemeyer, K. Ackermann, Chem. Ber. **115** (1982) 3851 – 3859.

(261) U. Schubert, D. Neugebauer, P. Hofmann, B. E. R. Schilling, H. Fischer, A. Motsch, Chem. Ber. **114** (1981) 3349 – 3365.

(262) E. O. Fischer, W. Kleine, F. R. Kreissl, Angew. Chem. **88** (1976) 646 – 647; Angew. Chem. Int. Ed. Engl. **15** (1976) 616 – 617.

(263) E. O. Fischer, W. Kleine, F. R. Kreissl, H. Fischer, P. Friedrich, G. Huttner, J. Organomet. Chem. **128** (1977) C49 – C53.

(264) W. Kleine, Thesis, Technische Universität München, 1978.

(265) D. Wittmann, Thesis, Technische Universität München, 1982.

(266) E. O. Fischer, D. Himmelreich, R. Cai, H. Fischer, U. Schubert, B. Zimmer-Gasser, Chem. Ber. **114** (1981) 3209 – 3219.

(267) H. Fischer, E. O. Fischer, R. Cai, D. Himmelreich, Chem. Ber. **116** (1983) 1009 – 1016.

(268) U. Schubert, E. O. Fischer, D. Wittmann, Angew. Chem. **92** (1980) 662 – 663; Angew. Chem. Int. Ed. Engl. **19** (1980) 643 – 644.

(269) E. O. Fischer, R. B. A. Pardy, U. Schubert, J. Organomet. Chem. **181** (1979) 37 – 45.

(270) H. Fischer, E. O. Fischer, R. Cai, Chem. Ber. **115** (1982) 2707 – 2713.

(271) E. O. Fischer, W. Kleine, J. Organomet. Chem. **208** (1981) C27 – C30.

(272) H. Fischer, unpublished results.

(273) A. J. Hartshorn, M. F. Lappert, J. Chem. Soc., Chem. Commun. (1976) 761 – 762.

(274) H. Fischer, R. Märkl, unpublished results.

(275) E. O. Fischer, D. Wittmann, D. Himmelreich, D. Neugebauer, Angew. Chem. **94** (1982) 451 – 452; Angew. Chem. Int. Ed. Engl. **21** (1982) 444 – 445.

(276) H. Fischer, A. Motsch, W. Kleine, Angew. Chem. **90** (1978) 914 – 915; Angew. Chem. Int. Ed. Engl. **17** (1978) 842 – 843.

(277) E. O. Fischer, H. Fischer, U. Schubert, R. B. A. Pardy, Angew. Chem. **91** (1979) 929 – 930; Angew. Chem. Int. Ed. Engl. **18** (1979) 871.

(278) H. Fischer, J. Organomet. Chem. **195** (1980) 55 – 61.

(279) H. Fischer, E. O. Fischer, D. Himmelreich, R. Cai, U. Schubert, K. Ackermann, Chem. Ber. **114** (1981) 3220 – 3232.

(280) E. O. Fischer, D. Himmelreich, R. Cai, Chem. Ber. **115** (1982) 84 – 89.

(281) E. O. Fischer, D. Wittmann, D. Himmelreich, U. Schubert, K. Ackermann, Chem. Ber. **115** (1982) 3141 – 3151.

(282) E. O. Fischer, K. Richter, Angew. Chem. **87** (1975) 359 – 360; Angew. Chem. Int. Ed. Engl. **14** (1975) 345 – 346.

(283) H. Fischer, A. Motsch, U. Schubert, D. Neugebauer, Angew. Chem. **93** (1981) 483 – 487; Angew. Chem. Int. Ed. Engl. **20** (1981) 463 – 464.

(284) E. O. Fischer, P. Rustemeyer, J. Organomet. Chem. **225** (1982) 265 – 277.

(285) J. H. Wengrovius, R. R. Schrock, M. R. Churchill, H. J. Wasserman, J. Am. Chem. Soc. **104** (1982) 1739 – 1740.

(286) S. J. Holmes, R. R. Schrock, J. Am. Chem. Soc. **103** (1981) 4599 – 4600.

(287) G. R. Clark, C. M. Cochrane, W. R. Roper, L. J. Wright, J. Organomet. Chem. **199** (1980) C35 – C38.

(288) H. Le Bozec, A. Gorgues, P. H. Dixneuf, J. Am. Chem. Soc. **100** (1978) 3946 – 3947.

(289) H. Le Bozec, A. Gorgues, P. H. Dixneuf, Inorg. Chem. **20** (1981) 2486 – 2489.

(290) H. Le Bozec, A. Gorgues, P. H. Dixneuf, J. Chem. Soc., Chem. Commun. (1978) 573 – 574.

(291) J. Y. LeMarouille, C. Lelay, A. Benoit, D. Grandjean, D. Touchard, H. Le Bozec, P. Dixneuf, J. Organomet. Chem. **191** (1980) 133 – 142.

(292) C. C. Frazier, N. D. Magnussen, L. N. Osuji, K. O. Parker, Organometallics **1** (1982) 903 – 906.

(293) T. J. Collins, K. R. Grundy, W. R. Roper, S. F. Wong, J. Organomet. Chem. **107** (1976) C37 – C39.

(294) D. H. Farrar, R. O. Harris, A. Walker, J. Organomet. Chem. **124** (1977) 125 – 129.

(295) M. Brookhart, G. O. Nelson, J. Am. Chem. Soc. **99** (1977) 6099 – 6101.

(296) P. W. Jolly, R. Pettit, J. Am. Chem. Soc. **88** (1966) 5044 – 5045.

(297) M. L. H. Green, M. Ishaq, R. N. Whiteley, J. Chem. Soc. A (1967) 1508 – 1515.

(298) P. E. Riley, C. E. Capshew, R. Pettit, R. E. Davis, Inorg. Chem. **17** (1978) 408 – 414.

(299) M. Brookhart, J. R. Tucker, T. C. Flood, J. Jensen, J. Am. Chem. Soc. **102** (1980) 1203 – 1205.

(300) A. R. Cutler, J. Am. Chem. Soc. **101** (1979) 604 – 606.

(301) A. Sanders, L. Cohen, W. P. Giering, D. Kenedy, C. V. Magatti, J. Am. Chem. Soc. **95** (1973) 5430 – 5431.

(302) A. Sanders, T. Bauch, C. V. Magatti, C. Lorenc, W. P. Giering, J. Organomet. Chem. **107** (1976) 359 – 375.

(303) W.-K. Wong, W. Tam, J. A. Gladysz, J. Am. Chem. Soc. **101** (1979) 5440 – 5442.

(304) A. C. Constable, J. A. Gladysz, J. Organomet. Chem. **202** (1981) C21 – C24.

(305) S. E. Kegley, M. Brookhart, G. R. Husk, Organometallics **1** (1982) 760 – 762.

(306) J. A. Marsella, K. Folting, J. C. Huffman, K. G. Caulton, J. Am. Chem. Soc. **103** (1981) 5596 – 5598.

(307) F. B. McCormick, R. J. Angelici, R. A. Pickering, R. E. Wagner, R. A. Jacobson, Inorg. Chem. **20** (1981) 4108 – 4111.

(308) E. Lindner, J. Organomet. Chem. **94** (1975) 229 – 234.

(309) D. L. Reger, M. D. Dukes, J. Organomet. Chem. **153** (1978) 67 – 72.

(310) G. R. Clark, S. V. Hoskins, W. R. Roper, J. Organomet. Chem. **234** (1982) C9 – C12.

(311) N. T. Allison, Y. Kawada, W. M. Jones, J. Am. Chem. Soc. **100** (1978) 5224 – 5226.

(312) P. E. Riley, R. E. Davis, N. T. Allison, W. M. Jones, J. Am. Chem. Soc. **102** (1980) 2458 – 2460.

(313) F. J. Manganiello, M. D. Radcliffe, W. M. Jones, J. Organomet. Chem. **228** (1982) 273 – 279.

(314) R. R. Schrock, J. Am. Chem. Soc. **96** (1974) 6796 – 6797.

(315) R. R. Schrock, J. D. Fellmann, J. Am. Chem. Soc. **100** (1978) 3359 – 3370.

(316) R. R. Schrock, J. Am. Chem. Soc. **98** (1976) 5399 – 5400.

(317) R. A. Andersen, Inorg. Chem. **18** (1979) 3622 – 3623.

(318) J. D. Fellmann, R. R. Schrock, G. A. Rupprecht, J. Am. Chem. Soc. **103** (1981) 5752 – 5758.

(319) J. D. Fellmann, G. A. Rupprecht, C. D. Wood, R. R. Shrock, J. Am. Chem. Soc. **100** (1978) 5964 – 5966.

(320) R. A. Andersen, M. H. Chisholm, J. F. Gibson, W. W. Reichert, I. P. Rothwell, G. Wilkinson, Inorg. Chem. **20** (1981) 3934 – 3936.

(321) R. R. Schrock, J. Am. Chem. Soc. **97** (1975) 6577 – 6578.

(322) S. J. McLain, C. D. Wood, R. R. Schrock, J. Am. Chem. Soc. **99** (1977) 3519 – 3520.

(323) R. R. Schrock, L. W. Messerle, C. D. Wood, L. J. Guggenberger, J. Am. Chem. Soc. **100** (1978) 3793 – 3800.

(324) C. D. Wood, S. J. McLain, R. R. Schrock, J. Am. Chem. Soc. **101** (1979) 3210 – 3222.

(325) L. W. Messerle, P. Jennische, R. R. Schrock, G. Stucky, J. Am. Chem. Soc. **102** (1980) 6744 – 6752.

(326) G. A. Rupprecht, L. W. Messerle, J. D. Fellmann, R. R. Schrock, J. Am. Chem. Soc. **102** (1980) 6236 – 6244.

(327) A. J. Schultz, J. M. Williams, R. R. Schrock, G. A. Rupprecht, J. D. Fellmann, J. Am. Chem. Soc. **101** (1979) 1593 – 1595.

(328) D. N. Clark, R. R. Schrock, J. Am. Chem. Soc. **100** (1978) 6774 – 6776.

(329) J. D. Fellmann, R. R. Schrock, D. D. Traficante, Organometallics **1** (1982) 481 – 484.

(330) P. R. Sharp, D. Astruc, R. R. Schrock, J. Organomet. Chem. **182** (1979) 477 – 488.

(331) R. R. Schrock, P. R. Sharp, J. Am. Chem. Soc. **100** (1978) 2389 – 2399.

(332) M. F. Lappert, C. R. C. Milne, J. Chem. Soc., Chem. Commun. (1978) 925 – 926.

(333) K. Öfele, Angew. Chem. **80** (1968) 1032 – 1033; Angew. Chem. Int. Ed. Engl. **7** (1968) 950.

(334) R. Gompper, E. Bartmann, Angew. Chem. **90** (1978) 490 – 491; Angew. Chem. Int. Ed. Engl. **17** (1978) 456 – 457.

(335) K. Öfele, J. Organomet. Chem. **22** (1970) C9 – C11.

(336) R. Weiss, C. Priesner, Angew. Chem. **90** (1978) 491 – 492; Angew. Chem. Int. Ed. Engl. **17** (1978) 457 – 458.

(337) D. Mansuy, M. Lange, J.-C. Chottard, P. Guerin, P. Morliere, D. Brault, M. Rougee, J. Chem. Soc., Chem. Commun. (1977) 648 – 649.

(338) D. Mansuy, P. Guerin, J.-C. Chottard, J. Organomet. Chem. **171** (1979) 195 – 201.

(339) J.-P. Battioni, D. Mansuy, J.-C. Chottard, Inorg. Chem. **19** (1980) 791 – 792.

(340) P. Guerin, J.-P. Battioni, J.-C. Chottard, D. Mansuy, J. Organomet. Chem. **218** (1981) 201 – 209.

(341) D. Mansuy, J.-P. Battioni, J. Chem. Soc., Chem. Commun. (1982) 638 – 639.

(342) J.-P. Battioni, J.-C. Chottard, D. Mansuy, Inorg. Chem. **21** (1982) 2056 – 2062.

(343) D. Mansuy, M. Lange, J.-C. Chottard, J. Am. Chem. Soc. **100** (1978) 3213 – 3214.

(344) J.-P. Battioni, D. Dupre, D. Mansuy, J. Organomet. Chem. **214** (1981) 303 – 309.

(345) D. Mansuy, J.-P. Lecomte, J.-C. Chottard, J.-F. Bartoli, Inorg. Chem. **20** (1981) 3119 – 3121.

(346) B. Cetinkaya, M. F. Lappert, K. Turner, J. Chem. Soc., Chem. Commun. (1972) 851 – 852.

(347) B. Cetinkaya, M. F. Lappert, G. M. McLaughlin, K. Turner, J. Chem. Soc., Dalton Trans. (1974) 1591 – 1599.

(348) A. J. Hartshorn, M. F. Lappert, K. Turner, J. Chem. Soc., Chem. Commun. (1975) 929 – 930.

(349) A. J. Hartshorn, M. F. Lappert, K. Turner, J. Chem. Soc., Dalton Trans. (1978) 348 – 356.

(350) W. Petz, Angew. Chem. **87** (1975) 288; Angew. Chem. Int. Ed. Engl. **14** (1975) 367 – 368.

(351) P. J. Fraser, W. R. Roper, F. G. A. Stone, J. Chem. Soc., Dalton Trans. (1974) 760 – 764.

(352) M. Green, F. G. A. Stone, M. Underhill, J. Chem. Soc., Dalton Trans. (1975) 939 – 943.

(353) K. Öfele, E. Roos, M. Herberhold, Z. Naturforsch. **31b** (1976) 1070 – 1077.

(354) W. R. Roper, A. H. Wright, J. Organomet. Chem. **233** (1982) C59 – C63.

(355) G. R. Clark, K. Marsden, W. R. Roper, L. J. Wright, J. Am. Chem. Soc. **102** (1980) 1206 – 1207.

(356) G. R. Clark, W. R. Roper, A. H. Wright, J. Organomet. Chem. **236** (1982) C7 – C10.

(357) M. F. Lappert, A. J. Oliver, J. Chem. Soc., Dalton Trans. (1974) 65 – 68.

(358) P. B. Hitchcock, M. F. Lappert, G. M. McLaughlin, A. J. Oliver, J. Chem. Soc., Dalton Trans. (1974) 68 – 74.

(359) H. Brunner, J. Wachter, J. Organomet. Chem. **155** (1978) C29 – C33.

(360) H. Brunner, G. Kerkien, J. Wachter, J. Organomet. Chem. **224** (1982) 295 – 300.

(361) H. Brunner, G. Kerkien, J. Wachter, J. Organomet. Chem. **224** (1982) 301 – 304.

(362) C. W. Fong, G. Wilkinson, J. Chem. Soc., Dalton Trans. (1975) 1100 – 1104.

(363) R. P. Beatty, J. M. Maher, N. J. Cooper, J. Am. Chem. Soc. **103** (1981) 238 – 239.

(364) K. Öfele, J. Organomet. Chem. **12** (1968) P42 – P43.

(365) K. Öfele, Angew. Chem. **81** (1969) 936; Angew. Chem. Int. Ed. Engl. **8** (1969) 916.

(366) C. G. Kreiter, K. Öfele, G. W. Wieser, Chem. Ber. **109** (1976) 1749 – 1758.

(367) H.-W. Wanzlick, H.-J. Schönherr, Angew. Chem. **80** (1968) 154; Angew. Chem. Int. Ed. Engl. **7** (1968) 141 – 142.

(368) H.-J. Schönherr, H.-W. Wanzlick, Chem. Ber. **103** (1970) 1037 – 1046.

(369) R. J. Sundberg, R. F. Bryan, I. F. Taylor, Jr., H. Taube, J. Am. Chem. Soc. **96** (1974) 381 – 392.

(370) M. J. Clarke, H. Taube, J. Am. Chem. Soc. **97** (1975) 1397 – 1403.

(371) C. W. Rees, E. von Angerer, J. Chem. Soc., Chem. Commun. (1972) 420.

(372) W. Petz, J. Organomet. Chem. **172** (1979) 415 – 420.

(373) G. Dettlaf, P. Hübener, J. Klimes, E. Weiss, J. Organomet. Chem. **229** (1982) 63 – 75.

(374) J. Daub, U. Erhardt, J. Kappler, V. Trautz, J. Organomet. Chem. **69** (1974) 423 – 427.

(375) J. Daub, J. Kappler, J. Organomet. Chem. **80** (1974) C5 – C8.

(376) H. Hoberg, G. Burkhart, C. Krüger, Y.-H. Tsay, J. Organomet. Chem. **222** (1981) 343 – 352.

(377) W. A. Herrmann, Angew. Chem. **86** (1974) 556 – 557; Angew. Chem. Int. Ed. Engl. **13** (1974) 599.

(378) W. A. Herrmann, Chem. Ber. **108** (1975) 486 – 499.

(379) W. A. Herrmann, Chem. Ber. **108** (1975) 3412 – 3413.

(380) W. A. Herrmann, J. Plank, M. L. Ziegler, K. Weidenhammer, Angew. Chem. **90** (1978) 817 – 818; Angew. Chem. Int. Ed. Engl. **17** (1978) 777 – 778.

(381) P. R. Sharp, R. R. Schrock, J. Organomet. Chem. **171** (1979) 43 – 51.

(382) M. F. Lappert, D. B. Shaw, J. Chem. Soc., Chem. Commun. (1978) 146 – 147.

(383) H. G. Raubenheimer, H. E. Swanepoel, J. Organomet. Chem. **141** (1977) C21 – C22.

(384) D. J. Cardin, B. Cetinkaya, E. Cetinkaya, M. F. Lappert, L. J. Manojlović-Muir, K. W. Muir, J. Organomet. Chem. **44** (1972) C59 – C62.

(385) D. J. Cardin, M. J. Doyle, M. F. Lappert, J. Chem. Soc., Chem. Commun. (1972) 927 – 928.

(386) M. F. Lappert, P. L. Pye, J. Chem. Soc., Dalton Trans. (1977) 2172 – 2180.

(387) M. F. Lappert, P. L. Pye, J. Chem. Soc., Dalton Trans. (1978) 837 – 844.

(388) B. Cetinkaya, P. Dixneuf, M. F. Lappert, J. Chem. Soc., Dalton Trans. (1974) 1827 – 1833.

(389) P. B. Hitchcock, M. F. Lappert, P. L. Pye, J. Chem. Soc., Dalton Trans. (1978) 826 – 836.

(390) B. Cetinkaya, P. Dixneuf, M. F. Lappert, J. Chem. Soc., Chem. Commun. (1973) 206.

(391) D. J. Cardin, B. Cetinkaya, E. Cetinkaya, M. F. Lappert, J. Chem. Soc., Dalton Trans. (1973) 514 – 522.

(392) P. B. Hitchcock, M. F. Lappert, P. Terreros, K. P. Wainwright, J. Chem. Soc., Chem. Commun. (1980) 1180 – 1181.

(393) P. B. Hitchcock, M. F. Lappert, P. L. Pye, S. Thomas, J. Chem. Soc., Dalton Trans. (1979) 1929 – 1942.

(394) K. Hiraki, K. Sugino, J. Organomet. Chem. **201** (1980) 469 – 475.

(395) K. Hiraki, M. Onishi, K. Ohnuma, K. Sugino, J. Organomet. Chem. **216** (1981) 413 – 419.

(396) B. H. Edwards, M. D. Rausch, J. Organomet. Chem. **210** (1981) 91 – 96.

(397) R. Aumann, E. O. Fischer, Chem. Ber. **114** (1981) 1853 – 1857.

(398) K. Öfele, M. Herberhold, Angew. Chem. **82** (1970) 775 – 776; Angew. Chem. Int. Ed. Engl. **9** (1970) 739 – 740.

(399) C. P. Casey, R. L. Anderson, J. Chem. Soc., Chem. Commun. (1975) 895 – 896.

(400) M. Berry, J. Martin-Gil, J. A. K. Howard, F. G. A. Stone, J. Chem. Soc., Dalton Trans. (1980) 1625 – 1629.

(401) R. Schrock, S. Rocklage, J. Wengrovius, G. Rupprecht, J. Fellmann, J. Mol. Catal. **8** (1980) 73 – 83.

(402) J. H. Wengrovius, R. R. Schrock, M. R. Churchill, J. R. Missert, W. J. Youngs, J. Am. Chem. Soc. **102** (1980) 4515 – 4516.

(403) J. H. Wengrovius, R. R. Schrock, Organometallics **1** (1982) 148 – 155.

(404) U. Klabunde, E. O. Fischer, J. Am. Chem. Soc. **89** (1967) 7141 – 7142.

(405) E. O. Fischer, H.-J. Kollmeier, Chem. Ber. **104** (1971) 1339 – 1346.

(406) J. A. Connor, E. O. Fischer, J. Chem. Soc. A (1969) 578 – 584.

(407) E. O. Fischer, B. Heckl, H. Werner, J. Organomet. Chem. **28** (1971) 359 – 365.

(408) E. O. Fischer, M. Leupold, Chem. Ber. **105** (1972) 599 – 608.

(409) E. O. Fischer, S. Fontana, J. Organomet. Chem. **40** (1972) 367 – 372.

(410) C. P. Casey, A. J. Shusterman, N. W. Vollendorf, K. J. Haller, J. Am. Chem. Soc. **104** (1982) 2417 – 2423.

(411) K. Weiss, E. O. Fischer, Chem. Ber. **106** (1973) 1277 – 1284.

(412) K. Weiss, E. O. Fischer, Chem. Ber. **109** (1976) 1868 – 1886.

(413) M. H. Chisholm, H. C. Clark, W. S. Johns, J. E. H. Ward, K. Yasufuku, Inorg. Chem. **14** (1975) 900 – 905.

(414) E. O. Fischer, H. J. Kalder, J. Organomet. Chem. **131** (1977) 57 – 64.

(415) L. Knauss, E. O. Fischer, Chem. Ber. **103** (1970) 3744 – 3751.

(416) L. Knauss, E. O. Fischer, J. Organomet. Chem. **31** (1971) C68 – C70.

(417) E. O. Fischer, R. Aumann, Chem. Ber. **101** (1968) 963 – 968.

(418) E. O. Fischer, E. Louis, W. Bathelt, J. Organomet. Chem. **20** (1969) 147 – 152.

(419) F. R. Kreissl, E. O. Fischer, C. G. Kreiter, H. Fischer, Chem. Ber. **106** (1973) 1262 – 1276.

(420) E. O. Fischer, M. Leupold, C. G. Kreiter, J. Müller, Chem. Ber. **105** (1972) 150 – 161.

(421) C. T. Lam, C. V. Senoff, J. E. H. Ward, J. Organomet. Chem. **70** (1974) 273 – 281.

(422) E. O. Fischer, G. Kreis, F. R. Kreissl, C. G. Kreiter, J. Müller, Chem. Ber. **106** (1973) 3910 – 3919.

(423) E. O. Fischer, V. Kiener, Angew. Chem. **79** (1967) 982 – 983; Angew. Chem. Int. Ed. Engl. **6** (1967) 961.

(424) C. G. Kreiter, Angew. Chem. **80** (1968) 402; Angew. Chem. Int. Ed. Engl. **7** (1968) 390 – 391.

(425) C. P. Casey, A. J. Shusterman, J. Mol. Catal. **8** (1980) 1 – 13.

(426) E. O. Fischer, T. Selmayr, F. R. Kreissl, Monatsh. Chem. **108** (1977) 759 – 765.

(427) E. Moser, E. O. Fischer, J. Organomet. Chem. **15** (1968) 147 – 155.

(428) C. P. Casey, W. R. Brunsvold, J. Organomet. Chem. **77** (1974) 345 – 352.

(429) E. O. Fischer, W. Held, J. Organomet. Chem. **112** (1976) C59 – C62.

(430) C. P. Casey, R. L. Anderson, J. Am. Chem. Soc. **96** (1974) 1230 – 1231.

(431) C. P. Casey, R. L. Anderson, J. Organomet. Chem. **73** (1974) C28 – C30.

(432) C. P. Casey, W. R. Brunsvold, D. M. Scheck, Inorg. Chem. **16** (1977) 3059 – 3063.

(433) M. Rudler-Chauvin, H. Rudler, J. Organomet. Chem. **212** (1981) 203 – 210.

(434) C. P. Casey, W. R. Brunsvold, J. Organomet. Chem. **102** (1975) 175 – 183.

(435) C. P. Casey, R. A. Boggs, D. F. Marten, J. C. Calabrese, J. Chem. Soc., Chem. Commun. (1973) 243 – 244.

(436) J. Levisalles, H. Rudler, F. Dahan, Y. Jeannin, J. Organomet. Chem. **187** (1980) 233 – 242.

(437) C. P. Casey, S. W. Polichnowski, H. E. Tuinstra, L. D. Albin, J. C. Calabrese, Inorg. Chem. **17** (1978) 3045 – 3049.

(438) C. P. Casey, S. W. Polichnowski, J. Am. Chem. Soc. **99** (1977) 6097 – 6099.

(439) C. P. Casey, S. W. Polichnowski, A. J. Shusterman, C. R. Jones, J. Am. Chem. Soc. **101** (1979) 7282 – 7292.

(440) L. Knauss, E. O. Fischer, J. Organomet. Chem. **31** (1971) C71 – C73.

(441) E. O. Fischer, K. Weiss, C. G. Kreiter, Chem. Ber. **107** (1974) 3554 – 3561.

(442) E. O. Fischer, K. Weiss, Chem. Ber. **109** (1976) 1128 – 1139.

(443) K. Weiss, E. O. Fischer, J. Müller, Chem. Ber. **107** (1974) 3548 – 3553.

(444) K. H. Dötz, J. Organomet. Chem. **118** (1976) C13 – C15.

(445) F. B. McCormick, R. J. Angelici, Inorg. Chem. **20** (1981) 1118 – 1123.

(446) D. Mansuy, M. Lange, J. C. Chottard, J. F. Bartoli, Tetrahedron Letters (1978) 3027 – 3030.

(447) E. O. Fischer, W. Kleine, F. R. Kreissl, J. Organomet. Chem. **107** (1976) C23 – C25.

(448) Y. Ito, T. Hirao, T. Saegusa, J. Organomet. Chem. **131** (1977) 121 – 131.

(449) K. H. Dötz, C. G. Kreiter, J. Organomet. Chem. **99** (1975) 309 – 314.

(450) K. H. Dötz, Chem. Ber. **110** (1977) 78 – 85.

(451) K. H. Dötz, I. Pruskil, J. Organomet. Chem. **132** (1977) 115 – 120.

(452) K. H. Dötz, I. Pruskil, Chem. Ber. **111** (1978) 2059 – 2063.

(453) H. Fischer, K. H. Dötz, Chem. Ber. **113** (1980) 193 – 202.

(454) K. H. Dötz, C. G. Kreiter, Chem. Ber. **109** (1976) 2026 – 2032.

(455) H. Fischer, U. Schubert, Angew. Chem. **93** (1981) 482 – 483; Angew. Chem. Int. Ed. Engl. **20** (1981) 461 – 463.

(456) H. Fischer, U. Schubert, R. Märkl, Chem. Ber. **114** (1981) 3412 – 3420.

(457) H. Fischer, S. Zeuner, J. Organomet. Chem., in the press.

(458) H. Fischer, S. Zeuner, unpublished results.

(459) H. Fischer, J. Organomet. Chem. **197** (1980) 303 – 313.

(460) R. Aumann, E. O. Fischer, Chem. Ber. **101** (1968) 954 – 962.

(461) C. G. Kreiter, R. Aumann, Chem. Ber. **111** (1978) 1223 – 1227.

(462) K. H. Dötz, Angew. Chem. **87** (1975) 672 – 673; Angew. Chem. Int. Ed. Engl. **14** (1975) 644 – 645.

(463) J. Levisalles, H. Rudler, D. Villemin, J. Organomet. Chem. **146** (1978) 259 – 265.

(464) J. Levisalles, H. Rudler, D. Villemin, J. Daran, Y. Jeannin, L. Martin, J. Organomet. Chem. **155** (1978) C1 – C4.

(465) K. H. Dötz, I. Pruskil, Chem. Ber. **114** (1981) 1980 – 1982.

(466) U. Schubert, E. O. Fischer, Justus Liebigs Ann. Chem. (1975) 393 – 400.

(467) E. O. Fischer, U. Schubert, W. Kalbfus, C. G. Kreiter, Z. Anorg. Allg. Chem. **416** (1975) 135 – 151.

(468) M. F. Lappert, J. J. MacQuitty, P. L. Pye, J. Chem. Soc., Dalton Trans. (1981) 1583 – 1592.

(469) J. D. Fellmann, H. W. Turner, R. R. Schrock, J. Am. Chem. Soc. **102** (1980) 6608 – 6609.

(470) H. W. Turner, J. D. Fellmann, S. M. Rocklage, R. R. Schrock, M. R. Churchill, H. J. Wasserman, J. Am. Chem. Soc. **102** (1980) 7809 – 7811.

(471) D. J. Cardin, M. J. Doyle, M. F. Lappert, J. Organomet. Chem. **65** (1974) C13 – C16.

(472) B. Cetinkaya, E. Cetinkaya, M. F. Lappert, J. Chem. Soc., Dalton Trans. (1973) 906 – 912.

(473) P. Braunstein, E. Keller, H. Vahrenkamp, J. Organomet. Chem. **165** (1979) 233 – 242.

(474) R. D. Rieke, H. Kojima, K. Öfele, J. Am. Chem. Soc. **98** (1976) 6735 – 6737.

(475) R. D. Rieke, H. Kojima, K. Öfele, Angew. Chem. **92** (1980) 550 – 551; Angew. Chem. Int. Ed. Engl. **19** (1980) 538 – 540.

(476) A. L. Balch, J. Organomet. Chem. **37** (1972) C19 – C20.

Spectroscopic Properties of Transition Metal Carbene Complexes

By Helmut Fischer and Fritz R. Kreissl

The IR spectra of pentacarbonyl(carbene) complexes generally show three absorptions in the ν(CO) region, as expected for C_{4v} symmetry. When aryl-carbene complexes are employed, however, the E band is split in most cases and the formally IR-forbidden B_1-absorption is then observed. Compared to CO, the carbene ligand has a higher σ-donor/π-acceptor ratio. In $(CO)_5Cr$-$[C(OMe)C_6H_4X]$ complexes, a linear relationship between the CO force constant k_1 **(trans)** and the Hammett σ-constants for the ring substituents X is observed as long as electron-releasing groups are involved. The effect of electron-withdrawing groups is somewhat smaller than expected, probably due to increased electron donation from the OMe group. The linear correlation between k_1 and the Hammett σ-constants, both for *para*- and *meta*-substituted complexes, indicates that there is more or less free rotation about the carbene carbon-phenyl ring bond (1,2). In $(CO)_5M[C(R)X]$ complexes (M = Cr, W; R = aryl, alkyl), the σ-donor/π-acceptor ratio increases according to the sequence X = OMe < SMe < SeMe < NHMe (3). Analogously, the dipole moment increases in the same sequence (3).

Carbene complexes show a remarkable variation in colour. The first absorption band in their electronic spectra can be assigned to a $\pi - \pi^*$ transition. This band undergoes a bathochromic shift when the heteroatom of the stabilizing group X is changed in the order N, O, S (4).

The proton NMR spectrum (CD_3COCD_3) of pentacarbonyl(methyl-methoxycarbene)chromium at 35 °C exhibits two sharp singlets having the same intensity at 4.68 and 3.03 ppm (5). When the sample is cooled, the signals broaden, though not to the same extent. Finally, at -40 °C each signal splits into two signals because the complex exists in two isomeric forms: OCH_3 at 4.89 and 4.37 ppm and CCH_3 at 3.00 and 3.17 ppm. The isomerism of $(CO)_5Cr[C(OCH_3)CH_3]$ results from differing orientations of the two methyl groups in the planar carbene ligand. Although at low temperatures

$$(CO)_5Cr{=}C\overset{\overset{\displaystyle H_3C}{\diagdown}O}{\underset{\underset{\displaystyle CH_3}{}}{}} \rightleftharpoons (CO)_5Cr{=}C\overset{\overset{\displaystyle O{-}CH_3}{}}{\underset{\underset{\displaystyle CH_3}{}}{}}$$

"trans"   "cis"

separate signals from the *cis* and *trans* isomers can be observed, at higher temperatures the rapid conversion of one isomer into the other results in an averaged spectrum (3).

This temperature dependence of the signals enables a calculation of the free energy of activation for this *cis-trans* isomerisation to be made. For $(CO)_5Cr[C(OCH_3)R]$, with R = CH_3, C_2H_5, n-C_3H_7, $\Delta G^{\#}$ is of the order of

13.5 kcal/mol; for R = C_6H_5 a value of 11.9 kcal/mol has been reported. Kreiter (6) demonstrated that in carbene complexes of the Fischer type, the rotation barrier depends on the metal, the ligands and on the nature of the substituents attached to the carbene ligand. An increasing σ-donor/π-acceptor ratio of the ligands causes a decrease in the carbene carbon-oxygen π-bonding. In the alkoxy-, thio-, seleno- and aminocarbene complexes $(CO)_5M$-[C(R)XR'] (X = O, S, Se, NH), the isomerisation barriers range from about 13.5 kcal/mol for X = O to more than 25 kcal/mol when X = NH.

Carbene complexes have distinct ^{13}C NMR spectra as revealed by the earliest measurements on various pentacarbonylcarbene complexes of chromium and tungsten (7). The available shift data for carbene carbon atoms presently extend over a range of more than 200 ppm, starting with a value for the aminocarbene complex at about 200 ppm and going up to more than 400 ppm for some silicon carbene complexes (8).

Electron impact mass spectra of many carbonylcarbene complexes $(CO)_5M[C(X)Y]$ have indicated in most instances the presence of the molecular ion, followed by a successive loss of the carbonyl ligands. The further fragmentation depends strongly upon the metal, the ligands and the nature of the substituents X and Y. This has been demonstrated by J. Müller for a wide variety of carbene complexes of the Fischer type (9 – 12), although a standardized fragmentation scheme cannot yet be given.

The ionisation potentials of carbene complexes permit an estimation of the charge transfer from the carbene ligand to the metal. For $(CO)_5Cr[C(XR)R']$, with X = O, S, NH and R and R' = alkyl or aryl groups, very low IPs (7.02 to 7.46 eV) have been reported, indicating that the first electron is ionized from a filled d orbital of the metal (9).

References

(1) E. O. Fischer, C. G. Kreiter, H. J. Kollmeier, J. Müller, R. D. Fischer, J. Organomet. Chem. **28** (1971) 237 – 258.

(2) E. O. Fischer, H. J. Kollmeier, Chem. Ber. **104** (1971) 1339 – 1346.

(3) C. G. Kreiter, E. O. Fischer, in XXIIIrd International Congress of Pure and Applied Chemistry. Butterworths, London 1971, pp. 151 – 168.

(4) E. O. Fischer, Pure Appl. Chem. **24** (1970) 407 – 423.

(5) E. O. Fischer, A. Maasböl, Chem. Ber. **100** (1967) 2445 – 2456.

(6) C. G. Kreiter, Habilitationsschrift, Technische Universität München 1971.

(7) C. G. Kreiter, V. Formaček, Angew. Chem. **84** (1972) 155 – 156; Angew. Chem. Int. Ed. Engl. **11** (1972) 141.

(8) E. O. Fischer, H. Hollfelder, P. Friedrich, F. R. Kreissl, G. Huttner, Chem. Ber. **110** (1977) 3467 – 3480.

(9) J. Müller, J. A. Connor, Chem. Ber. **102** (1969) 1148 – 1160.

(10) J. Müller, Habilitationsschrift, Technische Universität München 1969.

(11) E. O. Fischer, M. Leupold, C. G. Kreiter, J. Müller, Chem. Ber. **105** (1972) 150 – 161.

(12) J. Müller, K. Öfele, G. Krebs, J. Organomet. Chem. **82** (1974) 383 – 395.

Solid State Structures of Carbene Complexes

By Ulrich Schubert

1 Introduction

This article is concerned with the results of about 150 structure analyses of carbene complexes, determined by X-ray and neutron diffraction, which are summarized in Table 1. Our literature search extends up to the end of September 1982. In many cases structure determinations have been carried out mainly or exclusively to establish or confirm the constitution of a specific complex. However, the results of any structure analysis contain much information about the bonding present, if one bears in mind that structural parameters reflect the electronic structure. The structure of a given compound in the crystalline state, *i.e.* its constitution, configuration and conformation, or, in other words, the sum of its bond lengths, bond angles and torsional angles, is determined by the interactive influence of electronic and steric intra- and intermolecular forces. In order to extract information about bonding, one has to consider how the various interactions influence the observed structural parameters. The first way to do this is to compare changes in the structural parameters of a series of closely related compounds, and the second is to use the results of chemical theory to evaluate structural data.

A combination of both will be featured in this article. Based on the chapter on the electronic structures of carbene complexes by Peter Hofmann, structural results will be discussed for some types of carbene complexes, deemed important and characteristic by the autor. The main emphasis is on transition metal carbonyl carbene complexes not only to honor Professor E. O. Fischer, but also because such complexes have been studied extensively and more systematically than others.

2 Octahedral $(CO)_5$M-Carbene and $L(CO)_4$M-Carbene Complexes

For several reasons $(CO)_5$Cr-carbene complexes are especially well suited for systematic investigations of the electronic and steric influences on the bonding parameters of the carbene ligand:

(1) Carbene complexes with a wide variety of substituents at the carbene carbon can be and have been prepared.

(2) Because of the high symmetry of the $(CO)_5$Cr fragment and the favorable steric properties of CO ligands, steric interactions between substituents at the carbene carbon and ligands at the metal are usually not a crucial issue.

(3) X-ray structures of first row transition metal compounds are better suited to detailed discussion, because the standard deviations of bond lengths and angles are inherently lower than in analogous complexes based on heavier metals.

In any carbene complex the three groups attached to the sp^2 hybridized carbene carbon atom (the metal complex moiety, L_nM, and the two "organic" substituents, X and Y) compete with each other for π-bonding to the formally empty p orbital on the carbene carbon. The degree of π-bonding of any of these groups will depend not only on its own π-donating ability, but also on the π-donor properties of the other groups (31). The proper description of bonding lies somewhere between (A), (B) and (C) in valence bond terms (Scheme 1).

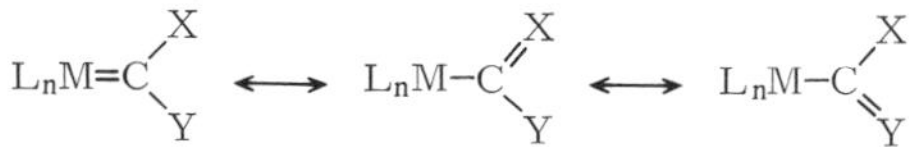

$$L_nM=C\begin{smallmatrix} X \\ \\ Y \end{smallmatrix} \longleftrightarrow L_nM-C\begin{smallmatrix} X \\ \\ Y \end{smallmatrix} \longleftrightarrow L_nM-C\begin{smallmatrix} X \\ \\ Y \end{smallmatrix}$$

$$\text{(A)} \qquad\qquad \text{(B)} \qquad\qquad \text{(C)}$$

Scheme 1.

In closely related compounds, bond lengths can roughly be correlated with bond orders. Accordingly, in the carbene complexes $(CO)_5CrC(X)Y$, the C(carbene)-Cr, -X and -Y distances should provide information about the bonding situation around the carbene carbon.

Inspection of Table 1 reveals that the Cr-C(carbene) bond lengths in such complexes are strongly influenced by the nature of the organic substituents attached. A word of caution on the accuracy of structural data is in order at this point. A reader not familiar with structural work might uncritically accept a certain value for a bond length or angle as a fixed quantity. However, each such number has an associated standard deviation as a measure of its reliability. As can be seen from Table 1, standard deviations for structural parameters may vary considerably, even within a series of closely similar complexes. Consequently, individual parameters from different structures may not be significantly different from a statistical point of view, even though their numerical values differ considerably. For example, the Cr-C(carbene) bond lengths in $(CO)_5CrC(OEt)C \equiv CPh$ [200(2)pm] (6) and $(CO)_5CrC(OH)$-Ph [205(1)pm] (8) are *not* significantly different. Nevertheless, trends can be established by comparison of a series of structural data from the same class of compounds.

The shortest Cr-C(carbene) distances (197 – 205 pm) are found when the $(CO)_5Cr$ fragment is competing *only* with a thio or oxy function (alkoxy, hydroxy or siloxy group), corresponding to a bonding situation between (A) and (B) (X = OR or SR). Replacement of these substituents by an amino group – which is a much better π-donor – reduces back bonding from the metal thereby lengthening the Cr-C(carbene) distance (208 – 215 pm). This means that the bonding is shifted closer towards (B) (X = NR_2). The high π-donating ability of an amino group is paralleled only by that of a phosphorus ylide function, -CR = PR'_3, as judged by the long Cr-C(carbene) distance of 213.7(7)pm in $(CO)_5CrC(OSiMe_3)CHPMe_3$ (11) (Fig. 1).

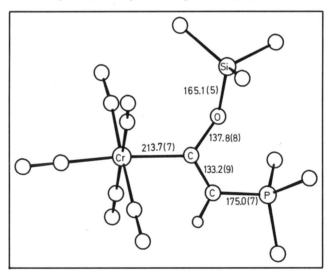

Figure 1: The "ylide"-carbene complex $(CO)_5CrC(OSiMe_3)CHPMe_3$ (11).

If a thio, oxy or amino substituent is engaged in π-bonding with the carbene carbon, the C(carbene)-S, -O or -N bond lengths should also reflect the degree of π-bonding between both atoms and should gradually become shorter as the π-bonding increases. In complexes in which these substituents do not have to compete with another π-donating organic substituent, the observed bond lengths (C-O 132 – 133 pm, C-N 128 – 133 pm, C-S 167 – 169 pm) are indeed shorter than the theoretical single bond distances [C(sp^2)-O 141 pm, C(sp^2)-N 145 pm, C(sp^2)-S 179 pm]. However, in a series of complexes, variations in the distance between the light atoms are usually small and in most cases get lost in the standard deviations. Therefore, one cannot rely on such differences alone to evaluate bonding modifications. For instance, the C(carbene)-O distance in $(CO)_5CrC(OSiMe_3)CHPMe_3$ [137.8(8)pm] (11) (Fig. 1) is longer than that in $(CO)_5CrC[OSi(SiMe_3)_3]C_4H_3O$ [132.1(9)pm] (10) (Fig. 2) and in $(CO)_5Cr$-alkoxycarbene complexes (132 – 133 pm). This suggests that in the former complex the siloxy substituent cannot (or must not) compete with the powerful ylide substituent for π-bonding. Since the observed difference between the

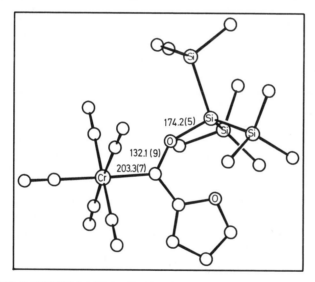

Figure 2: $(CO)_5CrC[OSi(SiMe_3)_3](1\text{-furyl})$ (10). The close approach of the furyl oxygen atom greatly distorts the coordination around the central silicon atom. In solution there is no free rotation about the O-Si bond.

C-O distances in both siloxycarbene complexes lies at the limit of significance, further evidence needs to be gained. As a matter of fact, this is contained in the other bond lengths around the carbene carbon (see Fig. 1), all of which are consistent with the interpretation that the vinyl-like form shown in Scheme 2 is the main contributor in a valence bond description of bonding in this complex.

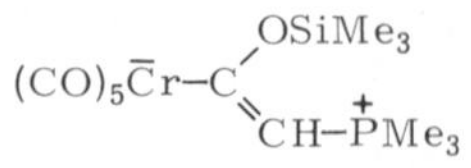

$$(CO)_5\bar{C}r-C\overset{\displaystyle OSiMe_3}{\underset{\displaystyle CH-\overset{+}{P}Me_3}{}}$$

Scheme 2.

As the carbene carbon is increasingly stabilized by π-bonding with the organic substituents, the carbene ligand as a whole becomes a weaker π-acceptor towards the metal. This results in increased back bonding to the carbonyl ligands, particularly the *trans*-CO ligand. However, significant shortening of the *trans*-Cr-C(CO) bond length is normally observed only if metal-C(carbene) back bonding is strongly decreased. Thus, in $(CO)_5CrC(O-SiMe_3)CHPMe_3$ (11) (Fig. 1) the mean Cr-C(CO,*cis*) distance is 189.1(7) pm whereas Cr-C(CO,*trans*) is 185.9(8). *Trans* effects of similar magnitude are reported for the aminocarbene complexes $(CO)_5CrC(NR_2)Y$. In the alkoxycarbene complexes $(CO)_5CrC(OR)Y$, where the carbene ligand is a stronger π-acceptor, shortening of the *trans*-Cr-C(CO) bond is less pronounced. Significant differences in the Cr-C(CO) distances in this case can be observed only if the standard deviations are very low. For $(CO)_5CrC(OEt)Me$ a high-precision structure determination has been performed (5) which shows the Cr-C(CO, *trans*) bond [189.4(1)pm] to be 1.4 pm shorter than the mean Cr-C(CO,*cis*) bond.

Additional information about participation of a carbene substituent in π-bonding can be gleaned from its orientation relative to the carbene plane (defined by the three σ-bonds between the carbene carbon and its substituents). Prerequisite for π-bonding is an orientation in which the orbitals concerned are (nearly) parallel to each other. If the occupied p orbital of sp^2 hybridized thio, alkoxy or amino substituents interacts with the empty p orbital at the carbene carbon, the plane of the (planar) amino group or the C(carbene)-O(or S)-C(alkyl) plane needs to be roughly coplanar with the carbene plane (see Fig. 3 for both the amino and the ethoxy group). If this is not the case, one can conclude that there will be no π-bond or at least no strong π-bond. Thus, in $(CO)_5CrC(OSiMe_3)CHPMe_3$ (11) (Fig. 1) only the ylidic C-P bond, but not the silicon atom, is coplanar with the carbene plane. Apart from the bond lengths, this fact provides additional evidence that in this complex the siloxy group plays no major role in the electronic stabilization of the carbene carbon.

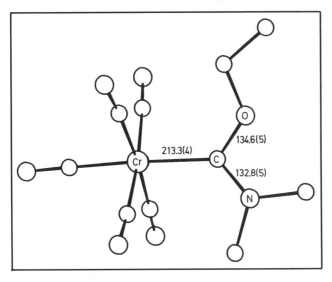

Figure 3: (CO)₅CrC(OEt)NMe₂ (12). All atoms of the carbene ligand are coplanar. The carbene plane bisects the Cr(CO)₅ fragment.

As a result of π-bonding between the carbene carbon and organic substituents *E/Z* isomers can be observed (Scheme 3). These arise from the differing arrangements of substituents along the partial C(carbene)-X double bond.

$$L_nM \!=\! C \!\!\!\overset{\displaystyle X-R}{\underset{\displaystyle R'}{}} \qquad L_nM \!=\! C \!\!\!\overset{\displaystyle \overset{R}{|}\; X}{\underset{\displaystyle R'}{}} \qquad (X = O, S, Se, NH)$$

E isomer   Z isomer

Scheme 3.

¹H-NMR spectroscopy has shown that the energies required for the thermal isomerization of (CO)₅CrC(OMe)R (R = alkyl or aryl) are typically in the range 11.5-14 kcal/mole (147). For aminocarbene complexes they are large enough [> 25 kcal/mole for (CO)₅CrC(NMe₂)Me (148)] to permit simultaneous observation of both isomers at room temperature. Which isomer of a given complex will be energetically favored is hard to predict and depends on the steric properties of the groups R and R′ and on the *cis* ligands attached to the metal. In the absence of steric influences, the *Z* isomer has been shown to prevail in (CO)₅MC(OR)R′ (M = Cr,Mo,W) (148). In reality, however, there will be steric influences between R and *cis*-CO ligands in the *Z* isomer and bet-

ween R and R' in the *E* isomer. Because of these interactions bulky substituents R tend to reinforce the *E* isomer [as in $(CO)_5CrC[OSi(SiMe_3)_3]C_4H_3O$ (10) (Fig. 2), $(CO)_5CrC(OSiMe_3)CHPMe_3$ (11) (Fig. 1), $(CO)_5CrC(SePh)$-NEt_2 (21) and the three complexes $(CO)_5CrC(SR)R'$ (23, 24), listed in Table 1]. In solution, the *E* isomer of $(CO)_5CrC(OSiMe_3)CHPMe_3$ can be photochemically converted to the *Z* isomer, which thermally re-isomerizes rapidly (11). In crystal structure analyses of methoxy- and ethoxy-substituted octahedral carbene complexes, *Z* isomers have been observed about twice as frequently as *E* isomers.

The mutual steric influence of the carbene substituents is also reflected in the bond angles. The mean metal-C(carbene)-O bond angle in alkoxy-substituted octahedral monocarbene complexes is 118° (based on six cases) for *E* isomers but 132° (ten cases) for *Z* isomers and is not strongly influenced by the second organic substituent R'. Steric interactions between group R and *cis* ligands seem to be responsible for the opening of the bond angle in the *Z* isomer since the same mean angle of 132° (based on eleven cases) is observed in analogous dialkylaminocarbene complexes. A pair of complexes has been investigated which differs only in the position of the alkoxy group. In $(CO)_5Re(CO)_4ReC(OMe)SiPh_3$ (Fig. 4) the *Z* isomer [Re-C-O 133(1)°] is found whereas in $(CO)_5Re(CO)_4ReC(OEt)SiPh_3$ the *E* isomer [Re-C-O 120(3)°] is produced (107), these bond angles being in good agreement with the mean values. Bond lengths are not (significantly) affected by the isomerism.

The above numbers reveal that bond angles at the carbene carbon can easily be adjusted to meet steric requirements. In contrast to bond lengths, they reflect more or less the steric situation within the carbene plane (see however, Chapter 6).

An unusual kind of π-bonding to an organic substituent on the carbene carbon is observed in $(CO)_5CrC(NEt_2)N = C(OMe)Ph$ (22). From spectroscopic investigations π-interaction of both the amino and imino groups with the carbene carbon atom had been postulated. However, a coplanar orientation of both substituents within the carbene plane is not possible for steric reasons. X-ray structure analysis indicated that the dihedral angle between the plane of the substituents around the imino double bond and the carbene plane is 100.6°, which allows the non-bonding pair of electrons at the imino nitrogen atom to interact with the carbene carbon.

The orientation of the carbene plane relative to the metal complex moiety is subject to the same orbital overlap condition as applies for the orientation of organic substituents. However, the spatial arrangement of metal orbitals able to π-interact with the carbene carbon is in most cases not as clearcut as that for organic substituents. Fragment orbital considerations have shown (cf. the chapter in this book by Peter Hofmann) that in (CO)$_5$Cr-carbene complexes there is identical π-interaction between the carbene ligand and the (CO)$_5$Cr fragment for any conformation. In crystal structures of such complexes the carbene plane usually is staggered with the *cis* CO ligands (see Figures 1 – 3). Since there can be no electronic preference, interactions between the carbene substituents and *cis*-CO ligands would seem to be responsible for the observed conformations.

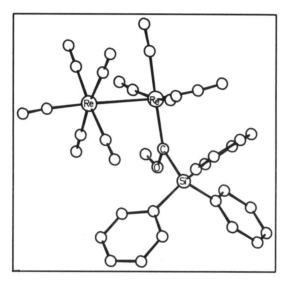

Figure 4: (CO)$_5$Re(CO)$_4$ReC(OMe)SiPh$_3$ (107). The Re-Re bond is approximately perpendicular to the carbene plane.

If a *cis*-CO ligand is replaced by another ligand, L, having different electronic properties, an electronically preferred orientation of the carbene plane is to be expected (149). In *cis*-L(CO)$_4$M-carbene complexes the carbene plane should be coplanar with two *cis* ligands. Depending on the σ- and π-bonding properties of the ligand L (relative to CO), this ligand L will be either within the carbene plane or perpendicular to it. The latter conformation is found in *cis*-(CO)$_5$M(CO)$_4$M-carbene complexes (M = Mn,Re) (see Fig. 4) in good agreement with MO arguments. The alternative orientation is complicated by the fact that both the carbene ligand and L should be located within the same plane. If L is bulky, severe steric problems will arise. This conformation can be realized without major distortion only if both the carbene carbon and L are members of the same chelate ring (28, 29, 63, 135, 136). In *cis*-Ph$_3$P(CO)$_4$CrC(OMe)Me (27), however, steric considerations result in the carbene plane being twisted away from the phosphane ligand by as much as 66°. Such a large departure from a geometry in which overlap between metal and carbene orbitals would be most effective, is possible only if the difference in energy between the various conformations is not too large. This problem will be discussed in more detail in Chapter 3.

Only two structures of *trans*-L(CO)$_4$M-carbene complexes are known: *trans*-(CO)$_4$Mo(CN(Me)CH$_2$CH$_2$NMe)$_2$ (69) and (CO)$_8$Re$_2$[C(OEt)SiPh$_3$]$_2$ (107). Problems arising from the bonding in such complexes have been studied in greater detail for *trans*-L(CO)$_4$CrCNEt$_2$ (149) in which the aminocarbyne ligand, CNEt$_2$, has marked carbene-like properties. In the dirhenium complex each rhenium atom bears one carbene ligand. At one rhenium atom the carbene ligand is equatorial while it is axial at the other. For the different *trans* ligands the Re-C(carbene) distances differ considerably [185(3) and 208(3) pm], the shorter bond length belonging to the carbene ligand which is *trans* to the second rhenium atom.

Biscarbene complexes are special cases, with L representing a second carbene ligand. Four examples have been structurally characterized in the (CO)$_4$M(carbene)$_2$ series (30, 68, 69); a fifth one, [(MeNC)$_4$FeC(NHMe)-NMe-C(NHMe)]$^{2+}$ (37), is closely related. Both *cis*- and *trans*-(CO)$_4$Mo(CN-(Me)CH$_2$CH$_2$NMe)$_2$ have been investigated (68, 69). In these complexes the carbene planes were found staggered with the *cis* ligands whereas in the *trans* isomer they are unexpectedly coplanar. However, because of the two nitrogen atoms attached to each carbene carbon, the degree of π-bonding in the Mo-C(carbene) bonds should be very low and the observed geometries should be largely influenced by steric factors. Complexes with carbene ligands of this kind have been regarded as metallated amidinium ions (150). The poor π-acceptor character of the imidazolidin-2-ylidene ligand relative to CO is also reflected in its *trans* effect. In the *cis* isomer the Mo-C(CO) distances *trans* to C(carbene) are 4.8 pm shorter than the mutually *trans* Mo-C(CO) distances, and the Mo-C(carbene) bond lengths in the *trans* isomer are 6.1 pm shorter than in the *cis* isomer (68, 69).

In (CO)$_4$CrC(OEt)-o-C$_6$H$_4$-COEt (30) and [(MeNC)$_4$FeC(NHMe)-NMe-C-NHMe]$^{2+}$ (37) both carbene units are forced into a coplanar arrangement by the nature of the chelating biscarbene ligand. A very interesting solid state structure is found for (CO)$_4$CrC(OEt)-CHPh-CHPh-COEt (29) (Fig. 5). In this particular biscarbene complex a "frozen" situation is observed, in which the carbene planes form differing torsional angles with the coordination plane of the metal octahedron (16° *vs.* 37°). Hence, the carbene p orbitals interact differently with the metal orbitals. As a result, varying Cr-C(carbene) bond lengths are found [197.6(13) and 202.4(10) pm], the shorter distance (*i.e.* the stronger π-bond) belonging to the carbene carbon with the smaller torsional angle. In spite of this, both carbene carbons are equally stabilized since the carbene carbon with the weaker Cr-C(carbene) π-bond gets more electron density from its ethoxy group than the other does [C(carbene)-O 127.3(13) *vs.* 133.8(12) pm].

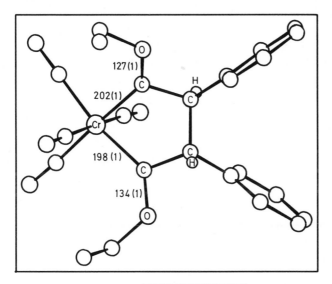

Figure 5: The biscarbene complex (CO)$_4$CrC(OEt)CHPhCHPhCOEt (30). The differing bond lengths in the two carbene units are caused by the different dihedral angles between the carbene planes and the metal octahedron.

3 π-Arene-substituted Carbene Complexes

A wide variety of carbene complexes has been prepared and structurally characterized in which π-bonded aromatic moieties, mainly the η^5-cyclopentadienly (Cp) or η^5-methylcyclopentadienyl (MeCp) groups, are present as co-ligands. The electronic nature of these ligands induces the metal complex fragments to become better π-donors compared to analogous $(CO)_n$M-fragments.

The effects of the increased metal-C(carbene) back-bonding have been studied by comparing the structures of $(CO)_5CrC(OMe)Ph$ (4) and $(\eta^6-C_6H_6)$-$(CO)_2CrC(OMe)Ph$ (31) in which the carbene ligand is kept constant. The main difference is a distinctly shorter Cr-C(carbene) distance [193.5(12) *vs.* 204(3) pm] and a somewhat longer C(carbene)-O distance [136.4(15) *vs.* 133(2) pm] in the η^6-benzene substituted complex. This is in accord with the arguments stated above: the metal complex fragment becomes a stronger competitor for π-bonding with the carbene carbon and thereby reduces the influence of the methoxy group. Inspection of Table 1 shows that shortening of the metal-C(carbene) bond by substitution of CO ligands for Cp is a general trend. For instance, the Re-C(carbene) distance is 209.4(7) pm in $(CO)_5Mn(CO)_4ReC(OMe)Me$ (106) or 209(2) in $(CO)_5Re(CO)_4ReC(OMe)$-$SiPh_3$ (107) (see Fig. 4), but 192(2) in $Cp(CO)_2ReC(H)SiPh_3$ (111). Similarly, the Mo-C(carbene) distance is 215(2) pm in $(CO)_5MoC(OEt)SiPh_3$ (7) but 206.2(11) pm in $Cp(CO)_2(Ph_3Ge)MoC(OEt)Ph$ (70) and the W-C(carbene) distance is 215(1) pm in $(CO)_5WCPh_2$ (132) but 205(2) pm in $Cp_2WC(H)Ph$ (140).

The most comprehensive data are available for Cp- or $MeCp(CO)_2Mn$-carbene complexes [see (55) for a comparison]. In these complexes, too, the π-aromatic ligand produces efficient metal-C(carbene) back-bonding. The most important practical consequence of the high Mn-C bond order is the fact that no π-donating organic substituent at the carbene carbon is necessary to obtain stable complexes. In fact, a number of $Cp(CO)_2Mn$-carbene complexes has been prepared and most of them have been structurally characterized. Such complexes would be non-existent or at least very unstable with poorer π-donating metal fragments, such as $(CO)_5Cr$ or $L_3Pt(II)$. Potentially π-donating organic substituents, which are nevertheless present, cannot compete or do so only to a minor extent with the $Cp(CO)_2Mn$ fragment. Accordingly, the C(carbene)-O bond lengths of oxy substituents in such complexes are longer than in the corresponding $(CO)_5CrC(OR)R'$ complexes (see Table 1). The amino-substituted complexes $Cp(CO)_2MnC(NR_2)R'$ have not yet been investigated.

Another consequence of this bonding situation is that the presence of an oxy substituent does not influence the Mn-C(carbene) distances significantly. The same bond lengths (187 – 189 pm) have been observed in Cp(CO)$_2$MnC(X)Y whether X or Y is bonded to C(carbene) *via* an oxygen atom or both X and Y are bonded *via* a carbon atom. The contrary would be expected in (CO)$_5$Cr-carbene and related complexes. However, constant bond lengths can be expected only as long as none of the organic substituents competes effectively with the metal complex fragment. In the ylide-substituted complexes MeCp(CO)$_2$MnC(OMe)C(Me)PMe$_3$ (58) and Cp(CO)$_2$MnC(CO$_2$-Me)CHPPh$_3$ (59), a distinct lengthening of the Mn-C(carbene) distances [199(1) and 198.5(3) pm] indicates that the organic substituent is now a much better π-donor than the (Me)Cp(CO)$_2$Mn fragment. Like (CO)$_5$CrC(OSi-Me$_3$)CHPMe$_3$, which has been discussed earlier, the "ylide-carbene" complexes of manganese are probably better regarded as vinyl complexes.

As in (CO)$_5$Cr-carbene complexes, a significant *trans* influence of the carbene ligand can be observed only if strongly π-donating organic substituents weaken the metal-C(carbene) π-bond. This is true in the case of MeCp-(CO)$_2$MnC(OMe)C(Me)PMe$_3$, where the Mn-C(Cp) distances *trans* to the carbene ligand are by 10 pm shorter than the *cis* distances (58).

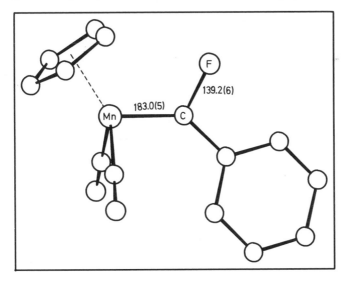

Figure 6: Cp(CO)$_2$MnC(F)Ph (57). The carbene plane bisects the Cp(CO)$_2$Mn fragment.

A special bonding situation is found in Cp(CO)₂MnC(F)Ph (57) (Fig. 6). The Mn-C(carbene) distance is shorter than in other carbene complexes of this type, the C(carbene)-F distance is longer than that expected for $C(sp^2)$-F and the Mn-C(carbene)-C(phenyl) angle is somewhat wider. Such changes indicate a hyperconjugative interaction between the electron-rich Cp(CO)₂Mn fragment and the C(carbene)-F bond. In valence bond terms Scheme 4 is valid.

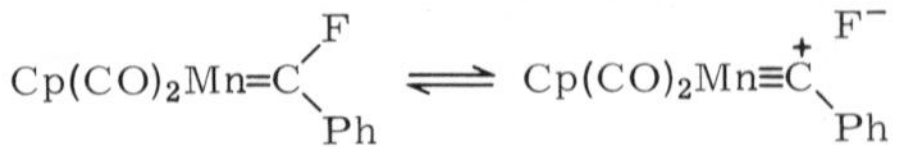

$$Cp(CO)_2Mn{=}C\underset{Ph}{\overset{F}{\big\langle}} \;\rightleftharpoons\; Cp(CO)_2Mn{\equiv}\overset{+}{C}\underset{Ph}{\overset{F^-}{\big\langle}}$$

Scheme 4.

Similar phenomena have been studied in detail for electron-deficient carbene complexes of Ta and W and will be discussed in Chapter 6.

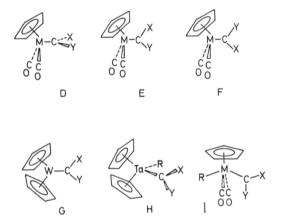

Figure 7: Idealized conformations of the cyclopentadienyl substituted carbene complexes.

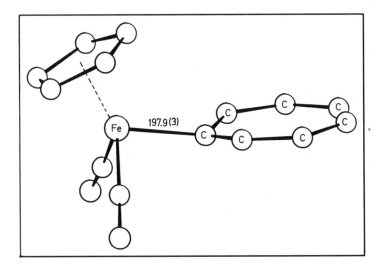

Figure 8: $[Cp(CO)_2FeC_7H_6]^+$ (39). Compared with Fig. 6, the carbene plane is rotated by 90°.

In (π-arene)(CO)$_2$MC(X)Y complexes having two different substituents X and Y there are three idealized conformations (illustrated as D, E and F in Figure 7) as far as the orientation of the carbene plane is concerned. MO calculations (151b) indicate that D should provide the best overlap between the carbene p orbital and the metal orbitals in (η^6-C$_6$H$_6$)(CO)$_2$Cr-carbene complexes. This configuration is found in crystalline (η^6-C$_6$H$_6$)(CO)$_2$CrC(OEt)Ph (31), the only Cr-carbene complex of this type whose structure is known. In Cp(CO)$_2$M-carbene (M = Mn,Re) and [Cp(CO)$_2$Fe-carbene]$^+$ complexes, the conformations E or F are likely to be favored and D should be second choice (151). For the majority of these complexes the conformations E or F have indeed been observed (see Fig. 6 for an example). However, for [Cp(CO)$_2$-FeCH$_2$]$^+$ the calculated difference in energy between E = F and D is only 6.2 kcal/mole (151a). Intra- or intermolecular steric requirements may therefore force the molecule to adopt the "wrong" conformation without noticeably affecting the metal-C(carbene) bond. In three out of the ten Cp(CO)$_2$Mn-carbene structures, namely in Cp(CO)$_2$MnC(OMe)menthyl (54), MeCp(CO)$_2$-MnC(OMe)C(Me)PMe$_3$ (58) and Cp(CO)$_2$MnC(CO$_2$Me)CHPPh$_3$ (59), and in both of the [Cp(CO)$_2$Fe-carbene]$^+$ complexes (39) investigated so far (see Fig. 8), the electronically unfavorable, yet sterically advantageous, conformation D is found. In ylide-substituted carbene complexes the low Mn-C bond order facilitates rotation about this bond. Cp(CO)$_2$ReC(H)SiPh$_3$ (111) adopts conformation E (X = H, Y = SiPh$_3$).

If the substituents X and Y differ, two conformations (E and F) are possible in which the carbene plane and the mirror plane of the $Cp(CO)_2M$ fragment are coplanar. Though one of them should be favored on electronic grounds (151b), the conformation in the crystalline state will also strongly depend on the steric properties of X and Y (55). In solution, different conformations can sometimes coexist and be observed by IR spectroscopy (152).

Relatively large barriers for carbene rotation have been calculated for Cp_2M-carbene, Cp_2LM-carbene and *trans* $Cp(CO)_2LM$-carbene complexes (153). Although comprehensive structural data are lacking, all available X-ray structures are consistent with theoretical predictions that $Cp_2WC(H)R$ [R = Ph (140) and R = $OZr(H)Cp_2^*$ (139)] will adopt conformation G (Fig. 7), $Cp_2(Me)TaCH_2$ (123), $Cp_2(Bz)TaCHPh$ (124) and $Cp_2(Cl)TaCHBu^t$ (125) will adopt conformation H (Fig. 7) and $Cp(CO)_2(Ph_3Ge)MoC(OEt)Ph$ (70), $Cp(CO)_2(I)Mo\text{-}\overline{C}OCH_2CH_2\overset{\frown}{C}H_2$ (71) and $Cp(CO)_2(Ph_3Sn)WCHTol$ (138) will adopt conformation I (Fig. 7). Of course, these conformational preferences will alter if the symmetry of the metal complex fragment is changed, *e.g.* a $Cp(CO)_2M$ fragment is replaced by $CpLL'M$ (112, 154).

4 Square-planar Carbene Complexes

X-ray structural analyses of square-planar carbene complexes were performed mainly on the Pt(II) series. Analysis of the observed bond lengths is
rendered difficult because the average standard deviations are twice as high as
those for chromium complexes and the co-ligands on the metal vary considerably. Apart from variation in the back-bonding ability of the L_3Pt fragment, dissimilar ligands may also induce changes in bonding through steric
interactions, particularly if the ligands *cis* to the carbene ligand are of differing size [see (149) for a related example]. A few general trends can be established:

(1) Pt-C(carbene) distances are shortest when (π-donating) chloride is *trans*
to the carbene ligand (182 – 201 pm). If the chloride is replaced by a
phosphane or methyl ligand, L_3Pt will become a weaker π-donor and the Pt-
C(carbene) distance will increase (200 – 213 pm).

(2) L_3Pt fragments are generally poor π-donors since, in alkoxy and amino
carbene complexes with only one π-donating organic substituent present, the
C(carbene)-O (122 – 132 pm) and C(carbene)-N distances (125 – 133 pm) are
very short.

The poor Pt-C(carbene) back-bonding is also reflected in the *trans*
influence of the carbene ligand, which can be determined accurately if
chloride is the *trans* ligand. Comparison of a series of complexes of the type
trans-ClL$_2$PtL′, revealed that the Pt-Cl distance decreased according to the
scheme L′ = alkyl > carbene ≈ phosphane > isonitrile > carbon monoxide (155). The same sequence is valid for octahedral Pt(IV) (105), Rh(III)
(115), and Ru(III) complexes (119). The position of carbene ligands between
alkyl and CO groups and their similarity to phosphane ligands indicates their
good σ-donor but poor π-acceptor properties in these particular complexes.

As in (CO)$_5$M-carbene complexes, the barrier for rotation about the M-
C(carbene) axis in square planar L_3M-carbene complexes is set by steric rather
than electronic considerations (156). In the crystalline state, most complexes
adopt a conformation in which the carbene plane is (roughly) perpendicular to
the coordination plane of the metal (see Fig. 9). However, there are two
square planar nickel complexes (78, 79), a palladium complex (87) and two
platinum complexes (95, 101) in which a coplanar arrangement of both planes
results from the carbene carbon being part of a chelating ligand.

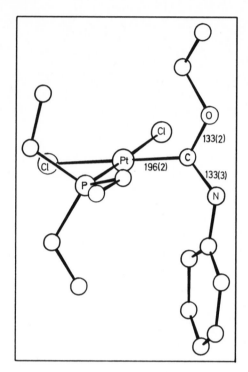

Figure 9: *Cis*-Cl$_2$(Et$_3$P)PtC(OEt)NHPh (92). The carbene plane is perpendicular to the coordination plane of the Pt.

A perpendicular orientation minimizes steric interactions between ligands on the metal and substituents on the carbene carbon and allows the latter to spread out into the space occupied by the fifth and sixth ligand in octahedral complexes. In open-chained alkoxy-, thio- or monoalkylamino-carbene ligands the alkyl group therefore adopts a *Z* configuration (see Fig. 9) for steric reasons. Only in (MeNC)$_2$Pt[C(NHMe)SEt]$_2^+$ (100) and Pt[C(NHMe)$_2$]$_4^{2+}$ (103), both of which have admidinium-like carbene ligands, one NHMe substituent in each carbene ligand adopts an *E*-conformation. As a second consequence, the M-C(carbene)-O and M-C(carbene)-N angles in square planar complexes are on average smaller than those in corresponding octahedral complexes. This provides additional evidence for steric intractions between ligands and carbene substituents in octahedral complexes.

5 Trigonal-bipyramidal Carbene Complexes

Only a few structures approximating to trigonal-bipyramidal carbene complexes are known [square-pyramidal geometry is found only in Np(Bu^{t}C)-(Me$_2$PCH$_2$CH$_2$PMe$_2$)WCHBu^t (141), with the carbyne ligand in the axial position]. Four complexes are of the type L(CO)$_3$M-carbene: [(3), (32) see Fig. 10, (33), (34)], L being an olefinic moiety in the latter two complexes. In all of these complexes the carbene ligand occupies an apical position and is oriented in such a way that the carbene plane is nearly eclipsed with one equatorial ligand.

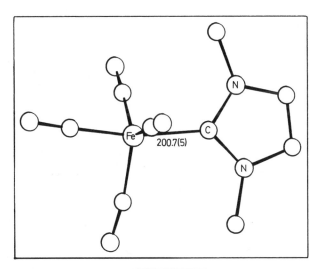

Figure 10: Trigonal bipyramidal (CO)$_4$FeCN(Me)CHCHNMe (32). The carbene ligand occupies an axial position.

The σ-donor character and the poorer π-acceptor ability of the carbene ligands (compared with CO or olefins) explain their preference for an axial site in carbonyl carbene complexes (157). The carbene carbon is stabilized mainly by π-donating organic substituents; the C(carbene)-O and C(carbene)-N distances are consequently rather short.

In a second type of trigonal-bipyramidal carbene complex, represented by the electron-deficient neopentylidene complexes (mesityl)(Me$_3$P)$_2$Ta(CHBu^t)$_2$ (128), [Np(Me$_3$P)$_2$TaCHBu^t]$_2$N$_2$ (129) and Cl$_2$(O)(Me$_3$P)WCHBu^t (144), the carbene ligand occupies an equatorial site and the carbene planes are coplanar with the trigonal plane. Neopentylidene and related ligands are strong π-acceptors since π-bonding with the metal is their only means of stabilization. They therefore prefer the equatorial sites with stronger π-bonding in these complexes.

6 Electron − deficient Carbene Complexes of Tantalum and Tungsten

As pointed out earlier, bond angles at the carbene carbon can easily be adjusted to meet steric requirements. Thus, the Ta-C(carbene)-C bond angle of 150.4(5)° in the 18-electron complex Cp$_2$(Cl)TaCHBu^t (125) is not at all surprising. However, in 14- or 16-electron complexes of tantalum and tungsten with CHR ligands [R = Bu^t (126, 128, 129, 141 − 145), Ph (127), Li (130), AlR$_3$ (146)], metal-C(carbene)-R angles of up to 170° are observed. At the same time, other structural parameters differ from the "normal" values. Two of these complexes have been investigated by neutron diffraction (126) in order to locate the crucial hydrogen in the neopentylidene ligand accurately.

From these investigations the carbene ligand appears to be pivotal in its position (see Fig. 11 and Scheme 5), *i.e.* while the C-C(carbene)-H angle within these ligands is virtually unaffected, the Ta-C(carbene)-H angle is strongly reduced [84.8(2)° and 78.1(3)° in the determined structures]. At the same time, the C-H bond is lengthened and the Ta-C(carbene) bond is shortened. Weakening of the C-H bond in these complexes is also indicated by the decreased C-H stretching frequencies and NMR coupling constants (158).

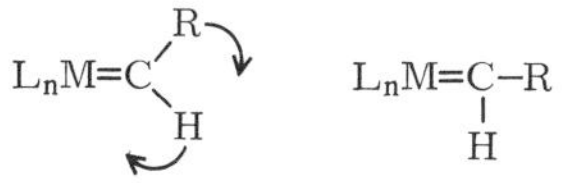

$$L_nM=C\underset{H}{\overset{R}{\diagdown}} \qquad L_nM=C{-}R \atop \qquad\qquad\quad | \atop \qquad\qquad\;\; H$$

Scheme 5.

Structural investigations on the 14- and 16-electron complexes of tungsten have shown that only in the absence of strongly π-donating ligands can extreme distortions be expected (145). MO analysis of such complexes (159) has traced these deformations back to electron deficiency at the metal. In this way an intramolecular electrophilic interaction between an empty acceptor orbital on the metal and both the lone pair and the σ(C-H) orbital of the carbene becomes possible. The magnitude of the distortion will depend on the actual electron deficiency. Accordingly, only small angular deformations will be observed if strong σ- and particularly π-donor ligands are present in formal 14- or 16-electron complexes.

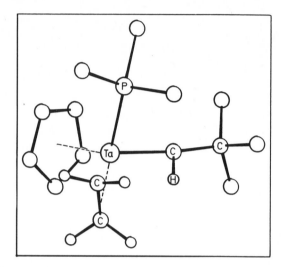

Figure 11. $(\eta^5\text{-}C_5Me_5)(\eta^2\text{-}C_2H_4)(Me_3P)TaCHBu^t$ (neutron diffraction) (126). The methyl groups on the Cp ring are omitted for clarity.

Table 1. Bond lengths [pm] from the carbene carbon in carbene complexes $L_nMC(X)Y$.

Complex	Metal to C(carbene)	C(carbene) to X and	Y	Ref.
$\{I_2Au[C(NHTol)_2]_2\}ClO_4$	207(2) 209(2)	N:129 – 135(2)		(1)
$ClAuC(NMe_2)Ph$	202(3)	N:127(4)	C:148(4)	(2)
$(Ph_3Ge)(CO)_3CoC(OEt)Et$	191.3(11	O:129(1)	C:153(2)	(3)
$(CO)_5CrC(OMe)Ph$	204(3)	O:133(2)	C:147(4)	(4)
$(CO)_5CrC(OEt)Me$	205.3(1)	O:131.4(1)	C:151.1(1)	(5)
$(CO)_5CrC(OEt)C\equiv CPh$	200(2)	O:132(2)	C:137(3)	(6)
$(CO)_5CrC(OEt)SiPh_3$	200(2)	O:133(2)	Si:200(2)	(7)
$(CO)_5CrC(OH)Ph$	205(1)	O:132(1)		(8)

	197.3(6)	O:136.0(7)	C:144.3(9)	(9)
(CO)₅CrC[OSi(SiMe₃)₃](1-furyl)				
	203.3(7)	O:132.1(9)	C:145.4(10)	(10)
(CO)₅CrC(OSiMe₃)CH = PMe₃				
	213.7(7)	O:137.8(8)	C:133.2(9)	(11)
(CO)₅CrC(OEt)NMe₂	213.3(4)	O:134.6(5)	N:132.8(5)	(12)
(CO)₅CrC(NHMe)Me	209	N:133	C:151	(13)
(CO)₅CrC(NEt₂)Me	216(1)	N:131(1)	C:150(1)	(14)
(CO)₅CrC(NEt₂)H	208.4(2)	N:130.0(3)	H:102(3)	(15)
(CO)₅CrC(NEt₂)Cl	211.0(5)	N:129.9(8)	Cl:178.0(6)	(16)
(CO)₅CrC(NEt₂)NCO	212(1)			(17)
(CO)₅CrC(NEt₂)NCS	211(1)			(17)
[(CO)₅CrC(NEt₂)]₂	219.0(7)	N:130.6(9)	C:148.0(9)	(18)
(CO)₅CrC(NH-c-C₆H₁₁)C(OMe) = CH₂				
	215(1)	N:132(2)	C:151(2)	(19)
(CO)₅CrC(NEt₂)SnPh₃	211.0(14)	N:133(2)	Sn:223.9(1)	(20)
(CO)₅CrC(NEt₂)SePh	217.1(8)	N:128(1)	Se:191.2(7)	(21)
(CO)₅CrC(NMe₂)N = C(OMe)Ph				
	213.5(4)	N:134.0(6), 134.2(5)		(22)
(CO)₅CrC(SPh)Me	202.0(3)	S:169.0(3)	C:149.0(4)	(23)
(CO)₅CrC[S-CH = C(OMe)Ph]Ph				
	202.9(10)	S:167.1(10)		(24)
(CO)₅CrC[SCH₂SCH₂SCH = C(OMe)Ph]Ph				
	202.7(5)	S:167.3(5)		(24)
(CO)₅CrC(1-furyl)(1-thienyl)				
	206(1)	C:145(1), 143(1)		(25)
(CO)₅CrC̄C(Ph)C̄Ph	205(1)	C:139, 140		(26)
cis-Ph₃P(CO)₄CrC(OMe)Me				
	200(2)	O:132(2)	C:153(3)	(27)

	200.4(12)	O:135(2)	C:149(2)	(28)

| | 202.5(4) | O:132.9(6) | C:147.3(5) | (28) |

$(CO)_4\overline{CrC(NEt_2)\text{-}CHMe\text{-}CPh}=NMe$

| | 208.0(11) | N:130(1) | C:147(1) | (29) |

$(CO)_4\overline{CrC(OEt)\text{-}o\text{-}C_6H_4\text{-}COEt}$

| | 200.4(7) | O:131.8(8) | C:146.2(9) | (30) |
| | 200.0(6) | 130.8(8) | 149.7(9) | |

$(CO)_4\overline{CrC(OEt)\text{-}CHPh\text{-}CHPh\text{-}COEt}$

| | 197.6(13) | O:134(1) | C:152(2) | (30) |
| | 202.4(10) | 127(1) | 153(2) | |

$(\eta^6\text{-}C_6H_6)(CO)_2CrC(OMe)Ph$

| | 193.5(12) | O:136(2) | C:150(2) | (31) |

$(CO)_4\overline{FeCNMe\text{-}CH}=CH\text{-}NMe$

| | 200.7(5) | N:134.9(6), 136.2(6) | | (32) |

| | 194(2) | O:130(2) | N:139(2) | (33) |

| | 181.9(3) | O:129.5(4) | C:142.0(4) | (34) |

| | 183.4(2) | C:140.7(4) | | (35) |

$[(CO)_2(Me_2NCS_2)\overline{FeC(NMe_2)SC(NMe_2)}]PF_6$

| | 195.4(8) | N:130.5(9) | S:176.0(8) | (36) |
| | 197.4(8) | 130.0(9) | 176.0(8) | |

$[(MeNC)_4\overline{FeC(NHMe)\text{-}NMe\text{-}C(NHMe)}](PF_6)_2$

| | 202(4) | N:126(5), 134 (5) | | (37) |
| | 203(4) | 131(5), 139(5) | | |

$[(MeNC)_4\overline{FeC(NHMe)\text{-}NH\text{-}CMe}=N\text{-}C=NHMe](PF_6)_2$

| | 195(1) | N:131(2), 132(2) | | (38) |

$[Cp(CO)_2FeC_7H_6]PF_6$	197.9(3)	C:139.3(5), 140.6(5)	(39)
$[Cp(CO) FeC_{11}H_8]PF_6$	199.6(2)	C:139.5(3), 140.7(3)	(39)
$[Cp(CO)_2Fe\!-\!\langle\rangle\!=\!Fe(CO)_2Cp]\,ClO_4$ (H Ph / Ph substituents)	192(3)	C_{sp^2}:137(4) C_{sp^3}:152(4)	(41)
	190(3)	141(4) 151(4)	
$Cp(CO)\overline{Fe}C(OEt)\text{-}NMe\text{-}C = CPh_2$	187.7(6)	N:132.0(7) O:133.9(7)	(40)
$Cp(CO)\overline{Fe}C(H)\text{-}NMe\text{-}BH_2\text{-}NMe\text{-}\dot{C}H$	186.3(15) 190.1(14)	N:129(2) 134(2)	(42)
$Cp(CO)Fe$ carbene (Ph, C≡NH aryl)	191.6(2)	N:132.6(2) C:148.8(2)	(43)
$Cp(CO)Fe$ carbene (NHBu^t, NHC$_6$H$_{11}$, NC$_6$H$_{11}$)	193.2(3)	N:133.6(4) C:140.7(4)	(44)
$(TTP)FeCCl_2$	183(3)	Cl:176(3)	(45)
$\{Hg[\overline{CN(Ph)\text{-}CH = CH\text{-}NPh}]_2\}(ClO_4)_2$	206(1)	N:133(1), 134(1)	(46)
$[Cp(I)(Ph_3P)IrC(SMe)Me]I$	203(3)	S:156(3) C:149(5)	(47)
Bu_2^tP…$Cl\!-\!Ir\!=\!C$…Bu_2^tP	200.6(4)	C:150.4(6), 150.2(6)	(48)
$Cl_3(Ph_3P)_2IrCCl_2$	187.2(7)	Cl:172.1(5)	(49)
$Cp(CO)_2MnCMe_2$	186.8(10)	C:151.6(11)	(50)
$Cp(CO)_2MnC(Ph)COPh$	188(2)	C:147(2), 149(3)	(51)
$Cp(CO)_2MnCPh_2$	188.5(2)	C:149.2(3), 148.5(3)	(52)

MeCp(CO)$_2$Mn=⟨structure⟩ ⟨structure⟩=Mn(CO)$_2$MeCp   187.8(8)   C:153(1), 146(2)   (53)

Cp(CO)$_2$MnC(OMe)menthyl

	189(2)	O:133(2)	C:153(2)	(54)
Cp(CO)$_2$MnC(OEt)Ph	186.5(14)	O:136(2)	C:154(2)	(55)
[Cp(CO)$_2$MnC(Ph)]$_2$O	185(2)	O:139(3)	C:150(3)	(56)
	188(2)	137(2)	149(2)	
Cp(CO)$_2$MnC(F)Ph	183.0(5)	F:139.2(6)	C:146.9(8)	(57)

MeCp(CO)$_2$MnC(OMe)C(Me)PMe$_3$

	198.7(10)	O:140(1)	C:139(1)	(58)

Cp(CO)$_2$MnC(CO$_2$Me)CHPPh$_3$

	198.5(3)	C:148.6(4), 136.1(4)	(59)

Cp(CO)[(MeO)$_3$P]MnC̅-S-C(CO$_2$Me)=C(CO$_2$Me)-S̅

	187.6(2)	S:175.3(2)173.2(2)	(60)

⟨structure⟩   196(3)   O:125(4)   C:152(4)   (61)

cis-(CO)$_5$Mn(CO)$_4$MnC(OEt)Ph

	195.0(5)	O:131.5(6)	C:148.3(7)	(62)

⟨structure⟩ (CO)$_4$Mn=C   199   O:132.2(7)   C:136   (63)

cis-Cl(CO)$_4$MnC̅OCH$_2$CH$_2$O̅

	196(1)	O:131(2), 133(2)	(64)

cis-(S)-(-)-[1-naphthyl(Me)(Ph)P](CO)$_4$MnC(OEt)Me

	195(2)	O:131(3)	C:152(3)	(65)

cis-(CO)$_4$MnC̅(Me)O-AlBr$_2$-Br

	195(2)	O:130(2)	C:155(2)	(66)

[Ph$_3$P(CO)$_3$MnC(SMe)NEt$_2$]BF$_4$

	184.3(17)	N:128(2)	S:178(2)	(67)

(CO)$_5$MoC(OEt)SiPh$_3$	215(2)	O:135(2)	Si:194(2)	(67)

cis-(CO)$_4$Mo(C̅N(Me)CH$_2$CH$_2$N̅Me)$_2$

	229.3(3)	N:134.1(3), 133.6(3)	(68)
	229.3(3)	134.2(3), 134.3(3)	

trans-(CO)$_4$Mo(C̅N(Me)CH$_2$CH$_2$N̅Me)$_2$

223.2(2)	N:133.3(3), 134.1(3)		(69)

trans-Cp(CO)$_2$(Ph$_3$Ge)MoC(OEt)Ph

206.2(11)	O:138(2)	C:150(1)	(70)

trans-Cp(CO)$_2$(I)Mo$\overline{\text{COCH}_2\text{CH}_2\text{CH}_2}$

208.6(8)	O:134(1)	C:145(1)	(71)

(R = Me)	203(3)	O:138(3)	C:147(4)	(72)
(R = H)	209.2(12)	O:136(2)	C:139(2)	(73)

Cp(CO)(I)$\overline{\text{MoC(NMe}_2)\text{C(Me)}=\text{NMe}}$

202.7(5)	N:130.5(6)	C:145.5(6)	(74)

Cp[(MeO)$_3$P](I)$\overline{\text{MoC(Np)P(OMe)}_2=\text{O}}$

200.6(4)	P:174.3(4)	(75)

Cp[Cp(CO)$_3$Mo](Tol$_2$CN$_2$)MoCTol$_2$

198(1)	(76)

(η^5-C$_9$H$_7$)[(MeO)$_3$P]$_2\overline{\text{MoC(SiMe}_3)\text{CH}_2}$

195.7(3)	Si:185.0(4)	C:143.6(5)	(77)

181.9(1)	N:133.5(4)	(78)

[(Me$_2$NCS)$\overline{\text{NiC(NMe}_2)\text{SC(NMe}_2)\text{S}}$]BPh$_4$

190.9(10)	N:133.2(11)	S:178.3(10)	(79)

[*trans*-(Ph$_3$P)$_2$(Cl$_5$C$_6$)Ni$\overline{\text{CCH}_2\text{CH}_2\text{CH}_2\text{O}}$]BF$_4$

183.7(7)	O:129.7(8)	C:148.2(10)	(80)

Cl$_2$(CO)(Ph$_3$P)$_2$OsCCl$_2$ (81)

H(CO)(Ph$_3$P)$_2\overline{\text{OsC(NMeTol)SC}=\text{S}}$

208(4)	N:130(5)	S:170(4)	(82)

(CO)$_9$Os$_3$[C(OMe)Me](μ-H)(μ-COMe)

199(2)	O:134(2)	C:151(3)	(83)

---

cis-Cl$_2$Pd[C(OMe)NHMe]$_2$	195.3(8)	O:131.6(9)	N:132(1)	(84)
	197.2(10)	134.3(11)	129(1)	

cis-Cl$_2$(Bu$^n_3$P)Pd$\overline{\text{CC(NMe}_2)\text{CNMe}_2}$

196.1(3)   C:138.5(5), 138.0(4)   (85)

trans-Cl(Et₃P)₂PdC̄N(Tol)C(CHCO₂Et)C(O)ĊH

196.1(3) — $C: 138.5(5), 138.0(4)$ — (85)

trans-Cl($Et_3P)_2$PdC̄N(Tol)C(CHCO₂Et)C(O)ĊH	196.1(3)	C:138.5(5), 138.0(4)	(85)
Cl_2P̄dC(NHMe)NHNHĊNHMe	194.2(4)	N:140.9(5) C:138.4(6)	(86)
Cl[Et_2N̄CH₂CH₂C(O)]PdC(NEt_2)NHMe	194.8(5)	N:132.7(7), 130.9(6)	(87)
(BuᵗNC)[ŌC($CF_3)_2$OC($CF_2)_2$]PdC(NEt_2)NHBuᵗ	198.8(6)	N:133.9(6), 134.0(6)	(88)
	207.4(9)	N:129(1), 134(1)	(89)

cis-Cl_2(Me_2PhP)PtC(OEt)CH₂Ph	192.0(9)	O:128(1)	C:149(2)	(90)
cis-Cl_2(Ph_3P)PtC(NMe_2)H 196(1)		N:125(1)		(91)
cis-Cl_2(Et_3P)PtC(NHPh)OEt	196.2(18)	N:133(3)	O:133(2)	(92)
cis-Cl_2Pt[C(OPrⁱ)Me]₂ 192(2)		O:124(2)	C:158(3)	(93)
	197(2)	122(1)	151(3)	
cis-Cl_2(Et_3P)PtC̄N(Ph)CH₂CH₂N̄Ph	200.9(13)	N:132.7(11)		(94)
[Cl_2PtC(OPrⁱ)Np]₂ 182(6)		O:132(7)	C:151(8)	(93)
[*trans*-Cl($Me_2PhP)_2$PtC(NMe_2)CH₂CH₂OH]PF₆	197.8(12)	N:129(2)	C:152(2)	(96)
trans-Cl_2(Et_3P)PtC̄N(Ph)CH₂CH₂N̄Ph	202.0(16)	N:133(2), 137(2)		(93)

$$\left[\begin{array}{c} \text{Ph}_2 \\ \text{P} \\ \text{Pt}\!-\!\overset{\text{NHTol}}{\underset{\text{NH}}{\text{C}}} \\ \text{P} \\ \text{Ph}_2 \end{array}\right] \text{ClO}_4$$

204.4(14)   N:129(2), 131(2)   (95)

[*trans*-Me($Me_2PhP)_2$PtC(OMe)Me]PF₆	213(2)	O:133	C:143	(97)
[*trans*-Me(Me_2PhP)PtC̄OCH₂CH₂ĊH₂]PF₆	200(2)	N:126(2)	C:153(2)	(98)
[*trans*-Me($Me_2PhP)_2$PtC(NMe_2)Me]PF₆	207.9(13)	N:127(2)	C:153(2)	(99)
{*trans*-(MeNC)₂Pt[C(NHMe)SEt]₂}(PF₆)₂	206(1)	N:130(1)	S:168(1)	(100)
[*cis*-(MeNC)₂P̄tC(NHMe)N(Me)=Ċ(NHMe)]BPh₄	195(2)	N:136(3), 143(2)		(101)
trans-[Cp(CO)₂Mo]₂(c-C₆H₁₁NC)PtC(OEt)NHc-C₆H₁₁				

	202(1)	N:134(1)	O:132(2)	(102)
[Pt(C(NHMe)$_2$)$_4$](PF$_6$)$_2$	204.7(7)	N:131.0(3)		(103)

(CO)$_4$Mn(μ-I)(Bu$_2^t$MeP)Pt$\overline{\text{COCH}_2\text{CH}_2\text{CH}_2}$

	188.9(8)	O:132.(1)	C:146(1)	(104)

	197.3(11)	N:133(1), 135(2)		(105)

(CO)$_5$Mn(CO)$_4$ReC(OMe)Me

	209.4(7)	O:129.9(8)	C:149(1)	(106)

(CO)$_5$Re(CO)$_4$ReC(OMe)SiPh$_3$

	209(2)	O:129(2)	Si:193(2)	(107)

(CO)$_5$Re(CO)$_4$ReC(OEt)SiPh$_3$

	209(4)	O:131(6)	Si:203(5)	(107)
(CO)$_8$Re$_2$[C(OEt)SiPh$_3$]$_2$	208(3)	O:133(4)	Si:195(3)	(107)
	185(3)	142(4)	205(3)	
[(CO)$_4$ReC(Me)OGeMe$_2$]$_2$	214(3)	O:127(3)		(108)
[(CO)$_4$ReC(ORe(CO)$_5$)O]$_2$	220(2)	O:125(2), 128(2)		(109)

(CO)$_7$[μ-C(SiPh$_3$)CO$_2$Et]Re$_2$C(OEt)SiPh$_3$

	195(2)	O:139(2)	Si:201(2)	(110)
Cp(CO)$_2$ReC(H)SiPh$_3$	192(2)	Si:191(2)		(111)

[Cp(NO)(Ph$_3$P)ReC(H)Ph]PF$_6$

	194.9(6)	C:145.4(9)		(112)

[(CF$_3$)$_2$CN](Ph$_3$P)$_2$$\overline{\text{RhCN(Me)CH}_2\text{CH}_2\text{NMe}}$

	200.6(15)	N:131(2), 140(2)		(113)
Cl$_3$(Et$_3$P)$_2$RhC(H)NMe$_2$	196.1(11)	N:129(1)	H:114	(114)

(X = O, R^1 = R^2 = Ph)	193.0(6)	N:132.6(8)	S:173.7(7)	(115)
[X = S, R^1 = N(CO$_2$Et)$_2$, R^2 = OEt]				
	193.0(11)	N:134(3)	S:174.3(10)	(116)

$I_3(CO)\overline{RhC(Ph)N(Me)C(Ph)} = NMe$

| | 196.8(13) | N:133(2) | C:146(2) | (117) |

{*trans*-Cl(NH$_3$)$_4$Ru$\overline{CN(H)C(Me)C(Me)NH}$}(PF$_6$)$_2$

| | 212.8 | N:134.7, 135.6 | | (118) |

| 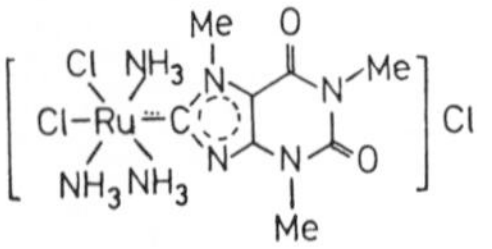 | 203(1) | N:131(1), 139(1) | | (119) |

$I_2(CO)(Ph_3P)(TolNC)RuC(NMeTol)H$

| | 204.6(15) | N:126(2) | H:129 | (120) |

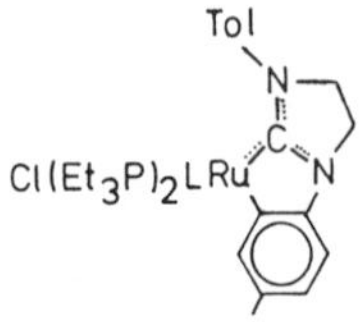

(L = hydrogen interaction)

| | 191.1(9) | N:134(1), 139(2) | | (121) |
| (L = CO) | 198.9(6) | N:133.5(7), 134.8(7) | | (121) |

trans-Cl$_2$Ru[$\overline{CN(Et)CH_2CH_2NEt}$]$_4$

	210.9(10)	N:134(1), 135(1)		(122)
	211.1(9)	136(1), 134(1)		
	210.3(9)	135(1), 134(1)		
	209.8(9)	137(1), 133(1)		

Cp$_2$(Me)TaCH$_2$	202.6(10)			(123)
Cp$_2$(Bz)TaCHPh	207(1)	C:147(2)		(124)
Cp$_2$(Cl)TaCHBu^t	203.0(6)	C:152.3(9)	H:82(6)	(125)
Cp*(C$_2$H$_4$)(Me$_3$P)TaCHBu^t(neutr.)				
	194.6(3)	C:151.8(3)	H:113.5(5)	(126)
Cp*(Bz)$_2$TaCHPh	188.3(14)	C:149(1)		(127)
(Me$_3$H$_2$C$_6$)(Me$_3$P)$_2$Ta(CHBu^t)$_2$				
	193.2(7)	C:152.7(11)		(128)
	195.5(7)	151.1(11)		
[Np(Me$_3$P)$_2$TaCHBu^t]$_2$N$_2$	193.2(9)	C:150(1)		(129)
	193.7(9)	151(1)		
[Cl$_3$(Me$_3$P)TaCHBu^t]$_2$ (neutr.)				
	189.8(2)	C:150(1)	H:113.1(3)	(126)

$Np_3TaC(Bu^t)Li(N,N'-Me_2N_2C_4H_8)$

	176(2)	C:153(3)	Li:219(3)	(130)

$(CO)_5WC(OMe)Ph$				(131)
$(CO)_5WCPh_2$	215(1)	C:145(2), 151(2)		(132)
$(CO)_5WC(OEt)C_5H_4RuCp$	223(2)	O:135(2)	C:143(3)	(133)
$(CO)_5WC(OEt)C_5H_8\text{-}CH=CPh_2$				
	218(1)	O:129(2)	C:152(2)	(134)

(R = OEt, X = CH$_2$)	213(1)	O:133(2)	C:151(2)	(135)
(R = Tol, X = NH)	220.6(7)	N:129.2(9)	C:148.4(8)	(136)
$Cp(CO)_2\overset{\frown}{W}C(CF_3)\overset{\frown}{C}(CF_3)C(O)SMe$				
	196	C:150, 144		(137)
trans-$Cp(CO)_2(Ph_3Sn)WCHTol$				
	203.2(7)			(138)
$Cp_2WCHOZr(H)Cp_2^*$	200.5(13)	O:135(2)		(139)
Cp_2WCHPh	205(2)	C:144(4)		(140)
$Np(Bu^tC)(Me_2PCH_2CH_2PMe_2)WCHBu^t$				
	194.2(9)	C:149(2)		(141)
$Cl_2(O)(Me_3P)_2WCHBu^t$	198.6(21)			(142)
$Cl_2(O)(Et_3P)_2WCHBu^t$				(143)
$Cl_2(O)(Et_3P)WCHBu^t$	188.2(14)			(144)
$Cl_2(CO)(Me_3P)_2WCHBu^t$	185.9(4)	H:105(4)		(145)
$Cl(Me_3P)_3WC(H)AlMe_{3-x}Cl_x$				
	180.7(6)	Al:211.3(6)		(146)

7 References

(1) L. Manojlović-Muir, J. Organomet. Chem. **73** (1974) C45 – C46.

(2) U. Schubert, K. Ackermann, R. Aumann, Cryst. Struct. Comm. **11** (1982) 591 – 594.

(3) F. Carre, G. Cerveau, E. Colomer, R. J. P. Corriu, J. C. Young, L. Ricard, R. Weiss, J. Organomet. Chem. **179** (1979) 215 – 226.

(4) O. S. Mills, A. D. Redhouse, J. Chem. Soc. (A) 1968, 642 – 647.

(5) C. Krüger, R. Goddard, Y.-H. Tsay, private communication.

(6) G. Huttner, H. Lorenz, Chem. Ber. **108** (1975) 1864 – 1870.

(7) E. O. Fischer, H. Hollfelder, P. Friedrich, F. R. Kreissl, G. Huttner, Chem. Ber. **110** (1977) 3467 – 3724.

(8) R. J. Klingler, J. C. Huffman, J. K. Kochi, Inorg. Chem. **20** (1981) 34 – 40.

(9) H. Berke, P. Härter, G. Huttner, J. v. Seyerl, J. Organomet. Chem. **219** (1981) 317 – 327.

(10) U. Schubert, M. Wiener, F. H. Köhler, Chem. Ber. **112** (1979) 708 – 716.

(11) S. Voran, H. Blau, W. Malisch, U. Schubert, J. Organomet. Chem. **232** (1982) C33 – C40.

(12) G. Huttner, B. Krieg, Chem. Ber. **105** (1972) 67 – 81.

(13) P. E. Bakie, E. O. Fischer, O. S. Mills, Chem. Comm. 1967, 1199 – 1200.

(14) J. A. Connor, O. S. Mills, J. Chem. Soc. (A) 1969, 334 – 341.

(15) A. Frank, Dissertation, Technische Universität München, 1978.

(16) G. Huttner, A. Frank, E. O. Fischer, W. Kleine, J. Organomet. Chem. **141** (1977) C17-C20.

(17) E. O. Fischer, W. Kleine, F. R. Kreissl, H. Fischer, P. Friedrich, G. Huttner, J. Organomet. Chem. **128** (1977) C49 – C53.

(18) E. O. Fischer, D. Wittmann, D. Himmelreich, D. Neugebauer, Angew. Chem. **94** (1982) 451 – 452; Angew. Chem. Int. Ed. Engl. **21** (1982) 444 – 445; Angew. Chem. Suppl. 1982, 1036 – 1049.

(19) G. Huttner, S. Lange, Chem. Ber. **103** (1970) 3149 – 3158.

(20) E. O. Fischer, R. B. A. Pardy, U. Schubert, J. Organomet. Chem. **181** (1979) 37 – 45.

(21) E. O. Fischer, D. Himmelreich, R. Cai, H. Fischer, U. Schubert, B. Zimmer-Gasser, Chem. Ber. **114** (1981) 3209 – 3219.

(22) H. Fischer, U. Schubert, R. Märkl, Chem. Ber. **114** (1981) 3412 – 3420.

(23) R. J. Hoare, O. S. Mills, J. Chem. Soc. Dalton 1972, 653 – 656.

(24) H. G. Raubenheimer, E. O. Fischer, U. Schubert, C. Krüger, Y.-H. Tsay, Angew. Chem. **93** (1981) 1103-1105; Angew. Chem. Int. Ed. Engl. **20** (1981) 1055 – 1056.

(25) E. O. Fischer, W. Held, F. R. Kreissl, A. Frank, G. Huttner, Chem. Ber. **110** (1977) 656 – 666.

(26) G. Huttner, S. Schelle, O. S. Mills, Angew. Chem. **81** (1969) 536; Angew. Chem. Int. Ed. Engl. **8** (1969) 515.

(27) O. S. Mills, A. D. Redhouse, J. Chem. Soc. (A) 1969, 1274 – 1279.

(28) G. J. Kruger, J. Coetzer, H. G. Raubenheimer, S. Lotz, J. Organomet. Chem. **142** (1977) 249 – 263.

(29) K. H. Dötz, B. Fügen-Köster, D. Neugebauer, J. Organomet. Chem. **182** (1979) 489 – 498.

(30) U. Schubert, K. Ackermann, N. H. Tran Huy, W. Röll, J. Organomet. Chem. **232** (1982) 155 – 162.

(31) U. Schubert, J. Organomet. Chem. **185** (1980) 373 – 384.

(32) G. Huttner, W. Gartzke, Chem. Ber. **105** (1972) 2714 – 2725.

(33) T. N. Sal'nikova, V. G. Andrianov, Koord. Khim. **3** (1977) 1607 – 1617.

(34) K. Nakatsu, T. Mitsudo, H. Nakanishi, Y. Watanabe, Y. Takegami, Chem. Letters 1977, 1447 – 1448.

(35) J. Klimes, E. Weiss, Angew. Chem. **94** (1982) 207; Angew. Chem. Int. Ed. Engl. **21** (1982) 205; Angew. Chem. Suppl. 1982, 477 – 482.

(36) W. K. Dean, D. G. Vanderveer, J. Organomet. Chem. **145** (1978) 49 – 55.

(37) J. Miller, A. L. Balch, J. H. Enemark, J. Amer. Chem. Soc. **93** (1971) 4613 – 4614.

(38) J. M. Castro, H. Hope, Inorg. Chem. **17** (1978) 1444 – 1447.

(39) P. E. Riley, R. E. Davis, N. T. Allison, W. M. Jones, Inorg. Chem. **21** (1982) 1321 – 1328.

(40) W. P. Fehlhammer, P. Hirschmann, H. Stolzenberg, J. Organomet. Chem. **224** (1982) 165 – 180.

(41) G. G. Aleksandrov, V. V. Skripkin, N. E. Kolobova, Yu. T. Struchkov, Koord. Khim. **5** (1979) 453 – 458.

(42) W. M. Butler, J. H. Enemark, J. Organomet. Chem. **49** (1973) 233 – 238.

(43) R. D. Adams, D. F. Chodosh, N. M. Golembeski, E. C. Weissman, J. Organomet. Chem. **172** (1979) 251 – 267.

(44) K. Aoki, Y. Yamamoto, Inorg. Chem. **15** (1976) 48 – 52.

(45) D. Mansuy, M. Lange, J. C. Chottard, J. F. Bartoli, B. Chevrier, R. Weiss, Angew. Chem. **90** (1978) 828 – 829; Angew. Chem. Int. Ed. Engl. **17** (1978) 781 – 782.

(46) P. Luger, G. Ruban, Acta Cryst., Sect. B **27** (1971) 2276 – 2279.

(47) G. Bombieri, F. Faraone, G. Bruno, G. Faraone, J. Organomet. Chem. **188** (1980) 379 – 387.

(48) H. D. Empsall, E. M. Hyde, R. Markham, W. S. McDonald, M. C. Norton, B. L. Shaw, B. Weeks, J. Chem. Soc. Chem. Comm. 1977, 589 – 590.

(49) G. R. Clark, W. R. Roper, A. H. Wright, J. Organomet. Chem. **236** (1982) C7 – C10.

(50) P. Friedrich, G. Besl, E. O. Fischer, G. Huttner, J. Organomet. Chem. **139** (1977) C68 – C72.

(51) A. D. Redhouse, J. Organomet. Chem. **99** (1975) C29 – C30.

(52) B. L. Haymore, G. Hillhouse, W. A. Herrmann, private communication.

(53) W. A. Herrmann, K. Weidenhammer, M. L. Ziegler, Z. Anorg. Allg. Chem. **460** (1980) 200 – 206.

(54) S. Fontana, U. Schubert, E. O. Fischer, J. Organomet. Chem. **146** (1978) 39 – 44.

(55) U. Schubert, Organometallics **1** (1982) 1085 – 1088.

(56) E. O. Fischer, J. Chen, U. Schubert, Z. Naturf., Sect. B, **37** (1982) 1284 – 1288.

(57) E. O. Fischer, W. Kleine, W. Schambeck, U. Schubert, Z. Naturf., Sect. B, **36** (1982) 1575 – 1579.

(58) a) W. Malisch, H. Blau, U. Schubert, Angew. Chem. **92** (1980) 1065 – 1066, **93** (1981) 134; Angew. Chem. Int. Ed. Engl. **19** (1980) 1020 – 1021, **20** (1981) 216. b) W. Malisch, H. Blau, U. Schubert, Chem. Ber. **116** (1983) 690 – 709.

(59) N. E. Kolobova, L. L. Ivanov, O. S. Zhvanko, I. N. Chechulina, A. S. Batsanov, Yu. T. Struchkov, J. Organomet. Chem. **238** (1982) 223 – 229.

(60) J. Y. Le Marouille, C. Lelay, A. Benoit, D. Grandjean, D. Touchard, H. Le Bozec, P. Dixneuf, J. Organomet. Chem. **191** (1980) 133 – 142.

(61) N. I. Pyshnograeva, V. N. Setkina, V. G. Andrianov, Yu. T. Struchkov, D. N. Kursanov, J. Organomet. Chem. **206** (1981) 169 – 176.

(62) G. Huttner, D. Regler, Chem. Ber. **105** (1972) 1230 – 1244.

(63) C. P. Casey, R. A. Boggs, D. F. Marten, J. C. Calabrese, J. Chem. Soc. Chem. Comm. 1973, 243 – 244.

(64) M. Green, J. R. Moss, I. W. Nowell, F. G. A. Stone, J. Chem. Soc. Chem. Comm. 1972, 1339 – 1340.

(65) F. Carré, G. Cerveau, E. Colomer, R. J. P. Corriu, J. Organomet. Chem. **229** (1982) 257 – 273.

(66) S. B. Butts, S. H. Strauss, E. M. Holt, R. E. Stimson, N. W. Alcock, D. F. Shriver, J. Amer. Chem. Soc. **102** (1980) 5093 – 5100.

(67) W. K. Dean, J. B. Wetherington, J. W. Moncrieff, Inorg. Chem. **15** (1976) 1566 – 1572.

(68) M. F. Lappert, P. L. Pye, G. M. McLaughlin, J. Chem. Soc. Dalton 1977, 1272 – 1282.

(69) M. F. Lappert, P. L. Pye, A. J. Rogers, G.M. McLaughlin, J. Chem. Soc. Dalton, 1981, 701 – 704.

(70) L. Y. Y. Chan, W. K. Dean, W. A. G. Graham, Inorg. Chem. **16** (1977) 1067 – 1071.

(71) N. A. Bailey, P. L. Chell, A. Mukhopadhyay, H. E. Tabbron, M. J. Winter, J. Chem. Soc. Chem. Comm. 1982, 215 – 217.

(72) C. K. Prout, T. S. Cameron, A. R. Gent, Acta Cryst., Sect. B, **28** (1972) 32 – 36.

(73) J. R. Knox, C. K. Prout, Acta Cryst., Sect. B, **25** (1969) 1952 – 1958.

(74) R. D. Adams, D. F. Chodosh, J. Amer. Chem. Soc. **99** (1977) 6544 – 6550.

(75) P. K. Baker, G. K. Barker, M. Green, A. J. Welch, J. Amer. Chem. Soc. **102** (1980) 7811 – 7812.

(76) L. Messerle, M. D. Curtis, J. Amer. Chem. Soc. **104** (1982) 889 – 891.

(77) S. R. Allen, M. Green, A. G. Orpen, I. D. Williams, J. Chem. Soc. Chem. Comm. 1982, 826 – 828.

(78) H. Hoberg, G. Burkhart, C. Krüger, Y.-H. Tsay, J. Organomet. Chem. **222** (1981) 343 – 352.

(79) W. K. Dean, R. S. Charles, D. G. Van Derveer, Inorg. Chem. **16** (1977) 3328 – 3333.

(80) K. Miki, H. Taniguchi, Y. Kai, N. Kasai, K. Nishiwaki, M. Wada, J. Chem. Soc. Chem. Comm. 1982, 1178 – 1180.

(81) G. R. Clark, K. Marsden, W. R. Roper, L. J. Wright, J. Amer. Chem. Soc. **102** (1980) 1206 – 1207.

(82) S. M. Boniface, G. R. Clark, J. Organomet. Chem. **208** (1981) 253 – 260.

(83) C. M. Jensen, T. J. Lynch, C. B. Knobler, H. D. Kaesz, J. Amer. Chem. Soc. **104** (1982) 4679 – 4680.

(84) P. Domiano, A. Musatti, M. Nardelli, G. Predieri, J. Chem. Soc. Dalton 1975, 2165 – 2168.

(85) R. D. Wilson, Y. Kamitori, H. Ogoshi, Z.-I. Yoshida, J. A. Ibers, J. Organomet. Chem. **173** (1979) 199 – 209.

(86) H. C. Clark, C. R. C. Milne, N. C. Payne, J. Amer. Chem. Soc. **100** (1978) 1164 – 1169.

(87) W. M. Butler, J. H. Enemark, Inorg. Chem. **10** (1971) 2416 – 2419.

108   *Ulrich Schubert*

(88) O. P. Anderson, A. B. Packard, Inorg. Chem. **17** (1978) 1333 – 1337.

(89) A. Modinos, P. Woodward, J. Chem. Soc. Dalton 1974, 2065 – 2069.

(90) G. K. Anderson, R. J. Cross, L. Manojlović-Muir, K. W. Muir, R. A. Wales, J. Chem. Soc. Dalton 1979, 684 – 689.

(91) E. K. Barefield, A. M. Carrier, D. J. Sepelak, D. G. Van Derveer, Organometallics **1** (1982) 103 – 110.

(92) E. M. Badley, K. W. Muir, G. A. Sim, J. Chem. Soc. Dalton 1976, 1930 – 1933.

(93) Lj. Manojlović-Muir, K. W. Muir, J. Chem. Soc. Dalton 1974, 2427 – 2433.

(94) Yu. T. Struchkov, G. G. Aleksandrov, V. B. Pukhnarevich, S. P. Sushchinskaya, M. G. Voronkov, J. Organomet. Chem. **172** (1979) 269 – 272.

(95) D. F. Christian, D. A. Clarke, H. C. Clark, D. H. Farrar, N. C. Payne, Can. J. Chem. **56** (1978) 2516 – 2525.

(96) R. F. Stepaniak, N. C. Payne, Can. J. Chem. **56** (1978) 1602 – 1609.

(97) R. F. Stepaniak, N. C. Payne, J. Organomet. Chem. **57** (1973) 213 – 223.

(98) R. F. Stepaniak, N. C. Payne, J. Organomet. Chem. **72** (1974) 453 – 464.

(99) R. F. Stepaniak, N. C. Payne, Inorg. Chem. **13** (1974) 797 – 801.

(100) W. M. Butler, J. H. Enemark, Inorg. Chem. **12** (1973) 540 – 544.

(101) W. M. Butler, J. H. Enemark, J. Parks, A. L. Balch, Inorg. Chem. **12** (1973) 451 – 457.

(102) P. Braunstein, E. Keller, H. Vahrenkamp, J. Organomet. Chem. **165** (1979) 233 – 242.

(103) S. Z. Goldberg, R. Eisenberg, J. S. Miller, Inorg. Chem. **16** (1977) 1502 – 1507.

(104) M. Berry, J. Martin-Gil, J. A. K. Howard, F. G. A. Stone, J. Chem. Soc. Dalton 1980, 1625 – 1629.

(105) R. Walker, K. W. Muir, J. Chem. Soc. Dalton 1975, 272 – 276.

(106) C. P. Casey, C. R. Cyr, R. L. Anderson, D. F. Marten, J. Amer. Chem. Soc. **97** (1975) 3053 – 3059.

(107) U. Schubert, K. Ackermann, P. Rustenmeyer, J. Organomet. Chem. **231** (1982) 323 – 334.

(108) M. J. Webb, M. J. Bennett, L. Y. Y. Chan, W. A. G. Graham, J. Amer. Chem. Soc. **96** (1974) 5931 – 5932.

(109) W. Beck, K. Raab, U. Nagel, M. Steinmann, Angew. Chem. **94** (1982) 556 – 557; Angew. Chem. Int. Ed. Engl. **21** (1982) 526 – 527.

(110) E. O. Fischer, P. Rustenmeyer, O. Orama, D. Neugebauer, U. Schubert, J. Organomet. Chem. **247** (1983) 7 – 19.

(111) E. O. Fischer, P. Rustenmeyer, D. Neugebauer, Z. Naturf. Sect. B, **35** (1980) 1083 – 1087.

(112) W. A. Kiel, G.-Y. Lin, A. G. Constable, F. B. McCormick, C. E. Strouse, O. Eisenstein, J. A. Gladysz, J. Amer. Chem. Soc. **104** (1982) 4865 – 4878.

(113) M. J. Doyle, M. F. Lappert, G. M. McLaughlin, J. McMeeking, J. Chem. Soc. Dalton 1974, 1494 – 1501.

(114) B. Cetinkaya, M. F. Lappert, G. M. McLaughlin, K. Turner, J. Chem. Soc. Dalton, 1974, 1591 – 1598.

(115) M. Cowie, J. A. Ibers, Inorg. Chem. **15** (1976) 552 – 557.

(116) K. Itoh, I. Matsuda, F. Ueda, Y. Ishii, J. A. Ibers, J. Amer. Chem. Soc. **99** (1977) 2118 – 2126.

(117) P. B. Hitchcock, M. F. Lappert, G. M. McLaughlin. A. J. Oliver, J. Chem. Soc. Dalton, 1974, 68 – 74.

(118) R. J. Sundberg, R. F. Bryan, I. F. Taylor, H. Taube, J. Amer. Chem. Soc. **96** (1974) 381 – 392.

(119) H. J. Krentzien, M. J. Clarke, H. Taube, Bioinorg. Chem. **4** (1975) 143 – 151.

(120) G. R. Clark. J. Organomet. Chem. **134** (1977) 51 – 65.

(121) P. B. Hitchcock, M. F. Lappert, P. L. Pye, S. Thomas, J. Chem. Soc. Dalton, 1979, 1929 – 1942.

(122) P. B. Hitchcock, M. F. Lappert, P. L. Pye, J. Chem. Soc. Dalton 1978, 826 – 836.

(123) L. J. Guggenberger, R. R. Schrock, J. Amer. Chem. Soc. **97** (1975) 6578 – 6579.

(124) R. R. Schrock, L. W. Messerle, C. D. Wood, L. J. Guggenberger, J. Amer. Chem. Soc. **100** (1978) 3793 – 3800.

(125) M. R. Churchill, F. J. Hollander, J. Amer. Chem. Soc. **17** (1978) 1957 – 1962.

(126) A. J. Schultz, R. K. Brown, J. M. Williams, R. R. Schrock, J. Amer. Chem. Soc. **103** (1981) 169 – 176.

(127) L. W. Messerle, P. Jennische, R. R. Schrock, G. Stucky, J. Amer. Chem. Soc. **102** (1980) 6744 – 6752.

(128) M. R. Churchill, W. J. Youngs, Inorg. Chem. **18** (1979) 1930 – 1935.

(129) M. R. Churchill, H. J. Wasserman, Inorg. Chem. **20** (1981) 2899 – 2904.

(130) L. J. Guggenberger, R. R. Schrock, J. Amer. Chem. Soc. **97** (1975) 2935.

(131) O. S. Mills, A. D. Redhouse, Angew. Chem. **77** (1965) 1142 – 1143; Angew. Chem. Int. Ed. Engl. **4** (1965) 1142.

(132) C. P. Casey, T. J. Burkhardt, C. A. Bunnell, J. C. Calabrese, J. Amer. Chem. Soc. **99** (1977) 2127 – 2134.

(133) E. O. Fischer, F. J. Gammel, J. O. Besenhard, A. Frank, D. Neu-gebauer, J. Organomet. Chem. **191** (1980) 261 – 282.

(134) J.-C. Daran, Y. Jeannin, Acta Cryst., Sect. B, **36** (1980) 1392 – 1395.

(135) C. A. Toledano, J. Levisalles, M. Rudler, H. Rudler, J.-C. Daran, Y. Jeannin, J. Organomet. Chem. **228** (1982) C7 – C11.

(136) C. P. Casey, A. J. Shusterman, N. W. Vollendorf, K. J. Haller, J. Amer. Chem. Soc. **104** (1982) 2417 – 2423.

(137) J. L. Davidson, M. Shiralian, Lj. Manojlovic-Muir, K. W. Muir, J. Chem. Soc. Chem. Comm. 1979, 30 – 32.

(138) G. A. Carriedo, D. Hodgson, J. A. K. Howard, K. Marsden, F. G. A. Stone, M. J. Went, P. Woodword, J. Chem. Soc. Chem. Comm. 1982, 1006 – 1008.

(139) P. T. Wolczanski, R. S. Threlkel, J. E. Bercaw, J. Amer. Chem. Soc. **101** (1979) 218 – 220.

(140) J. A. Marsella, K. Folting, J. C. Huffman, K. G. Caulton, J. Amer. Chem. Soc. **103** (1981) 5596 – 5598.

(141) M. R. Churchill, W. J. Youngs, Inorg. Chem. **18** (1979) 2454 – 2458.

(142) M. R. Churchill, A. L. Rheingold, Inorg. Chem. **21** (1982) 1357 – 1359.

(143) M. R. Churchill, A. L. Rheingold, W. J. Youngs, R. R. Schrock, J. H. Wengrovius, J. Organomet. Chem. **204** (1981) C17 – C20.

(144) M. R. Churchill, J. R. Missert, W. J. Youngs, Inorg. Chem. **20** (1981) 3388 – 3391.

(145) J. H. Wengrovius, R. R. Schrock, M. R. Churchill, H. J. Wasserman, J. Amer. Chem. Soc. **104** (1982) 1739 – 1740.

(146) M. R. Churchill, A. L. Rheingold, H. J. Wasserman, Inorg. Chem. **20** (1981) 3392 – 3399.

(147) E. O. Fischer, C. G. Kreiter, H. J. Kollmeier, J. Müller, R. D. Fischer, J. Organomet. Chem. **28** (1971) 237 – 258.

(148) C. G. Kreiter, E. O. Fischer, XXIIIrd Int. Congress of Pure and Appl. Chem. Vol. 6, p. 151 – 168, Butterworths, London, 1971.

(149) U. Schubert, D. Neugebauer, P. Hofmann, B. E. R. Schilling, A. Motsch, H. Fischer, Chem. Ber. **114** (1981) 3349 – 3365.

(150) J. Miller, A. L. Balch, Inorg. Chem. **11** (1972) 2069 – 2074.

(151) a) B. E. R. Schilling, R. Hoffmann, D. L. Lichtenberger, J. Amer. Chem. Soc. **101** (1979) 585 – 591; b) N. M. Kostić, R. F. Fenske, J. Amer. Chem. Soc. **104** (1982) 3879 – 3884.

(152) E. O. Fischer, E. W. Meineke, F. R. Kreissl, Chem. Ber. **110** (1977) 1140 – 1147.

(153) a) J. W. Lauher, R. Hoffmann, J. Amer. Chem. Soc. **98** (1976)

1729 – 1742; b) P. Kubáček, R. Hoffmann, Z. Havlas, Organometallics **1** (1982) 180 – 188.

(154) B. E. R. Schilling, R. Hoffmann, J. W. Faller, J. Amer. Chem. Soc. **101** (1979) 592 – 598.

(155) Lj. Manojlović-Muir, K. W. Muir, Inorg. Chim. Acta **10** (1974) 47 – 49.

(156) T. A. Albright, R. Hoffmann, J. C. Thibeault, D. L. Thorn, J. Amer. Chem. Soc. **101** (1979) 3801 – 3812.

(157) A. R. Rossi, R. Hoffmann, Inorg. Chem. **14** (1975) 365 – 374.

(158) R. R. Schrock, Acc. Chem. Res. **12** (1979) 98 – 104.

(159) R. J. Goddard, R. Hoffmann, E. D. Jemmis, J. Amer. Chem. Soc. **102** (1980) 7667 – 7676.

Electronic Structures of Transition Metal Carbene Complexes

By Peter Hofmann

1 Introduction

The initial report by Fischer and Maasböl (1) in 1964 of the targeted synthesis and structural characterization of the transition metal carbene complex, **1**, laid the foundations for a field of organometallic research which has since

$$(CO)_5W=C\underset{CH_3}{\overset{---OCH_3}{<}} \qquad \underline{\underline{1}}$$

attracted many active groups from virtually all branches of chemistry. Transition metal carbene (alkylidene) complexes have played and continue to play an important role in their own right in both synthetic and structural organometallic chemistry. Moreover, recent years have shown that these compounds may be of significance in organic syntheses and catalytic reactions. Mechanistic studies of transformations involving metal carbene species are now beginning to reveal the extensive range of applications offered by systems such as **2**, where a carbene unit C(X)(Y) ist bound to a transition metal ligand frag-

$$L_nM=C\underset{Y}{\overset{--X}{<}} \qquad \underline{\underline{2}}$$

ment ML_n. (The double bonds in **1** and **2** are drawn in only for convenience, as discussed later.)

From the large amount of experimental work carried out in the past by various research groups – most notably by E.O. Fischer and his collaborators in Munich, the "homebase" of transition metal carbene complexes – much insight has been gained into the bonding and electronic structure of these molecules. Synthetic efforts, including successes and failures alike, together with all kinds of spectroscopic investigations and many comparative molecular structure determinations have delineated both the limitations of stability and the basic reactivity patterns of carbene complexes. Improvement in experimental techniques has enabled the isolation of – or at least proof of the fleeting existence of – a number of hitherto elusive metal carbene species, ranging from matrix isolated species through gas phase molecular fragments to reactive intermediates in solution experiments. Thermochemical data are beginning to appear and the picture which emerges from experimental studies is becoming ever more complete.

From the mid-seventies, theoretical chemistry has paid increasing attention to such molecules. Throughout chemistry there exists the need to use theoretical models or quantitative calculations to generalize experimental findings and to accommodate them within some appropriate theoretical framework. The framework should not only be of value in the interpretation of experimental data, but should also permit extrapolations and predictions to be made.

The number of theoretical papers to date dealing with the electronic structure of transition metal carbene complexes, is not very large. Electronic structure theory here refers to the application of quantum-chemical calculations, in particular MO theory in its manifold stages of sophistication.

This chapter presents a brief, and admittedly limited, survey of theoretical work described in the literature insofar as it concerns mononuclear transition metal carbene systems with terminal, simple carbene ligands, $C(X)(Y)$ (see **2**).

Instead of reporting published results in detail, however, emphasis will be put upon a more general MO-theoretical interpretation of certain model carbene complexes. In the latter the numerical results of specific computations are of less importance than a qualitative understanding of the overall bonding situation and its implications for the structural and chemical properties of carbene complexes. We shall focus in particular on the class of "Fischer-type" transition metal carbene complexes and provide a detailed electronic structure analysis using qualitative MO theory for d^6 pentacarbonyl carbene systems, as exemplified by the prototype molecule $(CO)_5Cr = CH_2$, methylenepentacarbonylchromium(0). This approach is based not upon the author's preference, but upon the rather restricted capabilities of numerical computations, including *ab initio* calculations, in the field of transition metal organometallics (2), which are often of no greater use to the chemist than an accurate qualitative picture. Furthermore, this approach is intended to benefit those chemists not specialized in quantum chemistry and its methodology who possess a basic knowledge of applied molecular orbital theory.

After a general introduction to the problems associated with MO calculations for transition metal organometallics, and a short digression on carbene ligands per se, the basic molecular orbital structures of a few representative transition metal carbene complexes will be described. We shall start with $(CO)_5Cr = CH_2$ and discuss it in some detail before mentioning the interesting aspects of other systems. The simplest carbene, methylene, CH_2, and appropriate metal ligand fragments will form the basis of our electronic structure considerations. Where appropriate, reference will be made to related theoretical or experimental studies in the literature, and the implications concerning the predicted structural, reactivity and spectroscopic behavior of species will be elucidated. Structural variations within the carbene moiety, and to some extent those within the ML_n fragment, and their effect on electronic structures will be addressed in specific cases. As this report cannot possibly cover all aspects of electronic structure theory relating to transition metal carbene complexes, many will be mentioned only in passing. It is nevertheless hoped to provide a picture detailed and complete enough to be of some use to those interested in this field of organometallic chemistry.

2 Quantum-chemical Treatment of Transition Metal Carbene Complexes

Because molecular orbital theory provides a valuable quantum-chemical framework for the description of electronic structure in various transition metal carbene complexes, ist seems appropriate here to say a few words about its inherent limitations. Although recent years have seen an increasing number of *ab initio* treatments of "chemically interesting" (*e.g.* moderately sized) organometallic systems, our potential in performing numerically accurate electronic structure calculations is in fact still quite limited. The MO theory of transition metal organometallics has not yet reached the stage of electronic structure calculations performed on small organic (in general light atom) systems. For the latter, computations of molecular structures and properties using *ab initio* techniques at or beyond the Hartree-Fock limit are routinely performed nowadays and can be accomplished with an accuracy equivalent to or surpassing that attainable in experimental work. In addition, a number of highly sophisticated parametrized and rather accurate semi-empirical methods are available for light atom molecules. For transition metal systems this is not the case for obvious reasons.

The semi-empirical MO methods used today for heavy atom molecules, which are based on the Extended Hückel, CNDO, INDO, $X - \alpha$ or related formalisms, yield qualitative or at best semi-quantitative results. Such results can be of high value and utility, when only qualitative and comparative insights into electronic structure are sought. Moreover, they can help us to understand structural relationships and reactivity patterns when they form the basis of perturbation theory approaches and symmetry arguments. Indeed, the current goal of the electronic structure theory of organometallics is not so much the accurate numerical calculation of parameters for single molecules. It concerns rather the crucially important − especially in the area of experimental chemistry − devising of generally applicable, qualitatively reliable and transferable concepts to describe and organize the structural and reactivity patterns of organometallics. The numerous facets of experimental knowledge can thereby be unified and eventually utilized to predict new chemistry. This kind of approach, which exploits orbital interaction rules, frontier orbital theory and symmetry considerations, is well exemplified and has proved to be extremely valuable in the field of organic chemistry. It is again being pioneered, currently for inorganic and organometallic systems, in the work of

Hoffmann (3). Fascinating relationships and analogies between organic and inorganic molecules have been brought to light (4).

In parallel with this type of "applied" MO theory, the development of *ab initio* methods continues apace. Apart from the heavy computational costs involved, certain specific problems arise for heavy atom molecules, such as the transition metal carbene complexes. All-electron SCF-MO calculations, even at the single determinant level and approaching the Hartree-Fock limit, require first and foremost sufficiently large, flexible, balanced and adequate basis sets for the constituent atoms. Schäfer has recently summarized the problems and successes of the near Hartree-Fock theory of transition metal molecules (5). It is not only the basis sets employed that inevitably restrict the reliability of any *ab initio* treatment. Full geometrical optimization of organometallic systems is still often not feasible or − more accurately − not successful, even with very large basis sets, *e.g.* that for ferrocene (6). Frequently, calculations beyond the Hartree-Fock limit seem to be necessary to reproduce experimental findings correctly and therefore various approaches to compute partially correlated wave functions are being used. Relativistic effects become important for heavy atom systems, and in contrast to organic molecules, their closely bunched electronic states and sometimes fluxional nature, as well as the lack of experimental information about reaction pathways and intermediates, further complicate the matter.

Many variants of *ab initio* techniques using differing basis sets and varying approaches to get beyond the HF limit, utilizing all-electron or pseudopotential formalisms of both relativistic and non-relativistic type, are now being evolved by theoretical groups specializing in the calculation of the electronic structure of organometallics. Theory will undoubtedly devise ever more efficient and accurate methods. A complementary disadvantage of this development, however, seems to be that work of this type is becoming increasingly difficult to interpret and understand in terms of simple qualitative concepts of bonding. As soon as the much heralded millenium of the fully correlated "true" total wave function for organometallic molecules finally arrives, such an "understanding", in the opinion of some observers, may no longer be necessary and satisfaction may be obtained from the exact numbers being computed. As we are still far removed from this utopia, there remains the need for method-independent ways of describing the electronic structures of such systems. Often in a sense this is a prerequisite for discussing the results of calculations with chemists.

In the main we shall make use here of the fragment approach (3), i.e. the electronic structures, geometries *etc.* of metal carbene complexes will be described in terms of the molecular orbitals of appropriate fragments (building blocks), which may or may not exist in isolated form. The utilization of fragment MOs of well understood subunits enables us to employ standard rules for orbital interactions (intra- and intermolecular perturbation theory) (7), especially within the highly interesting valence (frontier) orbital region of the fragments and composite molecules in question.

For transition metal carbene complexes, a natural way of inspecting their electronic structure involves partitioning **2** into various carbene ligands $C(X)(Y)$ and metal ligand fragments ML_n. Let us consider the carbene ligands first.

3 The Carbene Ligands

For free carbenes, C(X)(Y), an extremely large body of experimental and theoretical information is now available. Many reviews have appeared, a recent one (8) providing an extensive and excellent overview of theoretical work undertaken to date.

Although the simplest carbene system, CH_2, has been found in only a few exceptional metal carbene complexes of type **2** (vide infra), we discuss this prototype first.

CH_2 is know to have a non-linear triplet groundstate structure, with an energy difference of about 9 kcal/mole separating it from the stronger bent singlet state. MO theory has played a crucial role here in guiding experimental progress to the present point, where both experimental and advanced theoretical determinations for the structure and energy of the triplet/singlet separation virtually converge (to within 1 to 2 kcal/mole). Stable transition metal carbene complexes generally exhibit a closed shell singlet structure with the carbene ligand geometry being approximately trigonal planar with respect to the carbene carbon atom. Accordingly, we can focus on the electronic and MO structure of a bent singlet CH_2 unit (C_{2v} symmetry). The actual level occupation pattern, singlet or triplet, is of no relevance here, since it is only the distribution of electrons within the MO manifold of the composite complexes that has significance. In terms of delocalized, symmetry-adapted (canonical) valence MOs, the bonding capability of a bent singlet CH_2 is dominated by its two frontier levels: the a_1 HOMO, an sp^2-type hybrid orbital on the carbon (with small bonding admixture of hydrogen 1s functions) and the LUMO, b_2, a pure carbon 2p atomic orbital. At a much lower energy, there is also a CH-bonding orbital, b_1, which will be seen lateron to play a certain role too. A diagram of these CH_2 wave functions is given in **3** (9).

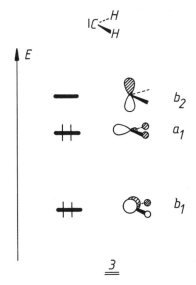

$$\underline{\underline{3}}$$

A CH_2 ligand thus provides a σ-donor orbital (a_1) for a metal center as well as a (single faced) empty p-type acceptor level (b_2) capable of forming π-bonds through back-donation from the metal. The final polarity and reactivity of a metal-methylene bond will depend upon the relative weighting of the synergistic interactions of the carbene valence levels with the ML_n fragment.

In substituted carbenes, CH(X) or C(X)(Y), the basic character of the HOMO and LUMO will still be preserved. The typical "Fischer-type" carbene complexes usually carry one or two groups (X and/or Y), which are π-donor substituents, such as $-NH_2$, -NHR, $-NR_2$, -OR or -SR. Free carbenes of this type have singlet groundstates and the qualitative reasons for this are to be seen in **4a** for a CH(X) carbene.

The LUMO is shifted to a higher energy relative to that for the CH_2 owing to a destabilizing π-interaction with the filled donor orbital of X. The HOMO experiences a stabilizing effect, due to the electronegativity of the substituent, which is usually greater than that of H (or alkyl). The π-system of the carbene CH(X) in **4a** can also be viewed as the strongly polarized π and π^* orbitals of a heteroethylene. Accordingly, the carbenes C(X)(Y) or $C(X)_2$ having two donor groups in an α-position will give rise to an allyl-type π-MO pattern with a total of 4 π electrons. The LUMO is thus the orbital given in **4b** and is located even higher in energy than for **4a**. As the HOMO-LUMO separation in cases **4a** and **4b** increases, a singlet groundstate will result. The destabilization of the LUMO will increase the more (the carbene is a weaker acceptor),

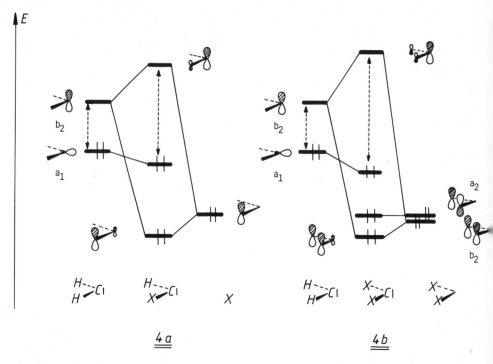

$\underline{\underline{4a}}$ $\qquad$ $\underline{\underline{4b}}$

the better donor groups X and/or Y are, consistent with the strategy of stabilizing singlet carbenes proposed early on (10). Even this simple picture indicates that the substituents X and/or Y will play an important role in determining the relative σ-donor and π-acceptor character of metal-bonded carbene ligands. As will become apparent, the carbene substituents generally provide an important electronic balancing mechanism for stabilizing carbene complexes and will compete with various attached metal fragments ML_n to utilize or satisfy the electronic needs of the carbene center. It should also be mentioned that attachment of substituents with p- or π-type orbitals causes especially the LUMO of a carbene C(X)(Y) to be less localized on the carbon atom, since the wave function then spreads out over the groups X and/or Y as well, resulting in a different overlap situation with the carbon atom.

The single-faced nature of the carbene's acceptor level will be responsible for the adoption of particular conformational options in metal complexes if the metal fragment ML_n is of the appropriate structure.

After this simple picture of the electronic structure of carbene ligands, we now turn to a description of the bonding in a prototype transition metal carbene complex.

4 Methylenepentacarbonylchromium(0), $(CO)_5Cr(CH_2)$, the Prototype of "Fischer-type" Carbene Complexes

As a lot of the fundamental work on typical tansition metal carbene complexes is concerned and still focuses on the d^6 metals Cr(0), Mo(0), W(0) and their carbonyl carbene derivatives – especially in Fischer's laboratories – we shall discuss a prototype molecule of this kind in some detail, as well as the implications of its electronic structure.

We shall take as a typical model the species $(CO)_5Cr(CH_2)$, which has not yet been characterized. Derivatives of this simple system, with aryl or aryl and alkyl groups replacing the hydrogens, are known for Cr and W (11). No Hartree-Fock level calculations for $(CO)_5Cr(CH_2)$ are available yet, but will undoubtedly appear in the near future.

The picture presented here is based on unpublished Extended Hückel calculations (12) and makes use of the aforementioned fragment approach.

A practicable approach to the molecular structure of $(CO)_5Cr(CH_2)$ employs both $Cr(CO)_5$ and CH_2 fragments and constructs the molecule from the orbital structure of these two subunits as indicated in **5**.

$$\underline{\underline{5}}$$

4.1 The $Cr(CO)_5$ Fragment

The $Cr(CO)_5$ fragment present in an idealized $(CO)_5Cr(CH_2)$ molecule (C_{2v} symmetry) is a typical d^6 ML_5 system having a square pyramidal structure, viz. an "octahedral" fragment with C_{4v} symmetry. This fragment – and $d^n ML_n$ transition metal ligand fragments in general – have been extensively described from an electronic standpoint in the literature (13). For $Cr(CO)_5$ in particular, a detailed *ab initio* calculation by Hay has been published (14).

The valence MOs of $Cr(CO)_5$, **6**, can easily be derived from the well-known orbital pattern of an octahedral $Cr(CO)_6$ by taking into account the removal of one CO ligand.

$$\underline{\underline{6}}$$

Figure 1 shows in qualitative terms what happens to the e_g and t_{2g} orbitals of a general $M(CO)_6$ system if transformation to a C_{4v} $M(CO)_5$ system occurs. The d-block valence MOs of $M(CO)_6$, e_g and t_{2g}, are depicted in an appropriate representation on the left of the diagram, displaying the relevant metal d AO and ligand contributions. (The coordinate system has been chosen with later diagrams for the composite molecule $(CO)_5Cr(CH_2)$ in mind.)

Removing one CO ligand causes a loss of anti-bonding for the z^2 member of e_g (15). This MO drops in energy and, because of the reduced symmetry, takes on some s and z character; the resulting MO is somewhat rehybridized toward the site of the missing ligand. The molecular orbital formed, a_1, is the LUMO of a C_{4v} $M(CO)_5$ system for a d^6 electron count. The $x^2 - y^2$ member of e_g remains unaffected to first order by the CO loss and becomes the b_1 of the $M(CO)_5$ at high energy. Similarly, the xy member of t_{2g} stays and is transformed into the b_2 of the $M(CO)_5$ as the wave function has no contribution from the CO ligand which has been removed. Both the xz and yz components of t_{2g} lose back-bonding when the CO group is taken off and a slight destabilization of these two levels occurs, though the resulting e set of $M(CO)_5$ is still energetically just above b_2. The subscripts s and a for the e orbital of $Cr(CO)_5$, the degenerate HOMO, denote its symmetric or antisymmetric nature with respect to the xz plane. On the right of Fig. 1 qualitative diagrams of the developing valence MOs are given. In the diagrams below utilizing the $Cr(CO)_5$ fragment only the metal contributions to the levels will be shown.

A d^6 $Cr(CO)_5$ molecular fragment thus provides four relevant valence MOs of predominantly d character: the LUMO a_1 (z^2, s, z), the two degenerate HOMOs, e_s (xz) and e_a (yz) and a slightly more stable occupied MO b_2 (xy). Contour plots to assist in visualizing the real appearance of these wave functions are available in the literature (16).

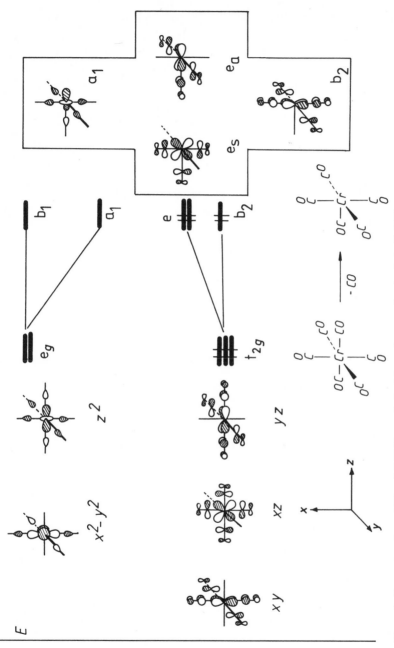

Figure 1: Fragment valence MOs of Cr(CO)$_5$, as derived from the t$_{2g}$ and e$_g$ levels in Cr(CO)$_6$. Metal and CO contributions are given.

4.2 The Electronic Structure of (CO)₅Cr(CH₂)

Now we are in a position to construct the MO interaction diagram for
(CO)₅Cr(CH₂), where the model carbene ligand CH₂ has replaced the original

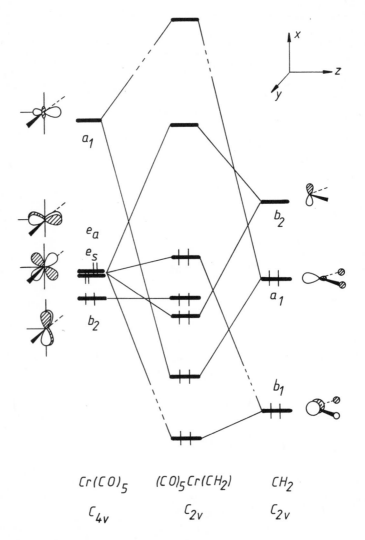

$$Cr(CO)_5 \qquad (CO)_5Cr(CH_2) \qquad CH_2$$
$$C_{4v} \qquad\qquad C_{2v} \qquad\qquad C_{2v}$$

Figure 2: Interaction diagram for (CO)₅Cr(CH₂). Only the metal contributions to the Cr(CO)₅
wave functions are shown. The highest and lowest MO levels are not to scale.

sixth CO group in $Cr(CO)_6$. Figure 2 builds the valence MOs of $(CO)_5Cr(CH_2)$ from the $Cr(CO)_5$ levels (on the left) and those of CH_2 (on the right).

For simplicity, the conformation chosen for the molecule has the two $C-H$ bonds eclipsing two of the four equatorial CO groups (see **5**), but, as will be seen, this conformational choice is irrelevant from the electronic viewpoint. The principal bonding relationships result from the interaction of the σ-acceptor orbital a_1 of $Cr(CO)_5$ and th σ-donor level a_1 of CH_2, and from the bonding interaction of the p-type acceptor orbital b_2 of CH_2 with the appropriate member e_s of the $Cr(CO)_5$ HOMO. The other component, e_a, is somewhat destabilized by a 2 orbital/4 electron interaction with the aforementioned low-lying, filled CH bonding orbital b_1 of the CH_2 ligand. The fragment MO of xy-type of $Cr(CO)_5$ (b_2 in C_{4v}) remains unaffected since it displays δ-symmetry toward the carbene part which has no orbitals of this symmetry available. The interesting valence orbital region of $(CO)_5Cr(CH_2)$ shown in Fig. 2 thus comprises five levels characterized in ascending order as follows:

(i)   The carbene to metal σ-bonding level, the bonding combination of the a_1 orbitals from both fragments;

(ii) The three levels emerging from the b_2 and e of $Cr(CO)_5$, mainly of metal d character. They correspond to t_{2g} of $Cr(CO)_6$, but now, because of the perturbed symmetry, they are split. The lowest of them, derived from e_s (xz), brings π-back-bonding to the carbene acceptor level. Note that xy (b_2) is back-bonding only to carbonyls, and that yz (e_a) is the least stable member of this MO triad, for it is destabilized by the CH bonds of CH_2, *i.e.* by b_1 of CH_2.

(iii) The LUMO of the complex $(CO)_5Cr(CH_2)$, which is the anti-bonding linear combination of e_s (xz) and the CH_2 b_2 acceptor level. This MO is heavily concentrated on the carbene center, a fact of some importance when we investigate the consequences of our MO picture.

The highest level shown in Fig. 2, the σ-anti-bonding level between the Cr and CH_2 and also lowest MO, C-H bonding within the CH_2 unit, will play no further role in our discussion.

4.3 The Implications of the $(CO)_5Cr(CH_2)$ MO Scheme

Using the molecular orbital scheme in Fig. 2 we now can draw a series of conclusions for our simple model system of "Fischer-type" carbene complexes and their derivatives.

4.3.1 Conformational and Structural Consequences

It is clear from the outset that, despite the formal presence of the metal to carbon "double bond" (σ-dative interaction and π-back-donation), no electronically imposed rotational barrier will exist for our model or for $(CO)_5CrC(X)(Y)$ carbene complexes. For any rotational orientation of the carbene ligand, the p-acceptor orbital of carbon will find an appropriate linear combination of both e HOMOs of the $Cr(CO)_5$ for back-bonding, leaving the bonding pattern shown in Fig. 2 essentially unchanged. Indeed, the observed orientations of carbene ligands in most cases are staggered toward the equatorial carbonyl ligands; the final conformations found in crystal structures are determined by steric requirements and crystal packing effects. For less symmetric metal fragments, ML_n, however, the situation is rather different.

In the overwhelming majority of carbene complexes, the hydrogen atoms of the CH_2 are replaced by heteroatomic substituents bearing lone pairs. These groups compete with the attached metal fragment as far as π-donation to the p-acceptor function of the carbene carbon is concerned. As shown in Section 3, carbene ligands of this type possess a high-lying, less carbon-centered LUMO and so back-donation from metal to carbene in the presence of π-donor substituents becomes less important. Especially for first row substituents, such as -OR or, to a greater extent, amino functions, π-interactions to the carbene carbon are quite strong. The overall situation for all three bonds emanating from the carbene carbon atom results from a mutual balance of these three groups. This is clearly reflected by the bond lengths to the carbon revealed from X-ray structure determinations of related carbene complexes of d^6 $M(CO)_5$ fragments, as discussed by Schubert in the preceding chapter. Generally, increased π-interaction of the carbene p-orbital with one of the attached groups will lessen conjugation with the others. The competition between donors ultimately serves to stabilize the carbene complex by reducing the carbon atom's electronic deficiency (due to σ-donation and the empty acceptor orbital of C) and by shifting the LUMO to a higher energy. Experimentally determined rotational barriers for carbon substituent bonds also reveal the mutual interplay of π-effects.

$(CO)_5Cr=C \begin{smallmatrix} NMe_2 \\ \\ Me \end{smallmatrix}$ $(CO)_5Cr=C \begin{smallmatrix} OEt \\ \\ NMe_2 \end{smallmatrix}$ $(CO)_5Cr=C \begin{smallmatrix} OMe \\ \\ Me \end{smallmatrix}$

$\underline{\underline{7}}$ $\underline{\underline{8}}$ $\underline{\underline{9}}$

The barrier for the $C-N$ bond in **7** (17) of 25 kcal/mole is reduced to 21 kcal/mole for **8** (17), the barrier for the $C-O$ bond in **8** is 8 kcal/mole as compared to about 14 kcal/mol in **9** (17). In each case the conformation of the $X-R$ groups is such as to keep R within the carbene plane, thereby making available the p-type lone pair on X for π-bonding to the carbene center, as shown in **10a** and **10b**, respectively, the E and Z isomeric forms of such systems.

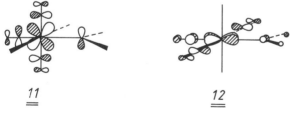

<u>10 a</u>   <u>10b</u>

Such stereoisomers can be observed, their relative stability and detailed steric arrangement being a consequence of the space-filling requirements superimposed upon the best conjugative situation (see previous chapter).

α-Heteroatom-stabilized carbene ligands are generally weaker π-acceptors than carbon monoxide. This has been shown by Fenske et al. in a detailed MO study of various substituted carbene ligands and their $Cr(CO)_5$ complexes (18). Based on the nodal pattern of the two valence MOs of $Cr(CO)_5$ (Fig. 1, e_s, e_a) or on those of a $(CO)_5CrC(X)(Y)$ complex, which "feel" the replacement of one carbonyl by a carbene (*i.e.* **11** *and* **12**, $e_s + p$ and e_a-$b_{1(CH_2)}$, respectively, in Fig. 2; xy is unaffected), the presence of the weak acceptor ligand $C(X)(Y)$ should exert a relative bond shortening effect on the *trans* metal-CO

<u>11</u>   <u>12</u>

bond compared to that for the *cis* carbonyls. For MO **11**, $(e_s + p)$, a weakly accepting carbene will leave more wave function density between the metal and the three carbonyls shown in **11**. For MO **12**, $(e_a$-$b_{1(CH_2)})$, the total lack of any suitable acceptor function on the carbon within the yz plane will do the same for the three COs in **12**, which mix into this level. The net effect for the eclipsed conformation due to **11** and **12** should thus be a stronger reinforcement of the back-bonding to the *trans*-CO (via **11** *and* **12**) than to both (different) pairs of *cis*-CO ligands. For the commonly encountered staggered rotamers, this conclusion will still hold, except that the four carbonyls will become more equivalent (totally equivalent in fact if a symmetric carbene CX_2 is present).

Of course, a σ-effect underlies the π-trends discussed so far if the carbene ligand is a better σ-donor than CO; both influences will be hard to separate out and to assess in quantitative terms. Structural studies have shown for weakly π-accepting carbenes in $(CO)_5MC(X)(Y)$ systems that the *trans* metal-CO bond is indeed shortened, whereas the *cis* shortening (again compared to $M(CO)_6$) is hardly noticeable. Clearcut observation of these effects is often hampered by a lack of accuracy in the structural data.

Apparently, however, the π-effect wins and IR studies also show unambiguously that the *trans* carbonyls are bound more strongly than the *cis* CO ligands.

So far we have concentrated on $Cr(CO)_5$ (or d^6 $M(CO)_5$) as the bonding partner for a carbene moiety. The degeneracy of the e HOMO of the $Cr(CO)_5$ fragment (xz, yz) is crucial here. In going to less symmetrically substituted $M(CO)_{5-x}L_x$ fragments instead of pentacarbonyls, it is easy to derive the likely consequences for conformational options. Comparisons of a number of $Cr(CO)_{5-x}L_x$ fragments with $Cr(CO)_5$ have been presented by us in the literature (19). It may be sufficient here to show the changes induced by a *cis* ligand L, which is a pure σ-donor (or a much weaker π-acceptor than CO). Changing from $Cr(CO)_5$ to $Cr(CO)_4L$ (C_s), **13**, for instance will destabilize the e_a and b_2

13

of $Cr(CO)_5$ because the back-bonding to one ligand site is lost or diminished. This is depicted qualitatively in Figure 3; e_s is unchanged.

In orther words, the degenerate HOMO (e_s, e_a) of $Cr(CO)_5$ is now split, "e_a" (yz) being higher in energy. According to the general rules for orbital

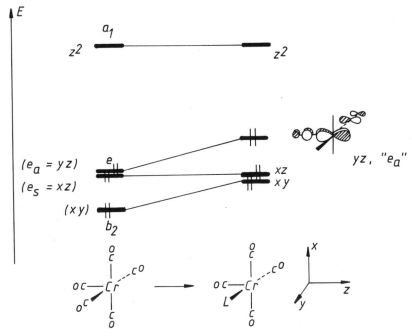

Figure 3: Evolution of the valence MOs of Cr(CO)₄L (C$_s$ symmetry, L = σ-donor) from those of Cr(CO)₅ (C$_{4v}$).

interactions, the carbene p-acceptor level will preferentially interact with this higher-lying (closer in energy) d level "e$_a$" (yz), and the preferred carbene orientation **14** will be the consequence. The carbene plane will be perpendicu-

14

lar to the M − L bond, thereby aligning the p-acceptor orbital with this bond. The rotational barrier has an electronic component now, superimposed upon the steric factors. Similarly, any other ligand substitution patterns within the ML$_5$ fragment can be analyzed. Examples for this type of preferred orientation are mentioned (and their steric limitations discussed) in the preceding chapter.

Let us now turn to some additional implications and experimental verifications of the electronic scheme outlined for $(CO)_5Cr(CH_2)$.

4.3.2 Photoelectron Spectra

UV-photoelectron spectroscopy provides a means of "inspecting" the highest occupied orbital manifold of molecules. Although the invalidity of Koopmans' theorem (large Koopmans' defects) often creates problems in the theoretical interpretation of PE data for transition metal organometallics, detailed studies of the photoelectron spectra of a series of nine $(CO)_5CrC(X)$-(Y) carbene complexes together with MO calculations by Block and Fenske (20) support the basic orbital pattern derived for $(CO)_5Cr(CH_2)$. Compared to the lowest energy band of $Cr(CO)_6$ (which results from a triply degenerate ionization of the t_{2g} orbital of the hexacarbonyl) the spectra of the $(CO)_5CrC(X)(Y)$ systems show split bands for the d ionizations and at higher energy the ionization of the metal carbene σ-bonding MO can be localized (see Fig. 2).

Jolly et al. have studied $(CO)_5CrC(OCH_3)(CH_3)$ by X-ray photoelectron spectroscopy (21). In terms of resonance theory, all three formulas in **15** appear to contribute significantly to the electronic structure, a result consistent with our MO description.

15

4.3.3 Electrochemistry and ESR

The chemical reactivity of molecular systems can often be deduced from the type and energetic location of their frontier MOs. For transition metal carbene complexes, the highest occupied orbitals are heavily metal-localized (d-type) levels. The lowest unoccupied orbital according to Fig. 2 for

$(CO)_5Cr(CH_2)$ not only has a rather low energy, well separated from the next orbitals above, but is also concentrated largely on the carbene moiety. This LUMO, shown in **16** for our model, is the anti-bonding combination of the e_s of $Cr(CO)_5$ and the p of CH_2 with the major weight in the wave function coming from the latter.

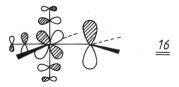

$$\underline{\underline{16}}$$

This appearance of the LUMO is consistent with ^{13}C-NMR and chemical studies, which have suggested an "inverse ylide" or metallo-carbenium ion nature for the carbene complexes $(CO)_5CrC(X)(Y)$ (see **15c**).

The nature of the LUMO **16** was verified experimentally by Krusić et al. (22), who showed that a series of $(CO)_5MC(X)(Y)$ complexes ($M = Cr$, Mo, W) could easily be reduced to their radical anions, the latter having been studied by ESR. A comparison of ESR data for **17** and the known organic radical **18** along with MO calculations is indicative of a singly occupied MO in

$$(CO)_5Cr - \overset{\cdot}{\underset{\ominus}{C}} \underset{C_6H_5}{\overset{OMe}{<}} \qquad H - \overset{\cdot}{C} \underset{C_6H_5}{\overset{OMe}{<}}$$

$$\underline{\underline{17}} \qquad\qquad\qquad \underline{\underline{18}}$$

17 localized only to about 5 % on the $(CO)_5Cr$ moiety and to 95 % on the carbene ligand. It is reasonable to assume that the SOMO (singly occupied molecular orbital) of radical anions like **17** is largely identical in character to the LUMO of neutral precursors. The degree of localization of the unpaired electron is variable within any series of complexes studied, though the basic feature of a carbene ligand centered LUMO (delocalized of course within the carbene fragment) always prevails. More studies of this type, expecially involving electrochemical work of quantitative nature, would be desirable.

If we relate the LUMO of the $(CO)_5Cr(CH_2)$ species, **16**, to the well-known LUMO of organic carbonyl compunds, **19**, the sometimes useful analogy of a $Cr(CO)_5$ group and an oxygen atom becomes clearer in MO terms. We shall come later to the implications for the reactivity.

$$\pi^*_{CO} \quad \text{⟨orbital diagram 19⟩}$$

$$O = C \text{⟨⟩}$$

Some information is available on the oxidation of metal carbene complexes to their radical cations from electrochemical studies (23a). Mass spectroscopic ionization potential measurements have demonstrated that electron loss occurs from the three highest levels of Fig. 2, and so electrochemical oxidation becomes easier with increasing donor capability of the carbene α-substituents. A theoretical description of the radical cation species (which might have to go beyond the single configuration level) is not available and currently there appears to be no further ESR work on such species. Very interesting results have been obtained from electrochemical, ESR, structural and theoretical studies for bis-carbene metal complexes (23b), although they will not be discussed here.

4.3.4 Electron Spectroscopy

From the MO scheme of Fig. 2, with the dominant d-character of the three highest occupied and nearly degenerate orbitals in sharp contrast to the carbene-centered and energetically isolated LUMO, one might expect an allowed metal to ligand charge transfer band as the transition of lowest energy in the electronic spectra of $(CO)_5MC(X)(Y)$ complexes. A full investigation of $(CO)_5CrC(R)(OCH_3)$ and $(CO)_5CrC(CH_3)(X)$ complexes revealed this to be the case (24). One relatively sharp and intense absorption band is found in the visible region around 24 000 cm^{-1}, which shifts in a predictable way as the carbene substitution pattern is varied. In passing we mention that recently photochemical *cis-trans* isomerization of a transition metal-carbene bond was observed for a complex of Re, the specific unsymmetry of the metal fragment causing the rotational barrier around the $Re = C$ linkage to be large enough for the isolation of two geometric isomers (25). Presumably, the excited state again has one electron promoted to the carbene (in this case: benzylidene) ligand, thus reducing the rotational barrier as in photoexcited olefins.

4.3.5 Reactivity

The electronic structure of $(CO)_5Cr(CH_2)$ suggests that for $(CO)_5MC(X)(Y)$ complexes the electron deficiency at the carbene carbon, in conjunction with the low-lying LUMO centered on the unique ligand, will dominate reactivity patterns in the groundstate, at least in competition with reaction channels opened by CO loss (thermal dissociation of the carbene ligand is not normally observed). Attack by nucleophiles will be a major feature for "Fischer-type" carbene complexes. Although we shall not report here on experimental studies of reactions of metal carbene complexes with nucleophiles, the general picture which emerges fits well with a frontier orbital controlled interaction of nucleophiles with the carbene center. Theoretical papers utilizing Fenske-Hall MO calculations (26) treat this aspect in detail. MO model calculations (27) for insertion reactions, as shown in **20** (28), also reveal a nucleophilic attack

$$(CO)_5Cr=C\overset{X}{\underset{Y}{\diagup}} \quad + \quad IN\equiv C-\bar{N}R_2 \quad \longrightarrow \quad (CO)_5Cr=C\overset{\diagup \bar{N}R_2}{\underset{N=C\overset{X}{\underset{Y}{\diagup}}}{\diagdown}}$$

$$\underline{\underline{20}}$$

on the carbene center to induce the insertion step. The calculations also indicate a stereochemistry of attack highly reminiscent of that for nucleophilic attack on carbonyl groups (29). An overall positive charge on a carbene complex of course enhances the rate of nucleophilic attack; the regioselectivity however results from the localization of the LUMO.

Similarly again to the situation in organic carbonyl compounds, $C-H$ bonds in α-position to the carbene carbon have been shown to be rather acidic. Anions such as **21** can easily be prepared and are useful synthetically (30).

$$(CO)_5Cr=C\overset{OR}{\underset{\underset{\ominus}{\bar{C}H_2}}{\diagup}} \quad \longleftrightarrow \quad (CO)_5\overset{\ominus}{\bar{C}}r-C\overset{OR}{\underset{CH_2}{\diagup}}$$

$$\underline{\underline{21}}$$

The limited scope of this chapter does not enable us to give more specific examples of experimental work to further elucidate the bonding and molecular orbital picture elaborated in some detail for one single group of metal carbene complexes using the parent molecule $(CO)_5Cr(CH_2)$ as a model. The qualitative conclusions from MO theory and the fragment MO approach, which can be obtained for related systems, are nicely reflected by experiment and have demonstrated the utility of such an approach.

5 Other Types
of Transition Metal Carbene Complexes

Metal carbene complexes derived directly from d^6 hexacarbonyls represent but one subclass of this type of organometallic compounds. In principle, the electronic structure of any ML_n carbene complex can be fully described and analyzed in terms of suitable fragment MOs as for $(CO)_5Cr(CH_2)$ above. For many representative prototypes of ML_n carbene systems this has been done or can be deduced from the literature. We consider it appropriate here to mention only some typical classes of carbene complexes together with citations of relevant published theoretical work.

5.1 $(\eta^5\text{-}C_5H_5)M(CO)_2$(Carbene) and Related Systems
(Pseudooctahedral Complexes)

A large number of carbene complexes in which three carbonyl groups of a d^6 hexacarbonyl or its derivatives are formally replaced by a six-electron donor arene ligand – usually $\eta^5\text{-}C_5H_5^-$, $\eta^5\text{-}C_5H_4(CH_3)^-$ or occasionally $\eta^6\text{-}C_6H_6$ – have been prepared and studied in some detail. The bonding capabilities and frontier levels of the general $CpML_2$ or $CpMLL'$ (Cp $= \eta^5\text{-}C_5H_5$) sixteen-electron fragment, **22**, have been extensively discussed by Hoffmann (31) and others (32).

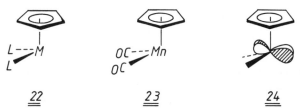

22   **23**   **24**

For d^6 fragments, exemplified by $CpMn(CO)_2$, **23**, or by $CpFe(CO)_2^+$, $CpFe(PR_3)_2^+$, etc., the three highest occupied MOs deriving from the "t_{2g}-like" set of pseudooctahedral complexes such as $CpMn(CO)_3$ are split in energy. As in *cis*-substituted chromiumpentacarbonyl carbene complexes, this leads to conformational preferences for a single-faced carbene group. Electronically, the dominant interaction between the p of the carbene and the HOMO a'' of **23**, depicted in **24** with its metal contributions, favors the conformation **25** rather than **26**, which would have steric advantages. Both types of conformation have been observed in X-ray studies. Again, steric *and*

25   26

electronic properties of the carbene ligands determine the conformational option adopted in the solid state. The $M = C(X)(Y)$ rotational barriers are rather low; in known cases experimental values compare well with calculated ones. Generally, arene-containing fragments are better donors toward carbenes than their isoelectronic and isolobal (4) metal carbonyl moieties. Complexes with carbene substituents [such as $CpMn(CO)_2(CMe_2)$] can be isolated which are much less stable when a pentacarbonyl fragment is bonded to the alkylidene ligand. Apart from the theoretical study by Hoffmann et al. (31) on $CpM(CO)_2$ complexes, outlining the electronic structure for $CpFe(CO)_2(CH_2)^+$, Fenske and Kostic also performed an interesting MO analysis of conformations of $CpMn(CO)_2$ and $(C_6H_6)Cr(CO)_2$ carbene complexes (33). A second study by these authors dealt *inter alia* with the electronic structure of $CpFe(PH_3)_2(CH_2)^+$ and its interaction with nucleophiles and electrophiles (26).

5.2 Trigonal Bipyramidal d⁸ (CO)₄M(Carbene) Complexes

A number of $(CO)_4MC(X)(Y)$ complexes have been prepared experimentally. If one takes $(CO)_4Fe(CH_2)$ as a model for such systems, the $Fe(CO)_4$ fragment in a trigonal bipyramidal molecule can be either of C_{3v} or of C_{2v} symmetry, as in **27** or **28** (where ligands are omitted for clarity). The methylene ligand is then held in either an apical or equatorial position in the trigonal bipyramid, as in **29** or **30/31**. While no sound numerical prediction is currently available for the parent system, $(CO)_4Fe(CH_2)$, to indicate whether **29** or **30** is of lower energy, **31** can be eliminated as a possible groundstate conformation on the basis of fragment MO considerations. The HOMO of a C_{2v}

27   28   29   30   31

$Fe(CO)_4$ fragment **28** is shown in **32** and readily explains why an equatorial carbene ligand would favor conformation **30** over **31**, analogously to the

32 High-energy good donor MO   33 Low-energy, poor donor MO

ethylene orientation in $(CO)_4Fe$(ethylene) complexes. For **31**, the carbene LUMO has to overlap with a lower-lying valence MO of C_{2v} $Fe(CO)_4$ and weaker back-donation results. A rotation barrier exists for **30**. For **29**, which has an apical carbene ligand, no electronically based rotational barrier for the $Fe=C$ bond is to be expected. Both the HOMO of **27** and the next two lower levels are e sets as is the HOMO in $Cr(CO)_5$. The lower-lying e set, shown on the metal in **33**, is mainly responsible for carbene metal π-interactions. Qualitatively, one would predict **30** as favored, though from the well-known fluxional nature of d^8 trigonal bypyramidal complexes, one must infer a rather small energy difference for **29** vs. **30**. The derivation of such small energetic differences, however, is beyond qualitative fragment analysis and perhaps even out of the ambit of *ab initio* approaches at the moment.

In all $(CO)_4Fe$(carbene) complexes for which structures are known from diffraction studies or spectroscopy, the carbene ligands are good donors but rather poor π-acceptors because of one or two α-substituents having lone pairs. The carbene ligand always occupies the apical position, as in models **34** (34) and **35** (35), with its orientation determined by steric factors. This is in accord with a theoretical analysis on site preferences for transition metal pentacoordination (36).

$$\underset{\underset{=}{34}}{} \qquad\qquad \underset{\underset{=}{35}}{}$$

A molecular orbital study (INDO, EH) for a series of $(CO)_4Fe(carbene)$ complexes along with photoelectron spectroscopic observations has recently appeared (37).

5.3 Tetrahedral $(CO)_3M(Carbene)$ Systems

The electronic structure of C_{3v} $M(CO)_3$ molecular fragments is well documented from MO studies (38) and has been amply used in discussing $M(CO)_3$ or ML_3 complexes having differing types of ligands (39). In regard to our topic, it is important to note that a very recent *ab initio*, near Hartree-Fock study of a realistic transition metal carbene complex has been performed for the model system $(CO)_3Ni(CH_2)$. This system is so elusive that it has not yet been prepared, although substituted derivatives are known (40). The work by Schaefer et al. (41) contains an interesting comparison of $(CO)_3Ni(CH_2)$ and $Ni(CO)_4$, *i.e.* of the CH_2 and CO ligands, and analyzes in detail the bonding situation in the model carbene complex. The quantitative results of this *ab initio* study of $(CO)_3Ni(CH_2)$ can be readily interpreted qualitatively on a fragment MO basis as $Ni(CO)_3$ interacting with CH_2. A predicted total (Mulliken) charge of $-0,58$ for the carbene carbon seems especially interesting. As $Ni(CO)_3$ offers an e-type π-donating HOMO, practically no barrier to internal rotation will exist.

5.4 Square Planar $L_3M(Carbene)$ Complexes

A molecular orbital analysis of such compounds has been published in connection with a study on the bonding, rotational barriers and conformational preferences of ethylene complexes (42). Generally speaking, steric requirements seem to establish the preferred geometry as **36** rather than **37** for square planar L_3M(carbene) systems.

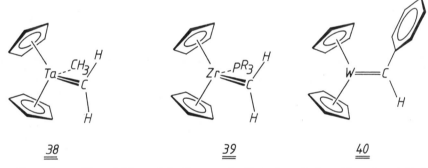

The well-known fragment MOs for a C_{2v} ML_3 unit can again serve as an appropriate basis for discussing variations in structural details as a function of the ligands L.

5.5 Cp₂M(Carbene) and Cp₂ML(Carbene) Complexes

Here we discuss only three complexes, namely **38**, **39** and **40** (43 – 45).

Complexes **38** and **39** contain the unsubstituted methylene ligand and their electronic and geometric structure is easily understandable in terms of an orbital structure of a bent Cp_2M or a "pyramidal" Cp_2ML fragment with a d^4 or d^2 electron count. The frontier orbital structure of these building blocks has been extensively studied from the theoretical standpoint (46). For a d^2 Cp_2ML fragment, such as $Cp_2Ta(CH_3)$ or $CpZr(PR_3)$, the HOMO, shown in **41** qualitatively and only for the metal center, is a high-lying donor orbital with π-symmetry pointing toward an incoming CH_2. For the d^4 Cp_2M fragment, **42**, the b_2 orbital as the HOMO presents the same nodal characteristic toward a carbene as the third ligand. In both cases, only the carbene orientation shown in **41** and **42** allows for good π-interaction and this is in fact a rather strong one. The benzylidene ligand conformation in **40**, which, as in **38**

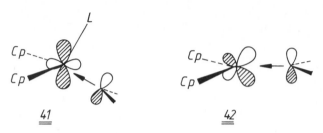

41   42

and **39** for CH$_2$, is contrary to steric expectations, thus becomes readily understandable. A strong metal-carbon double bond is formed: the rotational
barrier for the CH$_2$ ligand in **38** was found to be too high for observation in an
NMR experiment (47). Owing to the high donor capability of d^2 Cp$_2$ML and
d^4 Cp$_2$M fragments, carbene complexes such as **38 – 40** display nucleophilic
character at the carbene center, giving rise to fascinating chemistry. There is a
whole family of nucleophilic metal carbene complexes presently under active
investigation. Interestingly, this inverse, "ylide-like" polarity of the metal-
carbene double bond is found for electron-poor metal centers and is clearly in
sharp contrast to "Fischer-type" carbene complexes, where the carbene
ligands are electron-deficient even though the metal has a higher number of d
electrons. The issues of electrophilic versus nucleophilic nature and of total
charges for carbene ligands in various transition metal carbene systems have
been addresses in an MO study of unusual M-C-H angles in electron-deficient
carbene complexes of mainly tantalum (48), as synthesized in particular by
Schrock (49) ("Schrock-type" metal carbene complexes). The energy and
overlap match of the frontier orbitals of CH$_2$ or CHR ligands and metal fragments in complexes such as **43** result in a very high population for the p

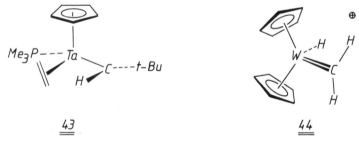

43   44

acceptor orbital in alkylidenes and therefore in a nucleophilic nature for such
carbene ligands. The LUMO here has a higher energy than that for electrophilic carbene complexes and is moreover much less localized on the carbene.

The electronic structure of the postulated reaction intermediate **44** (50), again a CH_2 complex, has also been briefly discussed in (48), as well as the carbene to carbyne α-hydrogen abstraction. **44** seems to have an electrophilic carbene center.

6 Other Theoretical Studies of Transition Metal Carbene Species

Ab initio MO theory at the single configuration SCF level has been employed in an extensive study of several of the electronic states of $MnCH_2$ (51). Another study using GVB calculations including configuration interaction has investigated the electronic structure of $NiCH_2$ (52). The latter system was also included in the MO study of $(CO)_3Ni(CH_2)$ by Schaefer mentioned earlier. Based on the mechanism of the olefin metathesis reaction, Goddard et al. carried out GVB *ab initio* calculations to elucidate the role of high-valent oxoalkylidene complexes (53). An *ab initio* study combined with EH calculations by Dedieu (54) has also treated such species from an electronic structure viewpoint.

7 Addendum

After this chapter had been completed, an *ab initio* SCF MO study of the hydroxycarbene complexes $(CO)_5Cr(CHOH)$ and $(CO)_4Fe(CHOH)$ was published (55). The results for the chromium complex agree very nicely with the picture drawn above.

For $(CO)_4Fe(CHOH)$ the apical position of the carbene ligand is predicted to be more stable than the equatorial one by 8 kcal/mole in accord with the rather weak acceptor capability of a hydroxycarbene ligand. We still expect the unsubstituted CH_2 ligand to prefer an equatorial site as in **30**. In contrast to results from EH or INDO calculations for a C_{3v} $Fe(CO)_4$ **27** (3, 4, 37), the 2e HOMO of this fragment is described as having mainly 4p character (55). It is not clear, however, to what extent this is a consequence of the choice of the basis set in this study.

Acknowledgement: Part of this review was written while the author held a visiting professorship at the Anorganisch-chemisches Institut der TU München. The hospitality of Prof. E.O. Fischer during this time and his support after the author's move to the present address is cordially acknowledged. The author would also like to thank Prof. Roald Hoffmann and Dr. M.M.L. Chen for making unpublished material available. Discussions with colleagues in Munich are highly appreciated, as is the skillfull help of Mrs. I. Rauscher in preparing the manuscript.

8 References

(1) E.O. Fischer, A. Maasböl: Angew. Chem. **76** (1964) 645; Angew. Chem. Int. Ed. Engl. **3** (1964) 580.

(2) P. Hofmann: Nachr. Chem. Techn. Lab. **30** (1982) 498 – 506.

(3) R. Hoffmann: Science **211** (1981) 995 – 1002; R. Hoffmann, T.A. Albright, D.L. Thorn: Pure Appl. Chem. **50** (1978) 1 – 9; for a recent detailed review see T.A. Albright: Tetrahedron **38** (1982) 1339 – 1388.

(4) R. Hoffmann: Angew. Chem. **94** (1982) 725 – 739; Angew. Chem. Int. Ed. Engl. **21** (1982) 711 – 724.

(5) H.F. Schaefer III: J. Mol. Struct. (Theochem) **76** (1981) 117 – 135.

(6) H.P. Lüthi, J. Ammeter, J. Almlöf, K. Korsell: Chem. Phys. Lett. **69** (1980) 540 – 542.

(7) See for example: R. Hoffmann: Acc. Chem. Res. **4** (1971) 1 – 9; J.K. Burdett: Molecular Shapes. Wiley, New York (1980); B.M. Gimarc: Molecular Structure and Bonding. Academic Press, New York 1979; E. Heilbronner, H. Bock: Das HMO Modell und seine Anwendung. Verlag Chemie, Weinheim 1968, vol. 1; I. Fleming: Frontier Orbitals and Organic Chemical Reactions. Wiley, London 1976, and references cited therein.

(8) E.R. Davidson: Singlet Triplet Energy Separation in Carbenes and Related Diradicals, in W.T. Borden (Ed.): Diradicals. Wiley, New York 1982.

(9) Three-dimensional contour plots of these MOs can be found in: W.L. Jorgensen, L. Salem: The Organic Chemist's Book of Orbitals. Academic Press, New York 1973.

(10) R. Gleiter, R. Hoffmann: J. Am. Chem. Soc. **90** (1968) 5457 – 5460.

(11) E.O. Fischer, W. Held, R.R. Kreissl, A. Frank, G. Huttner: Chem. Ber. **110** (1977) 656 – 666; C.P. Casey, T.J. Burkhardt: J. Am. Chem. Soc. **95** (1973) 5833 – 5834. C.P. Casey, S.W. Polichnowsky, A.J. Shusterman, C.R. Jones: J. Am. Chem. Soc. **101** (1979) 7282 – 7292.

(12) P. Hofmann: unpublished work; M.M.L. Chen: Ph.D. Thesis, Cornell University 1975. The author thanks Prof. R. Hoffmann and Dr. M.M.L. Chen for providing a copy of this work.

(13) M. Elian, R. Hoffmann: Inorg. Chem. **14** (1975) 1058 – 1076. See also references (3) and (4).

(14) P.J. Hay: J. Am. Chem. Soc. **100** (1978) 2411 – 2417.

(15) For simplicity z^2 is used instead of $3d_{z^2}$, z instead of $4p_z$ etc.

(16) See J.K. Burdett's book cited in reference (7), page 224.

(17) C.G. Kreiter, E.O. Fischer: XXIII International Congress of Pure and Applied Chemistry (Boston). Butterworths 1971, vol. 6; C.G. Kreiter, E.O. Fischer: Angew. Chem. **81** (1969) 780 – 781; Angew. Chem. Int. Ed. Engl. **8** (1976) 761 – 762.

(18) T.F. Block, R.F. Fenske: J. Organometal. Chem. **139** (1977) 235 – 296; T.F. Block, R.F. Fenske, C.P. Casey: J. Am. Chem. Soc. **98** (1976) 441 – 443.

(19) U. Schubert, D. Neugebauer, P. Hofmann, B.E.R. Schilling, H. Fischer, A. Motsch: Chem. Ber. **114** (1981) 3349 – 3365.

(20) T.F. Block, R.F. Fenske: J. Am. Chem. Soc. **99** (1977) 4321 – 4330.

(21) W.B. Perry, T.F. Schaaf, W.L. Jolly, L.J. Todd, D.L. Cronin: Inorg. Chem. **13** (1974) 2038 – 2039; For X-ray PES work on Pd and Pt carbene complexes see: P. Brant, J.E. Enemark, A.L. Balch: J. Organometal. Chem. **114** (1976) 99 – 106.

(22) P.J. Krusić, U. Klabunde, C.P. Casey, T.F. Block: J. Am. Chem. Soc. **98** (1976) 2015 – 2018.

(23) a) M.K. LLoyd, J.A. McCleverty, D.G. Orchard, J.A. Connor, M.B. Hall, I.H. Hillier, E.M. Jones, G.K. McEwen: J.C.S. Dalton **1973** 1743 – 1747.
b) R.D. Rieke, H. Kojima, K. Öfele: J. Am. Chem. Soc. **98** (1976) 6735 – 6737; R.D. Rieke, H. Kojima, K. Öfele: Angew. Chem. **98** (1980) 550 – 552; Angew. Chem. Int. Ed. Eng. **19** (1980) 538 – 540; M.F. Lappert, R.W. McCabe, J.J. McQuitty, P.L. Pye, P.I. Riley: J.C.S. Dalton **1980** 90 – 98; K. Ackermann, P. Hofmann, F.H. Köhler, H. Kratzer, H. Krist, K. Öfele, H.R. Schmidt, Z. Naturforsch. **39b** in press.

(24) E.O. Fischer, C.G. Kreiter, H.J. Kollmeier, J. Müller, R.D. Fischer: J. Organometal. Chem. **28** (1971) 237 – 258.

(25) F.B. McCormick, W.A. Kiel, J.A. Gladysz: Organometallics **1** (1982) 405 – 408; W.A. Kiel, G.-Y. Lin, J.A. Gladysz: J. Am. Chem. Soc. **102** (1980) 3299 – 3301; W.A. Kiel, G.-Y. Lin, A.G. Constable, F.B. McCormick, C.E. Strouse, O. Eisenstein, J.A. Gladysz: J. Am. Chem. Soc. **104** (1982) 4865 – 4878.

(26) N.M. Kostić, R.F. Fenske: Organometallics **1** (1982) 974 – 982.

(27) P. Hofmann, H. Fischer, R. Märkl: in preparation.

(28) H. Fischer, U. Schubert: Angew. Chem. **93** (1981) 482 – 483; Angew. Chem. Int. Ed. **20** (1981) 461 – 463.

(29) A.J. Stone, R.W. Erskine: J. Am. Chem. Soc. **102** (1980) 7185 – 7192 and references cited therein.

(30) See, for instance, C.P. Casey: Metal Carbene Complexes in Organic Synthesis, in H. Alper (Ed.): Transition Metal Organometallics in Organic Sythesis. Academic Press, New York 1976, vol. I. Also see the chapter by K.H. Dötz in this book.

(31) B.E.R. Schilling, R. Hoffmann, D.L. Lichtenberger: J. Am. Chem. Soc. **101** (1979) 585 – 591.

(32) P. Hofmann: Angew. Chem. **89** (1977) 551 – 552; Angew. Chem. Int. Ed. Engl. **16** (1977) 536 – 537.

(33) N.M. Kostić, R.F. Fenske: J. Am. Chem. Soc. **104** (1982) 3879 – 3884.

(34) K. Öfele, C.G. Kreiter: Chem. Ber. **105** (1972) 529 – 540; G. Huttner, W. Gartzke: Chem. Ber. **105** (1972) 2714 – 2725.

(35) J. Daub. G. Endress, U.Erhardt, K.H. Jogun, J. Kappler, A. Laumer, R. Pfiz, J.J. Stezowski: Chem. Ber. **115** (1982) 1787 – 1809.

(36) A. Rossi, R. Hoffmann: Inorg. Chem. **14** (1975) 365 – 374.

(37) M.C. Böhm, J. Daub, R. Gleiter, P. Hofmann, M.F. Lappert, K. Öfele: Chem. Ber. **113** (1980) 3629 – 3646.

(38) M. Elian, M.M.L. Chen, D.M.P. Mingos, R. Hoffmann: Inorg. Chem. **15** (1976) 1148 – 1155; J.K. Burdett: Inorg Chem. **14** (1975) 375 – 382; J.K. Burdett: J.C.S., Faraday Trans. 2 **70** (1974) 1599 – 1613.

(39) T.A. Albright, P. Hofmann, R. Hoffmann: J. Am. Chem. Soc. **99** (1977) 7546 – 7557; T.A. Albright, R. Hoffmann, P. Hofmann: Chem. Ber. **111** (1978) 1591 – 1602; T.A. Albright, P. Hofmann, R. Hoffmann, C.P. Lillya, P.A. Dobosh: J. Am. Chem. Soc. **105** (1983) 3396 – 3411; T.A. Albright, R. Hoffmann, Y. Tse, T. D'Ottavio: J. Am. Chem. Soc. **101** (1979) 3812 – 3821; P. Hofmann: Z. Naturforsch. **33b** (1978) 251 – 260; T. Ziegler, A. Rauk: Inorg. Chem. **18** (1979) 1755 – 1759.

(40) E.O. Fischer, F.R. Kreissl, E. Winkler, C.G. Kreiter: Chem. Ber. **105** (1972) 588 – 598.

(41) D. Spangler, J.J. Wendoloski, M. Dupuis, M.M.L. Chen, H.F. Schaefer III: J. Am. Chem. Soc. **103** (1981) 3985 – 3990.

(42) T.A. Albright, R. Hoffmann, J.C. Thibeault, D.L. Thorn; J. Am. Chen. Soc. **101** (1979) 3801 – 3812.

(43) R.R. Schrock: J. Am. Chem. Soc.**97** (1975) 6577 – 6578.

(44) J. Schwartz, K.I. Gell: J. Organometal. Chem. **184** (1980) C1 – C2.

(45) J.A. Marsella, K. Folting, F.C. Huffman, K.G. Caulton: J. Am. Chem. Soc. **103** (1981) 5596 – 5598.

(46) J.W. Lauher, R. Hoffmann: J. Am. Chem. Soc. **98** (1976) 1729 – 1742; P. Hofmann, P. Stauffert, N.E. Schore: Chem. Ber. **115** (1982) 2153 – 2174.

(47) L.J. Guggenberger, R.R. Schrock: J. Am. Chem. Soc. **97** (1975) 6578 – 6579.

(48) R.J. Goddard, R. Hoffmann, E.D. Jemmis: J. Am. Chem. Soc. **102** (1980) 7667 – 7676.

(49) Review: R.R. Schrock: Acc. Chem. Res. **12** (19797 98 – 104.

(50) N.J. Cooper, M.L.H. Green: J.C.S., Dalton Trans. **1979** 1121 – 1127.

(51) B.R. Brooks, H.F. Schaefer III: Mol. Phys. **34** (1977) 193 – 213.

(52) A.K. Rappé, W.A. Goddard III: J. Am. Chem. Soc. **99** (1977) 3966 – 3968.

(53) A.K. Rappé, W.A. Goddard III: J. Am. Chem. Soc. **104** (1982) 448 – 456.

(54) S. Nakamura, A. Dedieu: Nouv. J. Chim. **6** (1982) 23 – 30.

(55) H. Nakatsuji, J. Uskio, T. Yonezawa: J. Am. Chem. Soc. **105** (1983) 426 – 434.

Metal Complexes from Carbene Complexes: Selected Reactions

By Fritz R. Kreissl

1 Introduction

The first planned synthesis of a transition metal carbene complex by E. O. Fischer and A. Maasböl opened up a highly interesting research field in organometallic chemistry (1). Many new preparative routes were developed and a wide variety of carbene complexes were prepared (2 – 5). Moreover, transition metal carbene complexes have become ever more important as starting materials for the synthesis of organic compounds or as precursors for new organometallic complexes. This chapter presents selected examples of such applications and will concentrate on reactions of carbene complexes of the Fischer type:

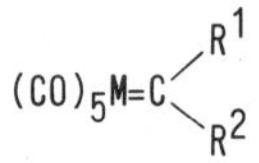

$$(CO)_5M=C{\diagup{R^1}\atop\diagdown{R^2}}$$

These carbene complexes may, in principle, react in several different ways as indicated in Scheme 1.

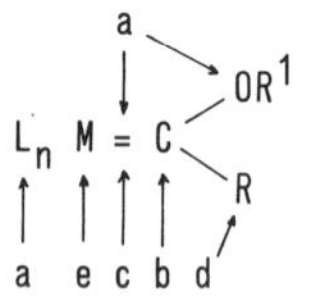

$$L_n\ M = C{\diagup OR^1 \atop \diagdown R}$$

Scheme 1. a) Substitution, b) addition of a nucleophile, c) insertion of a molecule, d) reaction at the carbene side chain, e) oxidation or reduction of the metal.

2 Reactions with Nucleophiles

According to ESCA measurements (6) and Mulliken population analyses (7, 8) of pentacarbonyl carbene complexes, the carbonyl ligands carry a greater positive charge than that on the carbene carbon atom. Nevertheless, it is well established that the carbene carbon atom shows a remarkable susceptibility to nucleophilic attack. This type of reaction is favoured by molecular orbital calculations which suggest that the nucleophilic attack may be better explained in terms of frontier orbital control. The lowest unoccupied molecular orbital in methoxymethylcarbene(pentacarbonyl)chromium has been demonstrated to be spatially and energetically localized on the carbene carbon atom.

2.1 Aminolysis and Related Reactions

2.1.1 Aminolysis

Alkoxycarbenepentacarbonyl complexes, as well as their phosphine- or phosphite-substituted analogues, can be regarded as esters in which the ketonic oxygen atom has been substituted by a $(CO)_4$ LM group (M = Cr, Mo, W; L = CO, PR_3, $P(OR)_3$). Accordingly, carbene complexes may react with ammonia or primary and uncrowded secondary amines to give amino-carbene complexes in good yields (9 – 16) (Scheme 2).

$$L(CO)_4M=C\begin{smallmatrix}OMe\\\\R\end{smallmatrix} \quad + \ HNR^1R^2 \quad \longrightarrow \quad L(CO)_4\,M=C\begin{smallmatrix}NR^1R^2\\\\R\end{smallmatrix} \quad + \ MeOH$$

Scheme 2. M = Cr, Mo, W; R = alkyl, aryl, H; L = CO, PR_3, $P(OR)_3$.

In general, primary amines react to give a mixture of EZ-isomers due to restricted rotation about the carbene carbon-nitrogen bond. Extensive proton nmr investigations by C. G. Kreiter have revealed that *E*-isomers predominate (17).

The aminolysis was also carried out with carbene complexes which are not of Fischer type (18) and used to synthesize optically active carbene complexes when chiral amine components were employed (19, 20). Furthermore, the mild reaction conditions of the aminolysis enabled the metal carbene moiety to be introduced as an amino protecting group in peptide synthesis (21) (Scheme 3).

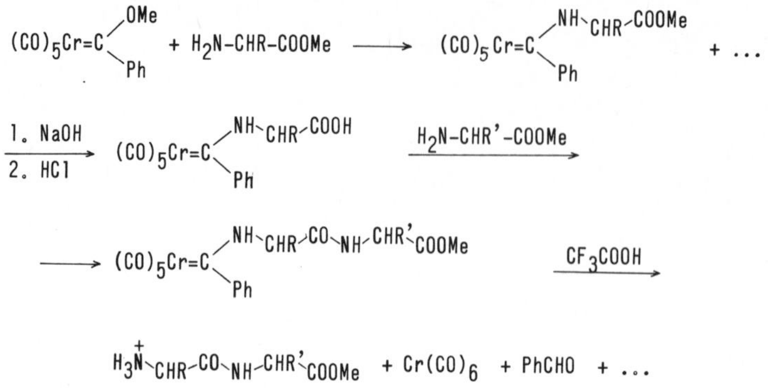

$$(CO)_5Cr=C\langle^{OMe}_{Ph} + H_2N-CHR-COOMe \longrightarrow (CO)_5Cr=C\langle^{NH-CHR-COOMe}_{Ph} + \ldots$$

$$\xrightarrow[2.\ HCl]{1.\ NaOH} (CO)_5Cr=C\langle^{NH-CHR-COOH}_{Ph} \xrightarrow{H_2N-CHR'-COOMe}$$

$$\longrightarrow (CO)_5Cr=C\langle^{NH-CHR-CO-NH-CHR'-COOMe}_{Ph} \xrightarrow{CF_3COOH}$$

$$\overset{+}{H_3N}-CHR-CO-NH-CHR'-COOMe + Cr(CO)_6 + PhCHO + \ldots$$

Scheme 3.

Similar reactions could be observed with bimetallic carbene complexes, for which significant differences in reaction rate were found depending on the position (eq or ax) of the pentacarbonylrhenio group (22) (Scheme 4).

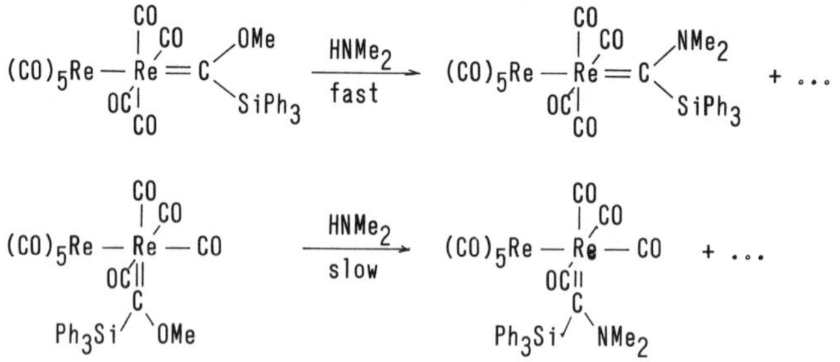

$$(CO)_5Re-\underset{\underset{CO}{\overset{|}{O\acute{C}|}}}{\overset{\overset{CO}{|}CO}{Re}}=C\langle^{OMe}_{SiPh_3} \xrightarrow[fast]{HNMe_2} (CO)_5Re-\underset{\underset{CO}{\overset{|}{O\acute{C}|}}}{\overset{\overset{CO}{|}CO}{Re}}=C\langle^{NMe_2}_{SiPh_3} + \ldots$$

$$(CO)_5Re-\underset{\underset{\underset{Ph_3Si}{}}{\overset{|}{O\acute{C}||}}}{\overset{\overset{CO}{|}CO}{Re}}-CO \xrightarrow[slow]{HNMe_2} (CO)_5Re-\underset{\underset{\underset{Ph_3Si}{}}{\overset{|}{O\acute{C}||}}}{\overset{\overset{CO}{|}CO}{Re}}-CO + \ldots$$
$$\quad\quad\quad\quad Ph_3Si\diagdown^{C}\diagup OMe \quad\quad\quad\quad\quad Ph_3Si\diagdown^{C}\diagup NMe_2$$

Scheme 4.

The aminolysis of methoxyphenylcarbene(pentacarbonyl)chromium has been studied in some detail and appears to involve a complicated reaction (23, 24). The rate law proved to be of third or fourth order, depending on the donicity of the solvent used.

On treating alkynylcarbene complexes of chromium and tungsten with dimethylamine, competition between aminolysis and conjugate addition was observed (25, 26) (Scheme 5).

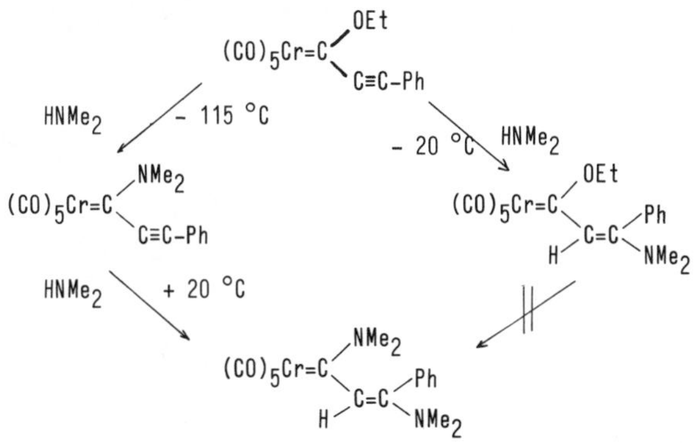

Scheme 5.

At very low temperatures ($-115°$ C), aminolysis is the favoured reaction while at $-20°$ C the conjugate addition reaction dominates. At intermediate temperatures both pathways are observed. The aminolysis at $-115°$ C may be followed by addition at 20° C to yield the amino(aminoallyl)carbene complex. The latter is not accessible via the initial conjugate addition product.

In contrast to silicon-containing carbene complexes (27), arylsiloxycarbene-(pentacarbonyl)tungsten compounds behave quite differently from their alkoxy analogues on treatment with dimethylamine (28) (Scheme 6).

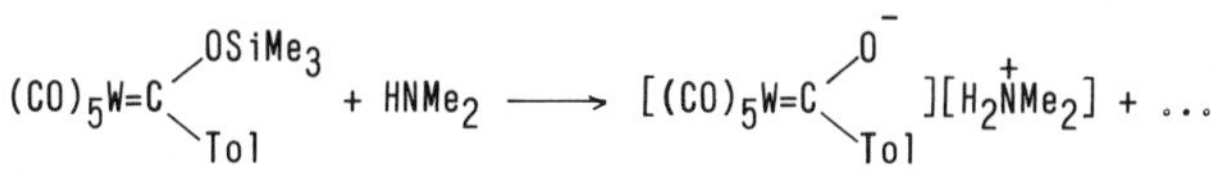

Scheme 6.

2.1.2 Reactions Related to Aminolysis

Thiols and methylselenol react with carbene complexes in a similar way to primary and secondary amines. Thus, sulfur (29, 30) and selenium (31) can also act as heteroatoms which stabilize the carbene ligands (Scheme 7). The thiolysis proceeds in two steps. The first leads to an ylide-like intermediate (30) which, after protonation, yields the corresponding thiocarbene complex.

$$(CO)_5M=C\diagup_{\displaystyle R}^{\displaystyle OMe} \quad + \text{ HSR'} \quad \longrightarrow \quad (CO)_5M=C\diagup_{\displaystyle R}^{\displaystyle SR'} \quad + \text{ MeOH}$$

$$(CO)_5Cr=C\diagup_{\displaystyle Ph}^{\displaystyle OMe} \quad + \text{ HSeMe} \quad \longrightarrow \quad (CO)_5Cr=C\diagup_{\displaystyle Ph}^{\displaystyle SeMe} \quad + \text{ MeOH}$$

Scheme 7. M = Cr, W; R,R′ = alkyl, aryl.

Enolate or alkoxide anions may also replace the alkoxy group in alkoxy-carbene complexes (32, 33). Carbene ligands can therefore be modified in a great variety of ways. In this connection it is important to note that the protons of the carbene-bonded methyl group in methoxymethylcarbene(pentacarbonyl)chromium display enhanced acidic character (34), attributable to the electron donor behaviour of the carbene ligand. Immediate and very fast deuteration catalysed by MeONa occurred in CH_3OD solution. A similar result was found for methylmethoxycarbene(dicarbonyl)η^5-methylcyclopenta-dienylmanganese. The transition state is presumably formed via proton abstraction by the base (Scheme 8).

$$L(CO)_2M=C\diagup_{\displaystyle CH_3}^{\displaystyle OMe} \quad \xrightarrow[\displaystyle CH_3ONa]{\displaystyle CH_3OD} \quad L(CO)_2M=C\diagup_{\displaystyle CD_3}^{\displaystyle OMe}$$

$$L(CO)_2M=C\diagup_{\displaystyle \underset{R}{\overset{|}{CH_2}}}^{\displaystyle OMe} \quad + \text{ B} \longrightarrow \left[L(CO)_2M=C\diagup_{\displaystyle \underset{R}{\overset{|}{CH^\ominus}}}^{\displaystyle OMe} \quad \longleftrightarrow \quad L(CO)_2M-\overset{\ominus}{C}\diagup_{\displaystyle \underset{R}{\overset{|}{CH}}}^{\displaystyle OMe} \right] + \text{ HB}$$

Scheme 8. M = Cr, Mn; L = (CO)$_3$, MeCp.

2.2 Ylide Complexes

Alkoxycarbene(pentacarbonyl)chromium complexes react with ammonia, primary and secondary amines, thiols and selenols by alcohol cleavage to yield the corresponding amino-, thio- and selenocarbene complexes. By way of contrast, dimethylphosphine merely added to methoxyphenylcarbene(penta-carbonyl)chromium or -tungsten in pentane at − 50° C without cleavage of either methanol or a carbonyl ligand to form an ylide complex (35) (Scheme 9).

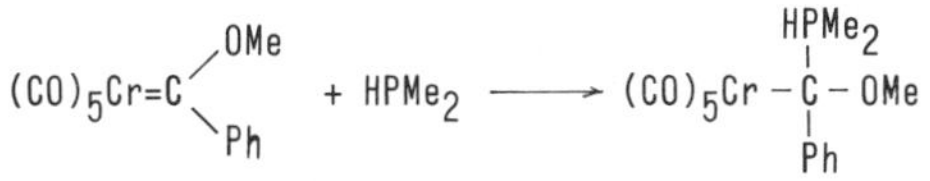

$$(CO)_5Cr=C\langle {}^{OMe}_{Ph} + HPMe_2 \longrightarrow (CO)_5Cr-\overset{\overset{\displaystyle HPMe_2}{|}}{\underset{\underset{\displaystyle Ph}{|}}{C}}-OMe$$

Scheme 9.

This type of reaction could easily be extended to related chromium, tungsten and rhenium complexes using a variety of carbene ligands and phosphines (Scheme 10).

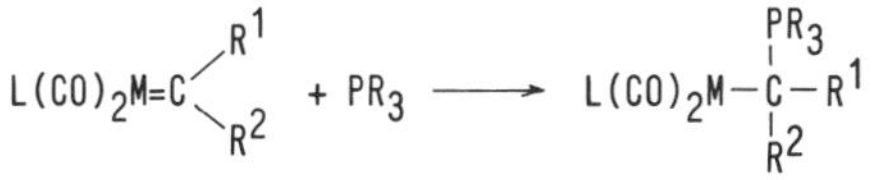

$$L(CO)_2M=C\langle {}^{R^1}_{R^2} + PR_3 \longrightarrow L(CO)_2M-\overset{\overset{\displaystyle PR_3}{|}}{\underset{\underset{\displaystyle R^2}{|}}{C}}-R^1$$

Scheme 10.

M	L	R^1	R^2	Ref.
Cr, W	$(CO)_3$	OMe	alkyl, aryl	(36)
W	$(CO)_3$	SMe	Me	(36)
Cr	$(CO)_3$	SeMe	Me	(31)
Cr	$(CO)_3$	OPh	aryl	(37)
Cr, W	$(CO)_3$	OSiMe₃	alkyl, aryl	(37)
Cr, W	$(CO)_3$	O(CO)R	Tol	(37)
W	$(CO)_3$	aryl	aryl	(38)
Re	Cp	Me	Ph	(39)

Thermodynamic investigations indicated that the formation of ylide complexes is a reversible reaction. The equilibrium constant depends on the nature of the phosphine, the metal and its ligands as well as on the solvent and the temperature. In most cases weakly basic triaryl or mixed alkyl-aryl phosphines yielded no stable ylide complexes (40). An exception to this generalisation is provided by $(CO)_5W$-$CH(C_6H_5)P(C_6H_5)_3$ (41), which is formed by treatment of the less stable phenylcarbene(pentacarbonyl)tungsten with triphenylphosphine.

[13]C nmr spectroscopy proved to be an excellent tool to observe the formation of the ylide complex. In going from the carbene to the ylide complex, a remarkable diamagnetic shift of the order of 200 to 320 ppm was recorded for the carbon atom coordinated to the metal (38).

In a similar reaction, carbene-ylide complexes of manganese, chromium and rhenium with trimethylphosphine gave cationic half ylide complexes (42, 43). The latter were used to synthesize free cationic half ylides via a heterolytic cleavage of the metal-carbon single bond (43) (Scheme 11).

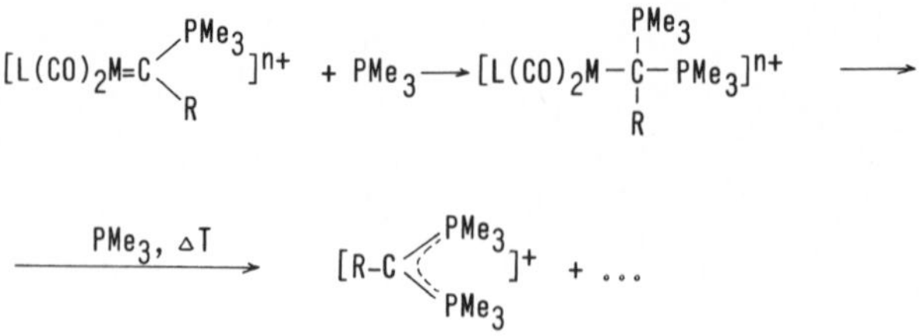

$$[L(CO)_2M=C<^{PMe_3}_{R}]^{n+} + PMe_3 \longrightarrow [L(CO)_2M-\underset{\underset{R}{|}}{\overset{\overset{PMe_3}{|}}{C}}-PMe_3]^{n+} \longrightarrow$$

$$\xrightarrow{PMe_3, \Delta T} [R-C<^{PMe_3}_{PMe_3}]^+ + \ldots$$

Scheme 11. M = Mn, Re; L = Cp; R = Ph; n = 1. M = Cr; L = Cl (CO)$_2$; R = Ph, Tol, SiPh$_3$; n = 0.

The aminolysis of carbene complexes with primary or secondary amines involves formation of a nitrogen ylide transition state complex (24). Closely related nitrogen ylide complexes were produced by treatment of the methoxy-phenylcarbene(pentacarbonyl) complexes of chromium and tungsten with "rigid" tertiary amines such as 1-azabicyclo[2.2.2]octane or 1,4-diazabicyc-lo[2.2.2]octane (44, 45) (Scheme 12).

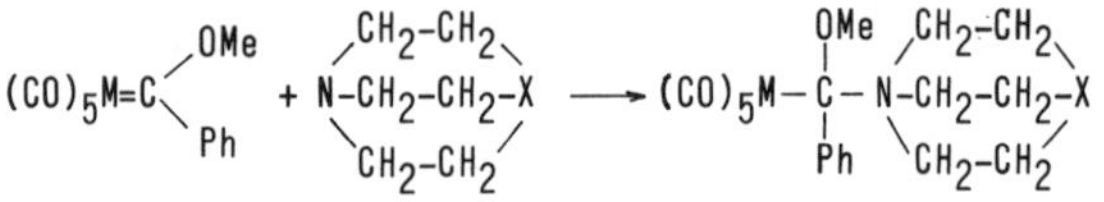

$$(CO)_5M=C<^{OMe}_{Ph} + \underset{\underset{CH_2-CH_2}{\diagdown}}{\overset{\overset{CH_2-CH_2}{\diagup}}{N}}-CH_2-CH_2-X \longrightarrow (CO)_5M-\underset{\underset{Ph}{|}}{\overset{\overset{OMe}{|}}{C}}-\underset{\underset{CH_2-CH_2}{\diagdown}}{\overset{\overset{CH_2-CH_2}{\diagup}}{N}}-CH_2-CH_2-X$$

Scheme 12. M = Cr, W; X = N, CH.

The amine addition was reversed if the nitrogen ylides were treated with KHSO$_4$ (Scheme 13). In contrast with phosphorus ylide complexes, a hydroxy-phenylcarbene(pentacarbonyl) complex – and not the starting carbene complex – was formed.

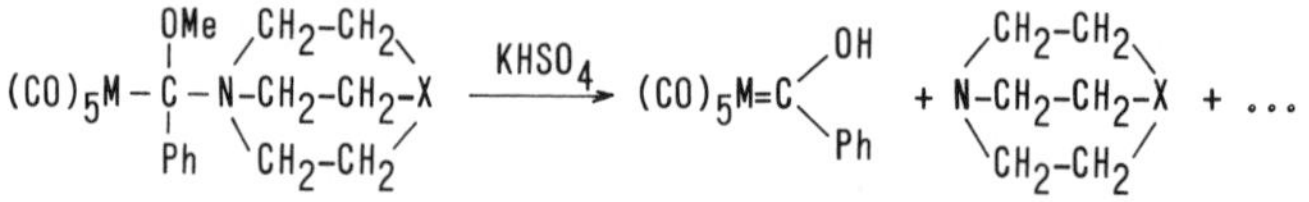

$$(CO)_5M-\underset{\underset{Ph}{|}}{\overset{\overset{OMe}{|}}{C}}-\underset{\underset{CH_2-CH_2}{\diagdown}}{\overset{\overset{CH_2-CH_2}{\diagup}}{N}}-CH_2-CH_2-X \xrightarrow{KHSO_4} (CO)_5M=C<^{OH}_{Ph} + \underset{\underset{CH_2-CH_2}{\diagdown}}{\overset{\overset{CH_2-CH_2}{\diagup}}{N}}-CH_2-CH_2-X + \ldots$$

Scheme 13.

In general, ylide complex formation proved to be a useful reaction for trapping unstable carbene complexes. For example, by making use of triphenyl- or tri-n-butylphosphine, the existence of a cationic rhenium methylene complex (46) and a tungsten phenylcarbene complex (41, 47) could be demonstrated. Additionally, the ylide reaction enables a distinction to be made between benzocyclobutenylidene- and possible π-benzocyclobuteneiron complexes (48) (Scheme 14).

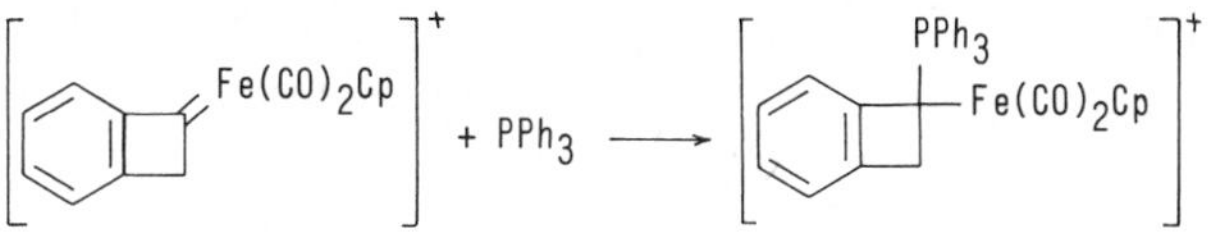

Scheme 14.

2.3 Addition of Carbon Nucleophiles

The distinctive electrophilic character of the carbene carbon atom (49) promotes the formation of ylide intermediates or ylide complexes when alkoxyorganylcarbene(pentacarbonyl) complexes of chromium, molybdenum or tungsten react with amines, thiols, selenols or phosphines. A similar type of reaction can be expected with metal organyls.

The treatment of methoxyphenylcarbene(pentacarbonyl)chromium with phenyllithium at 25° C yielded a persubstituted ethane derivative (50). This product can be explained by addition of the nucleophile at the carbene carbon atom, followed by a cleavage of the metal-carbon bond and subsequent dimerisation of the organic ligand (Scheme 15).

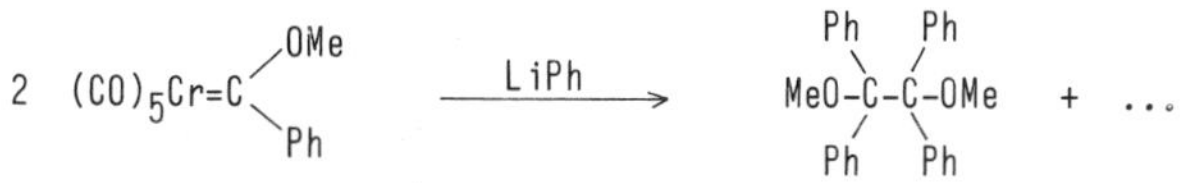

Scheme 15.

However, upon repeating this reaction at -78° C in ether, decomposition was avoided and the adduct could be isolated as a yellow solid. The corresponding bis(triphenylphosphine)imminium salts have been shown to be more stable than the lithium compounds (51, 52) (Scheme 16).

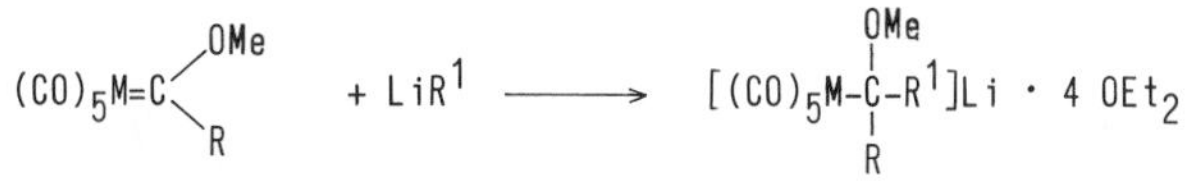

Scheme 16. M = Cr, W; R = Ph, Tol, $C_6H_4CF_3$-(4), OC_4H_3, SC_4H_3; R^1 = Me, Ph.

The conversion of the sp^2-carbene carbon in the carbene complex into an sp^3-hybridized metal-coordinated carbon atom in the adduct was paralleled by a significant diamagnetic shift of about 268 ppm.

Similar adducts were obtained on reacting pentacarbonyl(chloro)tungstate and lithium organyls (53). Such metallates could be used in the synthesis of new non-heteroatom-stabilized carbene complexes since, on treating them with acid at low temperature or with silica gel in pentane, intensely coloured carbene complexes were formed (52, 54, 55) (Scheme 17).

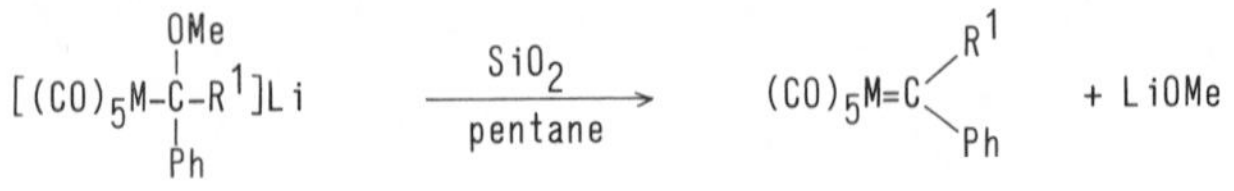

$$[(CO)_5M-\overset{OMe}{\underset{Ph}{\overset{|}{C}}}-R^1]Li \xrightarrow[\text{pentane}]{SiO_2} (CO)_5M=C\overset{R^1}{\underset{Ph}{\diagdown}} + LiOMe$$

Scheme 17. M = Cr, W; R^1 = Ph, Tol, $C_6H_4CF_3$-(4), OC_4H_3, SC_4H_3.

Attempts to obtain alkylarylcarbene complexes by this method have not been successful. On the contrary, rearrangement or decomposition occurred to produce the corresponding olefins or their π-complexes (56, 57) (Scheme 18).

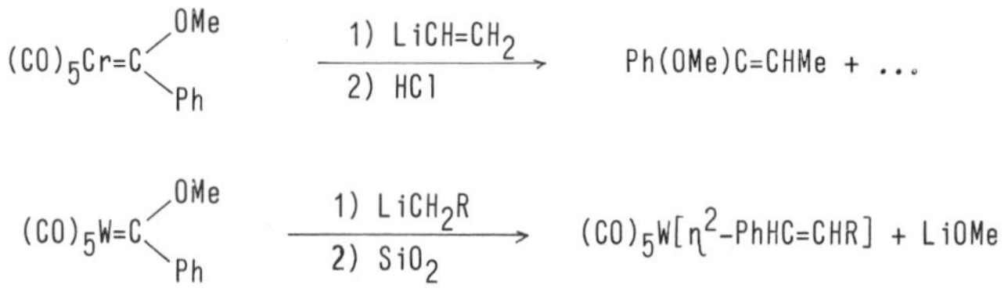

$$(CO)_5Cr=C\overset{OMe}{\underset{Ph}{\diagdown}} \xrightarrow[\text{2) HCl}]{\text{1) LiCH=CH}_2} Ph(OMe)C=CHMe + \ldots$$

$$(CO)_5W=C\overset{OMe}{\underset{Ph}{\diagdown}} \xrightarrow[\text{2) SiO}_2]{\text{1) LiCH}_2R} (CO)_5W[\eta^2\text{-PhHC=CHR}] + LiOMe$$

Scheme 18.

Like an alkoxy substituent, the chloride could function as the leaving group when chloro(dimethylamino)carbene(pentacarbonyl)chromium was reacted with potassium cyanide (58) (Scheme 19).

$$(CO)_5Cr=C\overset{Cl}{\underset{NMe_2}{\diagdown}} + KCN \longrightarrow (CO)_5Cr=C\overset{CN}{\underset{NMe_2}{\diagdown}} + KCl$$

Scheme 19.

In the comparable reaction of iodocyanide with methoxyphenylcarbene-(pentacarbonyl)chromium no metal complex could be isolated. The addition of the cyanide was immediately followed by a cleavage of the metal-carbon bond and dimerisation of the ligand (50).

3 Addition-Rearrangement Reactions

The interesting reaction capabilities of transition metal carbene complexes render the formation of new metal-heteroatom bonds, modified metal-carbene bonds or metal π-complexes feasible. Such results may be attained by reacting carbene complexes with systems containing heteroatoms or multiple bonds.

3.1 Metal-Heteroatom Bond Formation

A new metal-heteroatom bond was afforded by formal insertion of a molecule into the metal-carbene bond. In contrast to methylselenol, phenylselenol reacted with methoxymethylcarbene(pentacarbonyl)chromium to yield not a selenocarbene complex (31) but rather a selenol ether complex (59) (Scheme 20).

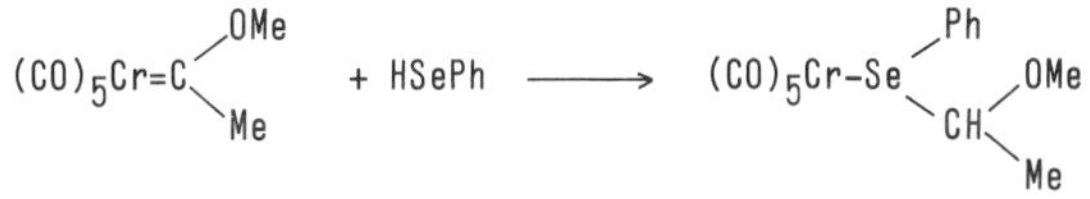

$$(CO)_5Cr=C\begin{smallmatrix}OMe\\\\Me\end{smallmatrix} \quad + HSePh \longrightarrow (CO)_5Cr-Se\begin{smallmatrix}Ph\\\\CH\begin{smallmatrix}OMe\\Me\end{smallmatrix}\end{smallmatrix}$$

Scheme 20.

This rearrangement was accompanied by hydrogen migration to the original carbene carbon.

Methoxyphenylcarbene(pentacarbonyl)chromium initially added at $-78°$ C one molecule of dimethylphosphine to give an ylide adduct (35). On warming to 25° C, a rearrangement, together with a hydrogen shift, took place to yield a phosphine complex (36) (Scheme 21).

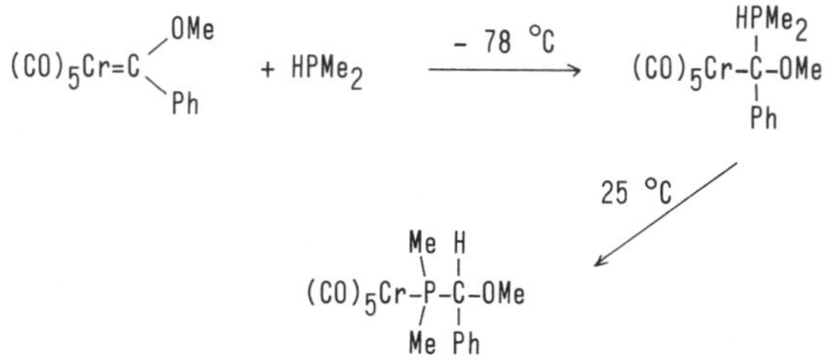

$$(CO)_5Cr=C\begin{smallmatrix}OMe\\\\Ph\end{smallmatrix} \quad + HPMe_2 \quad \xrightarrow{-78\ °C} \quad (CO)_5Cr-\overset{HPMe_2}{\underset{Ph}{C}}-OMe$$

$$25\ °C$$

$$(CO)_5Cr-\overset{Me\ \ H}{\underset{Me\ \ Ph}{P}}-\overset{}{C}-OMe$$

Scheme 21.

In a similar rearrangement, bis(thiomethyl)carbene(pentacarbonyl)tungsten reacted with diphenylphosphine to give a further phosphine complex (60) (Scheme 22). When a tertiary phosphine was used instead of a secondary phosphine, the reaction led to a thioether complex (60).

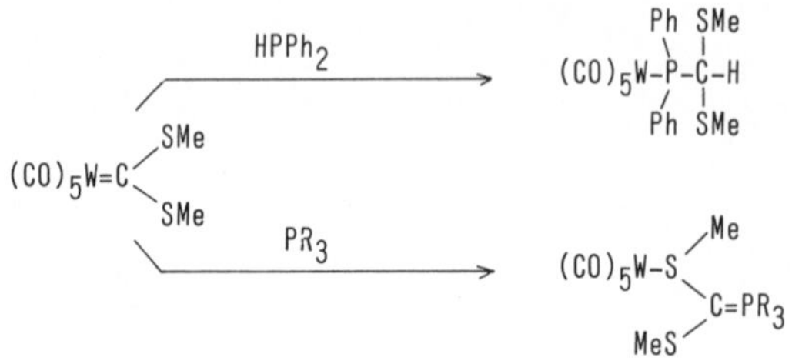

$$
(CO)_5W=C
\begin{array}{c}
SMe \\
SMe
\end{array}
$$

HPPh$_2$ →

$$
\begin{array}{c}
Ph \quad SMe \\
| \quad | \\
(CO)_5W-P-C-H \\
/ \quad | \\
Ph \quad SMe
\end{array}
$$

PR$_3$ →

$$
(CO)_5W-S
\begin{array}{c}
Me \\
C=PR_3 \\
MeS
\end{array}
$$

Scheme 22. PR$_3$ = PEt$_3$, PPhMe$_2$, PPh$_2$Me, P(OMe)$_3$.

The insertion of cyanide could be observed by treating methoxyphenylcarbene(pentacarbonyl)chromium with calcium cyanide (61) (Scheme 23).

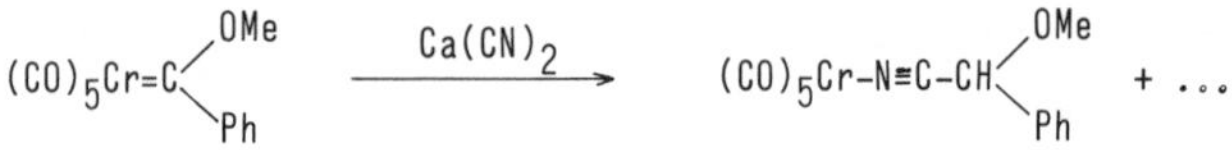

$$
(CO)_5Cr=C
\begin{array}{c}
OMe \\
Ph
\end{array}
$$

Ca(CN)$_2$ →

$$
(CO)_5Cr-N\equiv C-CH
\begin{array}{c}
OMe \\
Ph
\end{array}
+ \cdots
$$

Scheme 23.

Further examples of combined insertion and hydrogen migration can be seen in the reaction of other heteroatom-substituted carbene complexes with hydrogen halides at low temperatures. Thus, a chromium thiocarbene complex and hydrogen bromide yielded a thioether complex (62) (Scheme 24), and dialkylaminophenylcarbene complexes reacted similarly.

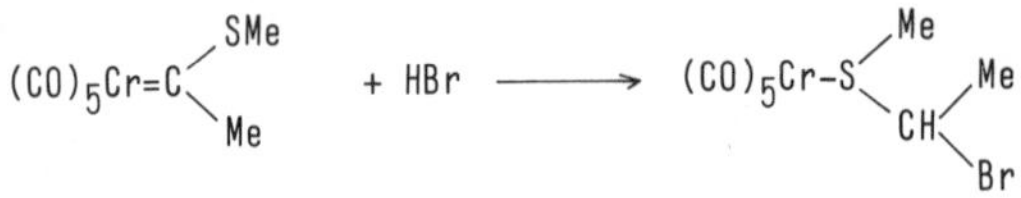

$$
(CO)_5Cr=C
\begin{array}{c}
SMe \\
Me
\end{array}
+ HBr \longrightarrow
(CO)_5Cr-S
\begin{array}{c}
Me \\
CH \\
Br
\end{array}
Me
$$

Scheme 24.

Stable immonium halogeno(pentacarbonyl)chromates were isolated (63) (Scheme 25) though α-halogenoamine complexes, which might be expected as intermediates, were not.

$$
(CO)_5Cr=C
\begin{array}{c}
NR_2 \\
Ph
\end{array}
+ HX \longrightarrow
[(CO)_5CrX]^-[R_2N=CHPh]^+
$$

Scheme 25.

Carbene substitution did occur, however, when HI reacted with a carbene complex such as methoxymethylcarbene(pentacarbonyl)chromium. The final product was the iodine-bridged iododecacarbonyldichromium anion. The first step was thought to involve cleavage of the carbene ligand with formation of the strong nucleophile [(CO)$_5$CrI]$^-$, which could then attack a second carbene complex molecule (64) (Scheme 26).

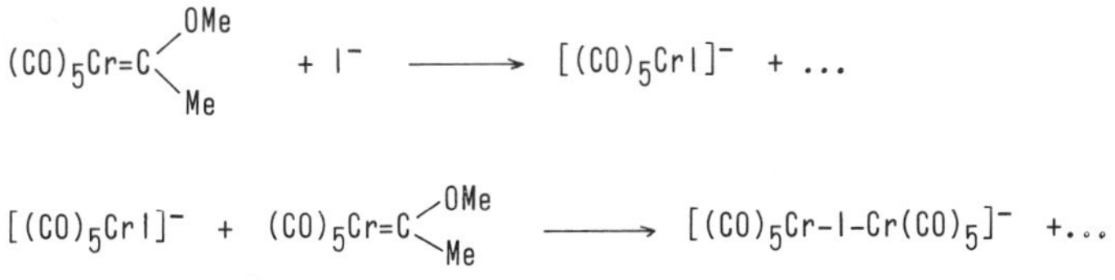

$$(CO)_5Cr=C\begin{cases} OMe \\ Me \end{cases} \quad + \quad I^- \quad \longrightarrow \quad [(CO)_5CrI]^- \quad + \ldots$$

$$[(CO)_5CrI]^- \quad + \quad (CO)_5Cr=C\begin{cases} OMe \\ Me \end{cases} \quad \longrightarrow \quad [(CO)_5Cr-I-Cr(CO)_5]^- \quad +\ldots$$

Scheme 26.

At low temperatures and in non basic solvents, methoxyphenylcarbene(pentacarbonyl)tungsten reacted with an excess of hydrogen halide HX (X = Cl, Br, I) with cleavage of the metal-carbene carbon bond to form pentacarbonyl(hydrogenhalide)tungsten complexes (Scheme 27), which proved to be strong acids (65).

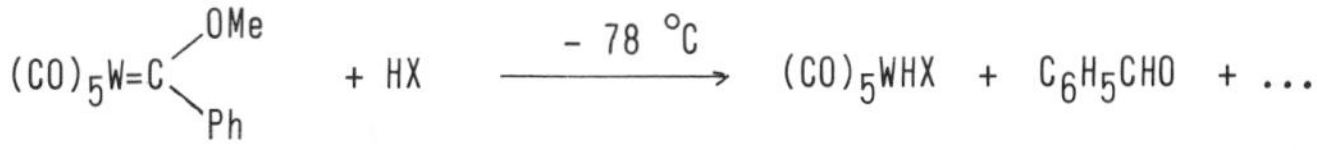

$$(CO)_5W=C\begin{cases} OMe \\ Ph \end{cases} \quad + \quad HX \quad \xrightarrow{\ -78\ °C\ } \quad (CO)_5WHX \quad + \quad C_6H_5CHO \quad + \ldots$$

Scheme 27.

On reacting methoxymethylcarbene(pentacarbonyl)chromium with hydroxylamine, formal insertion of an NH unit was observed to yield two isomeric nitrogen-coordinated acetimidic acid methyl ester complexes (66) (Scheme 28). In a similar reaction with dimethyl- or phenylhydrazine, cyanide complexes were obtained (66).

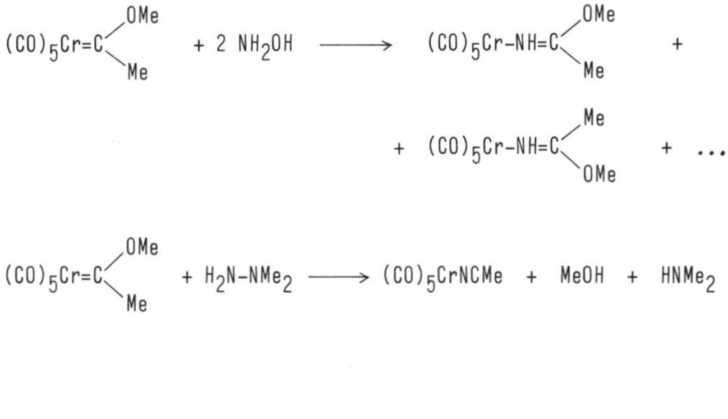

$$(CO)_5Cr=C\begin{cases} OMe \\ Me \end{cases} \quad + \quad 2\ NH_2OH \quad \longrightarrow \quad (CO)_5Cr-NH=C\begin{cases} OMe \\ Me \end{cases} \quad +$$

$$+ \quad (CO)_5Cr-NH=C\begin{cases} Me \\ OMe \end{cases} \quad + \ldots$$

$$(CO)_5Cr=C\begin{cases} OMe \\ Me \end{cases} \quad + \quad H_2N-NMe_2 \quad \longrightarrow \quad (CO)_5CrNCMe \quad + \quad MeOH \quad + \quad HNMe_2$$

Scheme 28.

Up to four products were isolated after treating a cyclic carbene complex of chromium with 1,1-dimethylhydrazine (67). In addition to aminolysis, the formation of two aldimine complexes and a thioether coordinated to the pentacarbonylchromium fragment has been reported (Scheme 29).

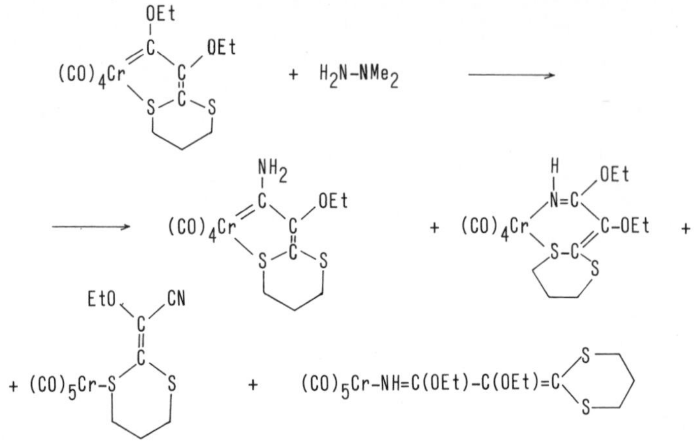

Scheme 29.

Reactions of the same type occurred with the oximes of aliphatic, alicyclic or aromatic ketones to form the corresponding ketimine(pentacarbonyl)chromium complexes. In the case of the aromatic oximes, alkylideneamine and cyanide complexes could be isolated as byproducts (68) (Scheme 30).

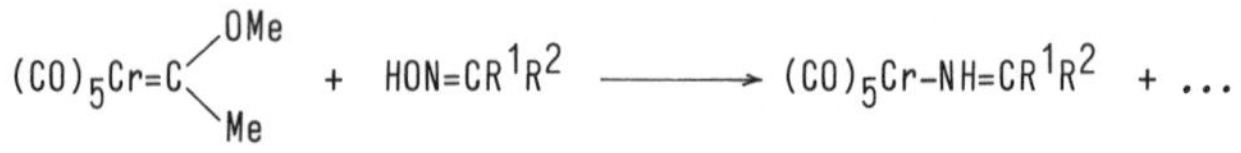

Scheme 30. R^1, R^2 = Me, Et, Ph.

Use of an aldoxime results in comparable reaction behaviour to give three products, *viz.* a cyanide, an aldimine and a modified carbene complex (69) (Scheme 31).

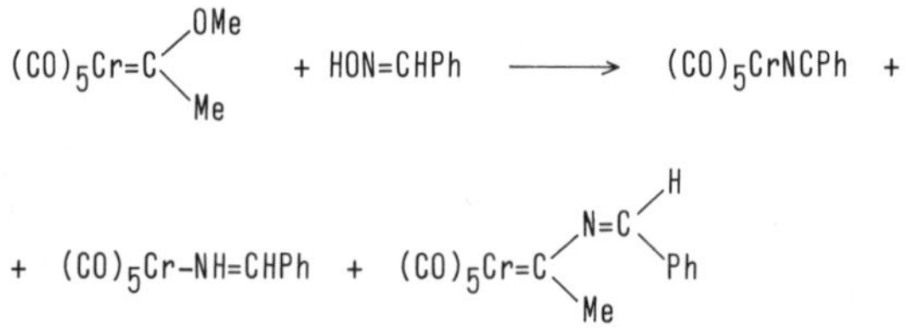

Scheme 31.

A further possibility for obtaining a cyanide from a carbene complex was via the reaction of acetoxycarbene complexes with hydrogen azide (70). Azidocarbene complexes have been postulated as intermediates. The loss of nitrogen initiates a rearrangement, the nature of which depends on the substituent R. If R acts as an electron donating group, a cyanide complex is favoured. Otherwise, if R is an electron withdrawing group, an isocyanide complex is formed (Scheme 32).

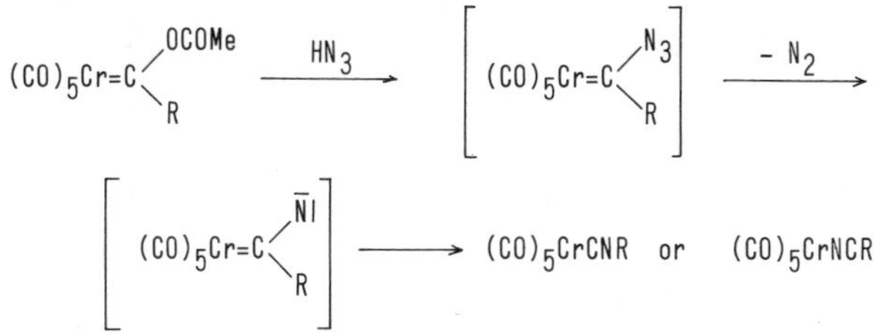

$$(CO)_5Cr=C\begin{array}{l}OCOMe\\R\end{array} \xrightarrow{HN_3} \left[(CO)_5Cr=C\begin{array}{l}N_3\\R\end{array}\right] \xrightarrow{-N_2}$$

$$\left[(CO)_5Cr=C\begin{array}{l}\overline{N}I\\R\end{array}\right] \longrightarrow (CO)_5CrCNR \quad or \quad (CO)_5CrNCR$$

Scheme 32.

Transition metal carbene complexes, e.g. methoxymethylcarbene(pentacarbonyl)chromium or -tungsten can add cyclohexyl isocyanide to give ketenimine complexes (71, 72) which either isomerize under the influence of a weak acid to an aminocarbene complex or add methanol or water to form other aminocarbene complexes (Scheme 33).

$$(CO)_5Cr=C\begin{array}{l}OMe\\Me\end{array} + C_6H_{11}NC \longrightarrow (CO)_5Cr-N(C_6H_{11})=C=C(Me)OMe$$

Scheme 33.

Arylphenylcarbene(pentacarbonyl)tungsten complexes react with elemental sulfur by insertion of one sulfur atom into the metal-carbene to yield arylphenylthioketone(pentacarbonyl)tungsten complexes (73).

$$(CO)_5W=C\begin{array}{l}Ph\\C_6H_4R-(4)\end{array} + S_8 \longrightarrow (CO)_5W-S=C\begin{array}{l}Ph\\C_6H_4R-(4)\end{array} +...$$

Scheme 34. R = H, OMe, CF$_3$.

Similar thioketone complexes may be obtained by reacting organyl isothiocyanates with arylphenylcarbene(pentacarbonyl) complexes of tungsten (74). Kinetic investigations indicate that the initial and rate determining step is nucleophilic attack of RNCS at the carbene carbon atom.

The adduct rearranges, mainly by elimination of the RNC unit, to give the thioketone complex and, to a lesser extent, by cleavage of thioketone to yield $(CO)_5WCNR$.

Ethoxyphenylethynecarbene(pentacarbonyl)tungsten and diazomethane display interesting conjugate addition to form a nitrogen-coordinated pyrazole complex (75) (Scheme 35). The initial step has been suggested as 1,3-dipolar addition of one molecule of diazomethane to the $C \equiv C$ triple bond followed by a hydrogen shift. An additional – and possibly preceding – step is the nucleophilic attack of another molecule of diazomethane on the carbene carbon atom to generate an ylide-like adduct. The final coordination of the nitrogen occurs perhaps by an inter- or an intramolecular reaction.

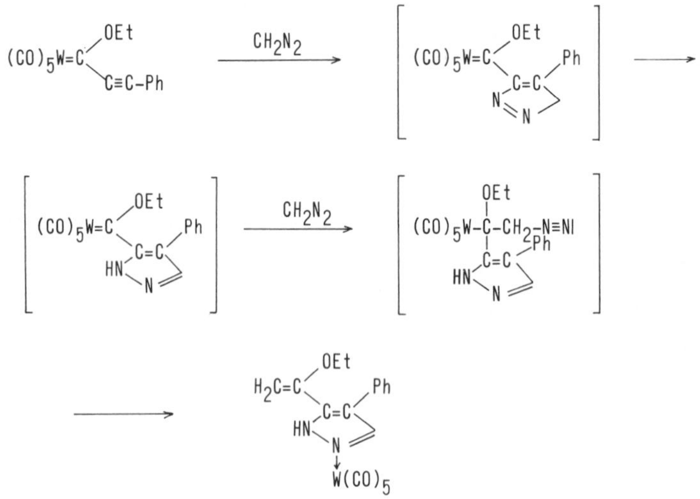

Scheme 35.

3.2 Modification of the Carbene Side Chain

Diaryl- or alkoxycarbene complexes of chromium, molybdenum and tungsten react with 1-aminoalkynes to yield alkenyl(amino)carbene complexes (Scheme 36). Kinetic investigations suggest that nucleophilic attack of the alkyne on the carbene carbon atom is followed by subsequent formation of a four-membered metallacycle and final cleavage of the original metal-carbene bond (76–78). This mechanism is highly stereoselective and favours the *E* isomers.

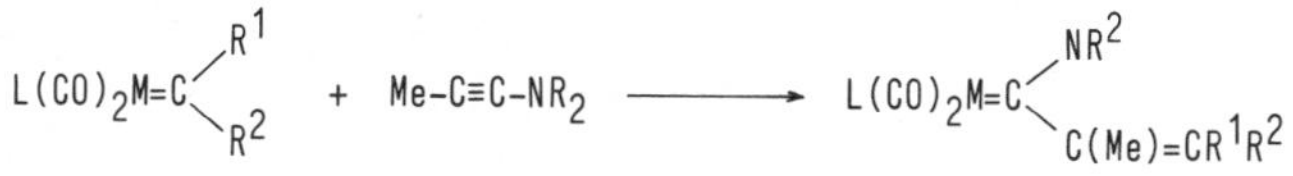

$$L(CO)_2M=C\begin{smallmatrix}R^1\\R^2\end{smallmatrix} \quad + \quad Me-C\equiv C-NR_2 \quad\longrightarrow\quad L(CO)_2M=C\begin{smallmatrix}NR^2\\C(Me)=CR^1R^2\end{smallmatrix}$$

Scheme 36. M = Cr, Mo, W, Mn; R = Me, Et; R^1 = Me, Ph; R^2 = Ph, OMe; L = (CO)$_3$, MeCp.

Dialkylaminocarbene complexes fail to react in the same fashion. However, monoalkylaminocarbene complexes insert an ynamine into the metal-carbene bond to yield a chelating aminocarbene ligand via substitution of one metal carbonyl ligand by the methylamino group (79) (Scheme 37).

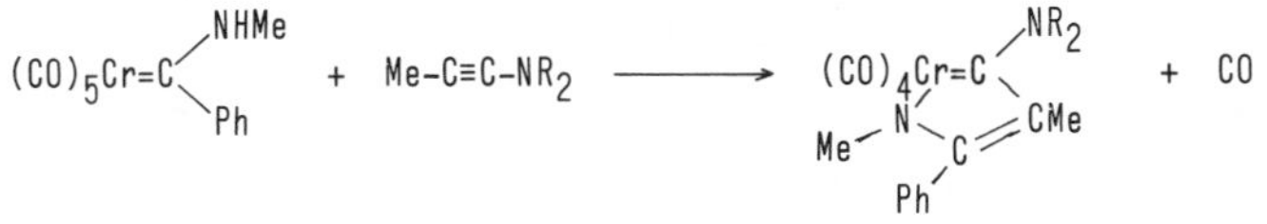

$$(CO)_5Cr=C\begin{smallmatrix}NHMe\\Ph\end{smallmatrix} \quad + \quad Me-C\equiv C-NR_2 \quad\longrightarrow\quad (CO)_4Cr=C\begin{smallmatrix}NR_2\\\\Me-N\\C=CMe\\Ph\end{smallmatrix} \quad + \quad CO$$

Scheme 37.

Finally, under identical conditions aminocarbene complexes yield alkylideneaminocarbene complexes via an addition-rearrangement reaction (80) (Scheme 38).

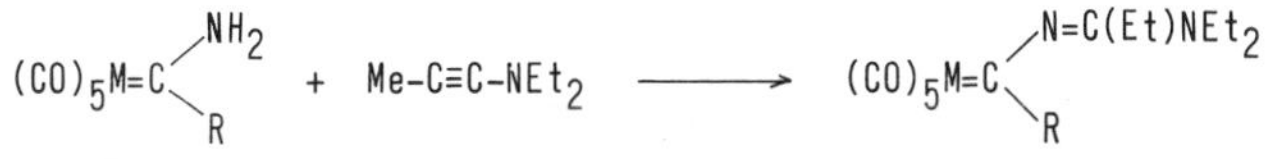

$$(CO)_5M=C\begin{smallmatrix}NH_2\\R\end{smallmatrix} \quad + \quad Me-C\equiv C-NEt_2 \quad\longrightarrow\quad (CO)_5M=C\begin{smallmatrix}N=C(Et)NEt_2\\R\end{smallmatrix}$$

Scheme 38. M = Cr, Mo, W.

Similar insertion reactions into the metal-carbene bond have been observed with ethoxyacetylene (47), cyclic enol ethers or enamines (81 – 83) and dimethylcyanamide (84) (Schemes 39 and 40).

$$(CO)_5W=C\begin{smallmatrix}R\\Ph\end{smallmatrix} \quad + \quad HC\equiv C-OEt \quad\longrightarrow\quad (CO)_5W=C\begin{smallmatrix}OEt\\CH=C(Ph)R\end{smallmatrix}$$

Scheme 39. M = Cr, W; R^1 = Ph, OMe; R^2 = Ph; R = OEt.

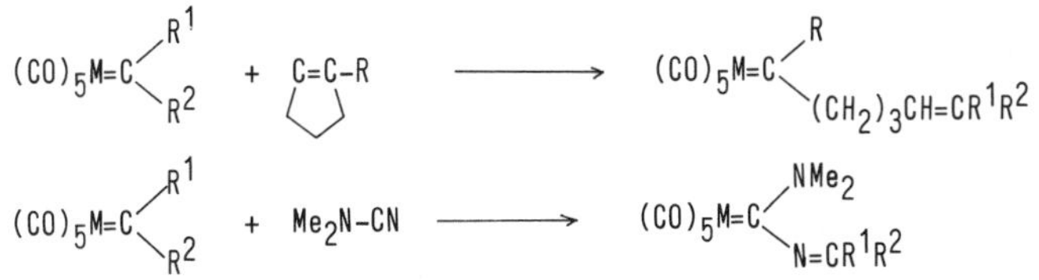

$$(CO)_5M=C\begin{smallmatrix}R^1\\R^2\end{smallmatrix} \quad + \quad C=C-R \quad\longrightarrow\quad (CO)_5M=C\begin{smallmatrix}R\\(CH_2)_3CH=CR^1R^2\end{smallmatrix}$$

$$(CO)_5M=C\begin{smallmatrix}R^1\\R^2\end{smallmatrix} \quad + \quad Me_2N-CN \quad\longrightarrow\quad (CO)_5M=C\begin{smallmatrix}NMe_2\\N=CR^1R^2\end{smallmatrix}$$

Scheme 40. M = Cr, W; R = OC$_2$H$_5$; R^1 = Me, Ph; R^2 = OMe, C$_6$H$_4$X-(4).

The synthesis of metal-coordinated indene derivatives was achieved by reacting methoxyphenylcarbene(pentacarbonyl)chromium with bis(diethyl-amino)acetylene. Depending on the temperature, the insertion-rearrangement could be divided into several steps. At room temperature, insertion of the acetylene into the metal-carbene bond occurs. On warming to 70° C, combined substitution of carbon monoxide and formation of a chelate ligand follows and finally at 125° C rearrangement to a π-bonded indene complex takes place (85, 86) (Scheme 41).

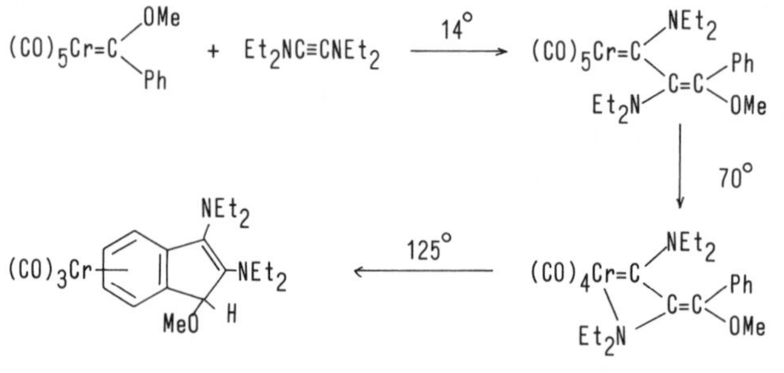

Scheme 41.

Hydroxycarbene complexes behave in differing ways when treated with dicyclohexylcarbodiimide (DCCD). The outcome depends mainly on the metal and the kind of carbene substituents in the particular molecule. Hydroxy-methylcarbene(pentacarbonyl)chromium yields, for instance, a cyclic four-membered aminocarbene complex (87), whereas the analogous phenyl complex reacts via an intermolecular condensation to produce a binuclear anhydride (88, 89) (Scheme 42).

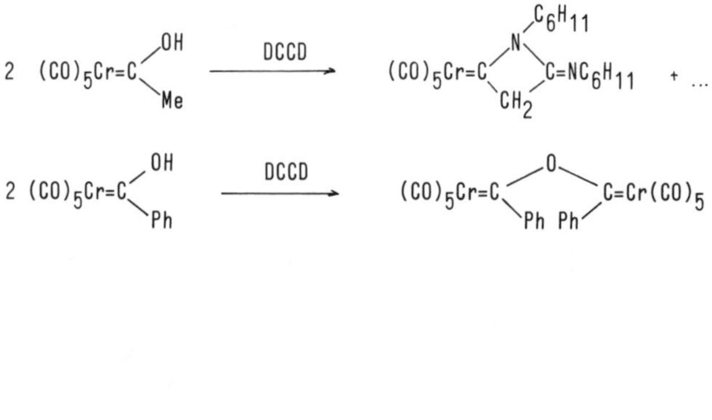

Scheme 42.

However, hydroxyphenylcarbene(pentacarbonyl)tungsten was transformed into a binuclear carbene-carbyne complex (88, 89) (Scheme 43).

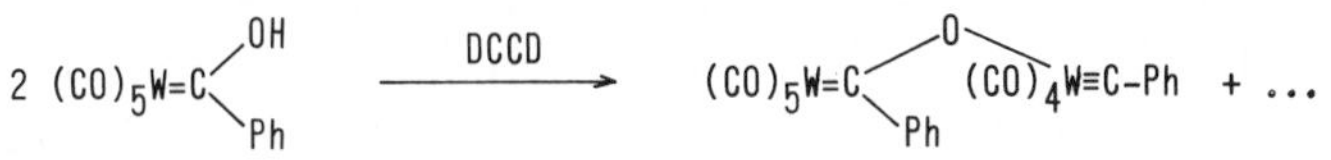

Scheme 43.

3.3 Formation of Metal π-Complexes

3.3.1 π-Aromatic Systems

Pentacarbonylchromium complexes with substituted phenyl-, naphthyl-, furyl-, thionyl- and cyclopentylcarbene ligands react with various alkynes in a strongly stereoselective way to give substituted naphthol, phenanthrene, benzofuran, benzothiophene and indene ligands coordinated to the tricarbonyl-chromium fragment (91 – 93). Schematically, these reactions could be reduced to a combination of the carbene ligand, a metal carbonyl ligand and the alkyne molecule used. The kinetic product could be converted to the more stable thermodynamic product simply by heating (Scheme 44).

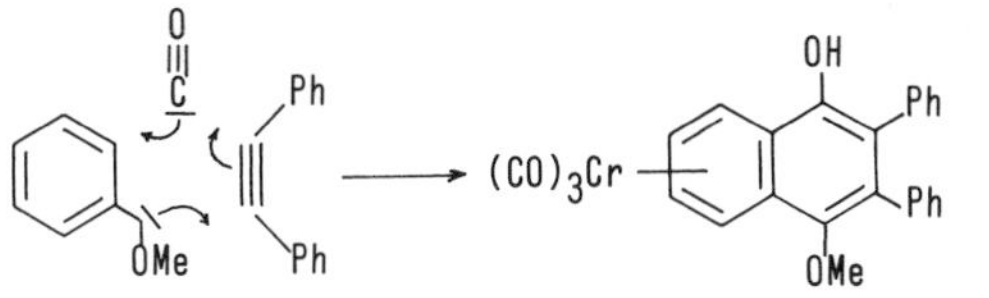

Scheme 44.

Heteroatom-substituted alkynes such as bis(diethylamino)acetylene or bis-(trimethylsilyl)acetylene react with methoxyphenylcarbene(pentacarbonyl)-chromium in comparable fashion to form either indeneπ-complexes or naphthol complexes with a stable silyl-substituted vinylketene group on the naphthol frame (94) (Scheme 45).

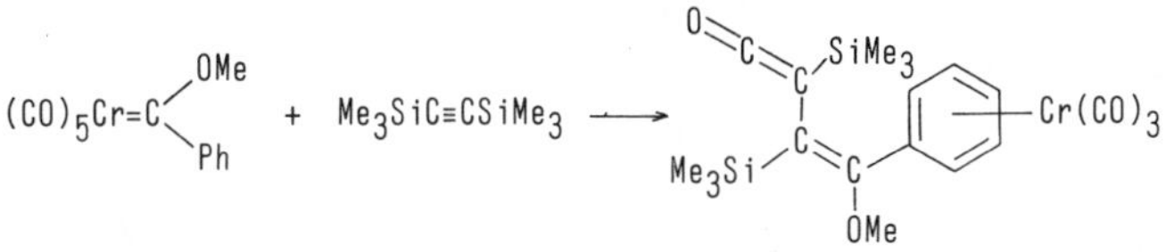

Scheme 45.

3.3.2 Formation of Ketene Complexes

Diphenylcarbene complexes of manganese can be converted to diphenyl-ketene complexes under conditions appropriate to high pressure carbonylation (95) (Scheme 46). Such carbonylation offers a new synthetic route to η^2-ketene complexes and may be regarded as preparative proof of the postulated carbene-ketene transformation in the coordination sphere of a transition metal carbonyl fragment (96). The carbene-ketene conversion is interesting as a model system for a partial Fischer-Tropsch synthesis (97).

$$R-C_5H_4(CO)_2Mn=C\begin{smallmatrix}Ph\\\\Ph\end{smallmatrix} \xrightarrow{CO} R-C_5H_4(CO)_2Mn-\overset{\overset{O}{\parallel}}{C}\begin{smallmatrix}C\\Ph \quad Ph\end{smallmatrix}$$

Scheme 46. R = H, Me.

In the reaction of η^3-vinylcarbene(tricarbonyl)iron with triphenylphosphine or carbon monoxide, the η^3-vinylcarbene ligand can be carbonylated to afford an apparant η^4-vinylketene complex (Scheme 47) which has the alternative structure $(\eta^3$-allyl $+ \eta^1)Fe(CO)_2L$ (98).

$$(CO)_3Fe=C\begin{smallmatrix}OMe\\\\C-CO_2Me\\H_2C\end{smallmatrix} \xrightarrow{L} \begin{smallmatrix}MeO_2C \quad\quad OMe\\C-C\\H_2C \quad\quad C=O\\Fe\\(CO)_2L\end{smallmatrix}$$

Scheme 47. L = CO, PPh$_3$.

4 Substitution Reactions

4.1 Substitution of a Non-Carbene Ligand

Ligand substitution reactions occur readily with most carbene complexes to yield a variety of new carbene complexes (2 – 5). The best studied substitution reaction for carbonyl ligands is the displacement by phosphines (24, 99 – 101) (Scheme 48) which can be brought about by mild heating or by a photochemically induced transformation. Study of the reaction mechanism provides information about the coordinatively unsaturated intermediates discussed in the olefin metathesis and cyclopropenation processes.

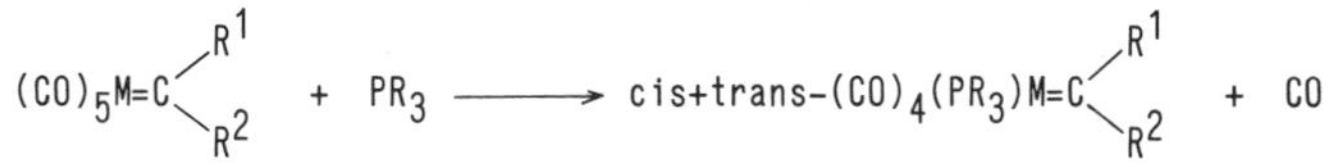

$$(CO)_5M=C\underset{R^2}{\overset{R^1}{<}} \quad + \quad PR_3 \quad \longrightarrow \quad cis+trans-(CO)_4(PR_3)M=C\underset{R^2}{\overset{R^1}{<}} \quad + \quad CO$$

Scheme 48. M = Cr, Mo, W; R^1 = OMe; R^2 = alkyl, aryl; R = alkyl, aryl, OR.

These substitution reactions can be extended to other metal carbene complexes and to other nucleophilic reagents such as arsines or stibines (102).

A detailed kinetic analysis of the reaction between the pentacarbonyl carbene complexes of chromium and tertiary phosphines by Werner (103) postulated an intermediate adduct, such as an ylide complex, which could be isolated on treating the carbene complexes with trimethylphosphine at low temperatures (36).

Attempts to replace all the carbonyl ligands in a Fischer-type carbene complex by other groups proved to be difficult. The photochemically induced reaction of methoxymethylcarbene(dicarbonyl)(η^5-methylcyclopentadienyl)-manganese with methylaminobis(difluorophosphine) yielded first carbene complex of this type free of carbonyl ligands (104) (Scheme 49).

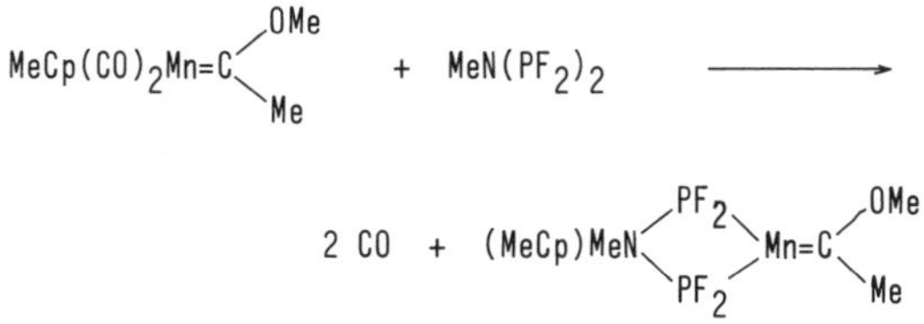

$$MeCp(CO)_2Mn=C\underset{Me}{\overset{OMe}{<}} \quad + \quad MeN(PF_2)_2 \quad \longrightarrow$$

$$2\ CO \quad + \quad (MeCp)MeN\underset{PF_2}{\overset{PF_2}{<}}Mn=C\underset{Me}{\overset{OMe}{<}}$$

Scheme 49 .

4.2 Substitution of the Carbene Ligand

The nucleophilic substitution of the carbene ligand in methoxymethylcarbene(pentacarbonyl) complexes of chromium, molybdenum and tungsten by phosphine (105), phosphorus bromide or iodide (106) offered an easy route to tetracarbonyl(bisphosphine)chromium and pentacarbonyl(trihalidephosphine) complexes (Scheme 50).

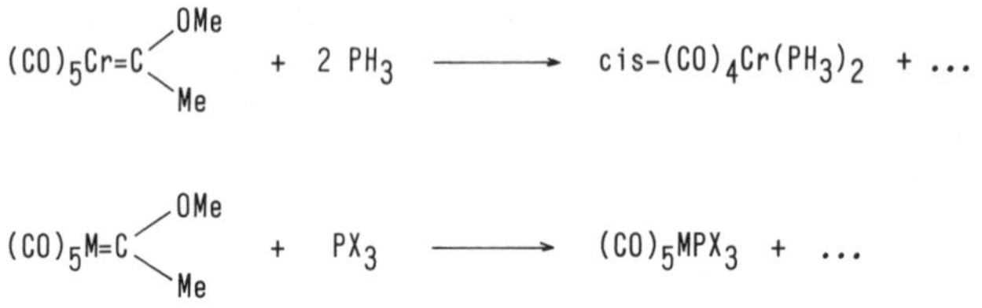

$$(CO)_5Cr=C\begin{smallmatrix}OMe\\Me\end{smallmatrix} \quad + \quad 2\ PH_3 \quad\longrightarrow\quad cis\text{-}(CO)_4Cr(PH_3)_2 \ + \ ...$$

$$(CO)_5M=C\begin{smallmatrix}OMe\\Me\end{smallmatrix} \quad + \quad PX_3 \quad\longrightarrow\quad (CO)_5MPX_3 \ + \ ...$$

Scheme 50. M = Cr, Mo, W; X = Br, I, C_6H_{11}(99).

Pyridine and related basic nitrogen compounds cause, in a similar way to phosphorus halides, a cleavage of the carbene ligand to form, for instance, pyridine complexes and enol ethers (107 – 109) (Scheme 51).

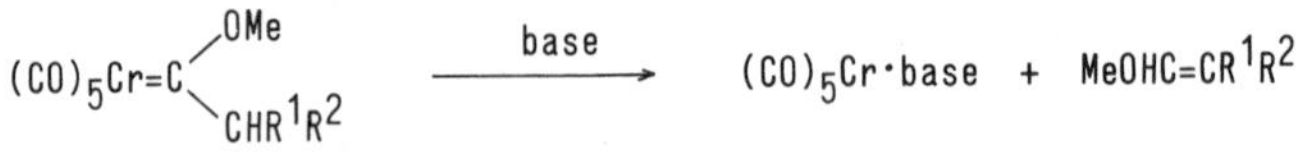

$$(CO)_5Cr=C\begin{smallmatrix}OMe\\CHR^1R^2\end{smallmatrix} \quad\xrightarrow{\text{base}}\quad (CO)_5Cr\cdot base \ + \ MeOHC=CR^1R^2$$

Scheme 51. R^1 = H, Me; R^2 = H, Me, Et.

When a thioether such as Et_2S is used, the same type of cleavage can be observed (110) (Scheme 52).

$$(CO)_5Cr=C\begin{smallmatrix}OEt\\Ph\end{smallmatrix} \quad + \quad Et_2S \quad\longrightarrow\quad (CO)_5CrSEt_2 \ + \ ...$$

Scheme 52.

In a few cases the transfer of the carbene ligand from one metal to another could be achieved. On irradiating a mixture of carbonyl(η^5-cyclopentadienyl)-nitrosylcarbene complexes of chromium, molybdenum or tungsten and pentacarbonyliron in benzene solution, methoxyphenylcarbene(tetracarbonyl)-iron was formed by a transfer reaction of the carbene (111 – 113). There is, however, no proof for the migration of a free carbene, since it is possible that a free tetracarbonyliron fragment could react with the initial carbene complex (Scheme 53).

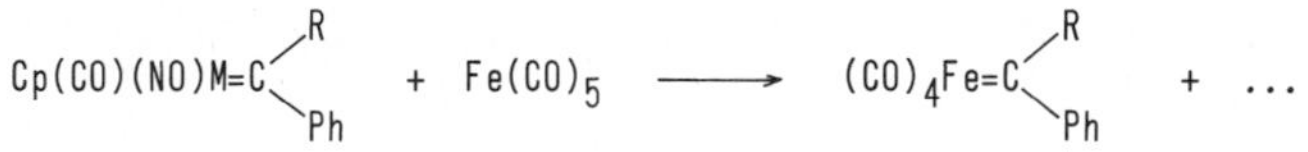

$$Cp(CO)(NO)M=C\overset{R}{\underset{Ph}{}} + Fe(CO)_5 \longrightarrow (CO)_4Fe=C\overset{R}{\underset{Ph}{}} + ...$$

Scheme 53. M = Cr, Mo, W; R = OMe, NMe$_2$.

The reaction of the molybdenum carbene complex with tetracarbonylnickel in tetrahydrofuran yielded, without irradiation, an extremely labile violet complex which is believed to be a trimer (111) (Scheme 54).

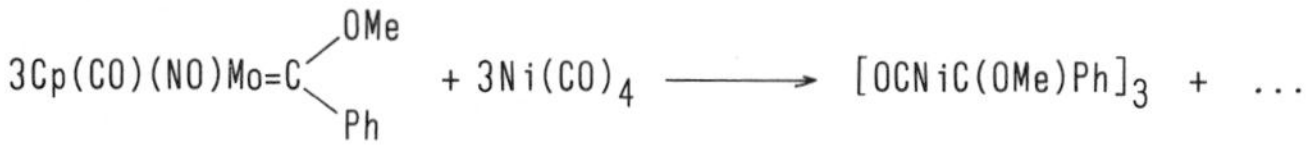

$$3Cp(CO)(NO)Mo=C\overset{OMe}{\underset{Ph}{}} + 3Ni(CO)_4 \longrightarrow [OCNiC(OMe)Ph]_3 + ...$$

Scheme 54.

A further thermally induced transfer of the carbene ligand was shown in the reaction of a chromium carbene complex with hexacarbonyltungsten (109) (Scheme 55).

$$(CO)_5Cr=C\overset{C-C}{\underset{O-C}{}} + W(CO)_6 \longrightarrow (CO)_5W=C\overset{C-C}{\underset{O-C}{}} + Cr(CO)_6$$

Scheme 55.

With electron-rich carbene ligands, thermal disproportionation has been found to yield biscarbene complexes (114 – 116) (Scheme 56).

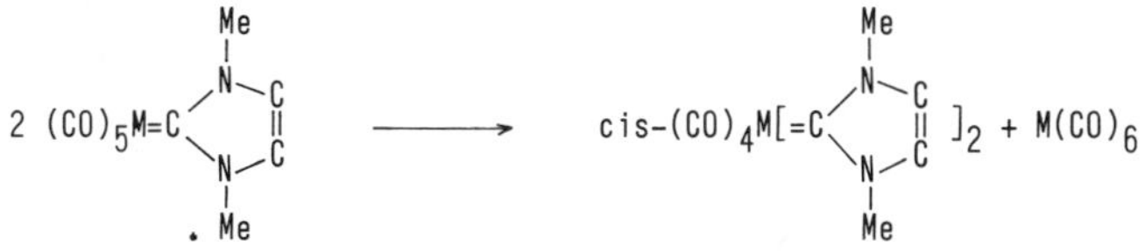

$$2 (CO)_5M=C\overset{Me}{\underset{Me}{\overset{N-C}{\underset{N-C}{}}}} \longrightarrow cis\text{-}(CO)_4M[=C\overset{Me}{\underset{Me}{\overset{N-C}{\underset{N-C}{}}}}]_2 + M(CO)_6$$

Scheme 56. M = Cr, Mo, W.

5 Reaction with Transition Metal Nucleophiles

The known reaction behaviour of nucleophilic d^8 or d^{10} metal complexes with alkenes and alkynes can be extended to other double bond systems such as those present in carbene complexes. The metal carbene carbon double bond acts like an olefin to give with d^8 and d^{10} metal complexes dimetal compounds containing a bridging carbene ligand. This reaction is applicable to many carbene complexes of chromium, molybdenum, tungsten and manganese, when nickel, palladium or platinum complexes are used (117 – 121) (Scheme 57).

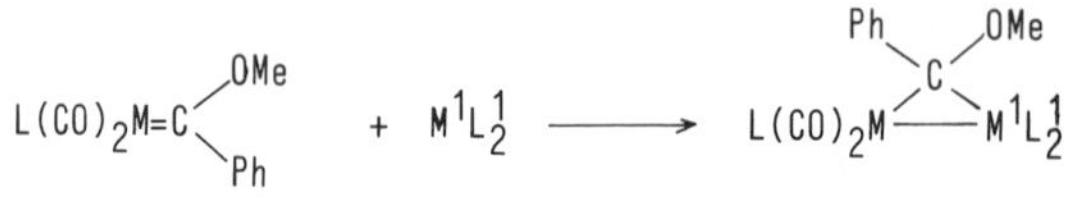

$$L(CO)_2M=C\overset{OMe}{\underset{Ph}{\big\langle}} \quad + \quad M^1L^1_2 \quad \longrightarrow \quad L(CO)_2M\overset{Ph\quad OMe}{\overset{\diagdown C \diagup}{\underset{}{\big|}}}M^1L^1_2$$

Scheme 57. M = Cr, Mo, W, Mn; L = (CO)$_3$, Cp; M^1 = Ni, Pd, Pt; L^1 = COD, PMe$_3$.

On treating enneacarbonyl[(2-oxacyclopentylidene)carbene]dimanganese with ethylene(bisphosphine)platinum, a mixture of a binuclear manganese-platinum complex and pentacarbonylmanganese hydride was isolated (117, 119) (Scheme 58).

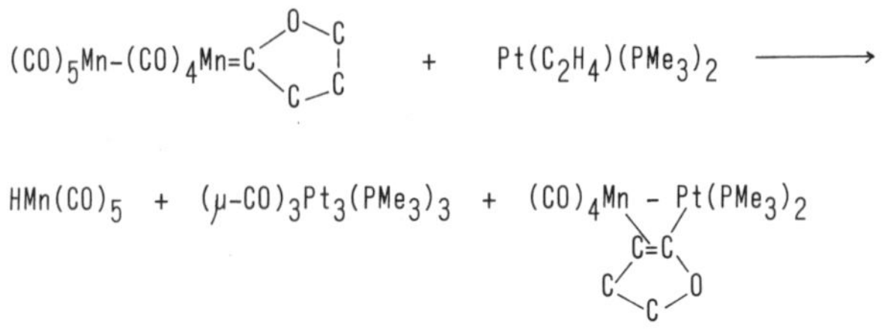

$$(CO)_5Mn-(CO)_4Mn=C\overset{O\diagdown C}{\underset{C-C}{\big\langle}} \quad + \quad Pt(C_2H_4)(PMe_3)_2 \quad \longrightarrow$$

$$HMn(CO)_5 \quad + \quad (\mu\text{-}CO)_3Pt_3(PMe_3)_3 \quad + \quad (CO)_4\overset{}{\underset{C=C}{Mn}} - \overset{}{\underset{C\diagdown C\diagup O}{Pt}}(PMe_3)_2$$

Scheme 58.

In the presence of bulky phosphine ligands, even trinuclear platinum compounds with μ^3-carbene ligands were formed (120). With I(CO)$_4$Mn = C(CH$_2$)$_3$O and diethylenephosphineplatinum total transfer of the carbene ligand was observed (121).

6 Reaction with Electrophilic Reagents

6.1 Synthesis of Neutral Transition Metal Carbyne Complexes

Carbyne complexes with terminally bonded C-alkyl, C-aryl or C-heteratom ligands were first synthesized in 1973 by E. O. Fischer and G. Kreis by reacting methoxyorganylcarbene complexes of chromium, molybdenum and tungsten with boron tribromide (122). Thus, carbene complexes proved to be an excellent source for the synthesis of different kinds of carbyne complexes when various trihalides of boron, aluminium or gallium were used. The alkoxy substituent and the *trans* carbonyl ligand were split off to yield *trans*-halogeno(tetracarbonyl)organylcarbyne complexes (122 – 129) (Scheme 59).

$$(CO)_5M=C\begin{matrix} R^1 \\ \\ R^2 \end{matrix} \quad + \text{ "}YX_3\text{"} \longrightarrow \quad trans\text{-}X(CO)_4M\equiv C\text{-}R^2 \quad + \quad CO \quad + ..$$

Scheme 59. M = Cr, Mo, W; R^1 = O-alkyl; R^2 = alkyl, aryl; Y = B, Al, Ga; X = Cl, Br, I.

This reaction turned out to be quite general and could be easily extended to carbene complexes having silyl (130), amino (131), vinylic (132), acetylenic (133) or metallocene (134 – 137) substituents. Instead of alkoxy groups, amino (138), thio (139), siloxy (140) or acetoxy groups (141) could be used as leaving groups in the reaction with Lewis acids.

With *cis*-phosphine, arsine and stibine substituted carbene complexes, the formation of *mer*-halogeno(phosphine or arsine or stibine)carbyne(tricarbonyl) complexes could be accomplished (142) (Scheme 60).

$$cis\text{-}(CO)_4(ZMe_3)Cr=C\begin{matrix} OMe \\ \\ Me \end{matrix} \quad + \quad BX_3 \longrightarrow mer\text{-}(CO)_3X(ZMe_3)Cr\equiv C\text{-}Me + CO$$

Scheme 60. Z = P, As, Sb.

Boron trifluoride reacted in a different way. Thus, a tetrafluoroborate unit rather than a fluoro ligand occupied the *trans* position at the metal (143 – 145) (Scheme 61).

$$cis\text{-}(CO)_4LM=C\begin{matrix} OMe \\ \\ R \end{matrix} \quad \xrightarrow{BF_3} \quad BF_4\begin{matrix} OC & L \\ \diagdown & | \\ -M\equiv C\text{-}R \\ \diagup & \diagdown \\ OC & CO \end{matrix} \quad + \quad CO \quad + \quad ...$$

Scheme 61. M = Cr, W; R = alkyl, aryl; L = CO, PMe$_3$.

6.2 Cationic Carbyne Complexes

Carbene complexes having a ligand in the *trans* position with a greater σ-donor/π-acceptor ratio than carbon monoxide (146), or a π-aromatic system (147 – 150) reacted with boron trihalides without evolution of carbon monoxide. On the contrary, a cationic carbyne complex was formed simply by removing the heteratom containing substituent from the carbene ligand (Scheme 62).

$$\text{trans-ZMe}_3(CO)_4Cr=C\begin{smallmatrix}OMe\\\\Me\end{smallmatrix} \xrightarrow{BX_3} [ZMe_3(CO)_4Cr\equiv C-Me][BX_4] + ..$$

$$\pi\text{-}C_nH_n(CO)_2M=C\begin{smallmatrix}OMe\\\\R\end{smallmatrix} \xrightarrow{BX_3} [\pi\text{-}C_nH_n(CO)_2M\equiv C-R][BX_4] + ...$$

Scheme 62. Z = P, As, Sb; X = F, Cl, Br.
M = Cr: n = 6; M = Mn, Re: n = 5.
R = alkyl, aryl, metallocene.

When two heteratom substituents are present in the carbene ligand, a unique type of cationic carbyne complex is obtained, either by reacting dialkylamino(ethoxy)carbene(pentacarbonyl) complexes of chromium (131, 151), molybdenum (152) or tungsten (153) with boron trihalides or by treating dimethylamino(chloro)carbene(pentacarbonyl)chromium with silver salts (151) (Scheme 63).

$$(CO)_5M=C\begin{smallmatrix}OEt\\\\NR_2\end{smallmatrix} \xrightarrow{BX_3} [(CO)_5M\equiv C-NR_2][BX_4] + ...$$

$$(CO)_5Cr=C\begin{smallmatrix}Cl\\\\NMe_2\end{smallmatrix} + AgA \longrightarrow [(CO)_5Cr\equiv C-NMe_2]A + ...$$

Scheme 63. M = Cr, Mo, W; R = alkyl; A = BF$_4$, PF$_6$, ClO$_4$.

The mechanism of the carbyne formation is believed to start with the addition of the Lewis acid to the heteroatom bound to the carbene carbon. The heteroatom containing substituent thereby becomes a better leaving group and is cleaved to yield a cationic carbyne complex. The positive charge, localized mainly on the heteroatom, leads to a decrease in the back-bonding to all ligands and especially to the *trans* ligand. If this ligand has a high σdonor/π-acceptor ratio, the ligand will remain attached to the metal to give, finally, a

cationic carbyne complex. In the opposite case of a low ratio, the *trans* ligand
will be replaced by one having a better donor/acceptor ratio (Scheme 64).

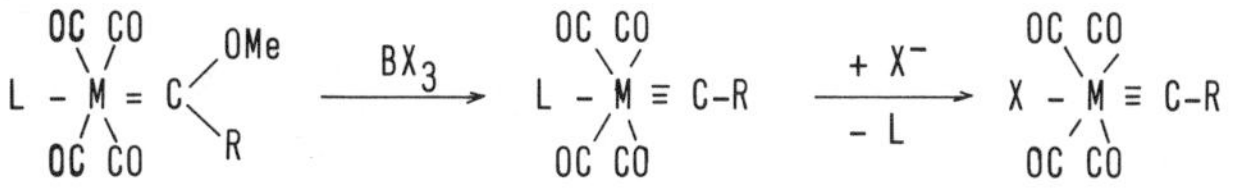

$$
\begin{array}{ccccc}
\text{OC} \; \text{CO} & & \text{OC} \; \text{CO} & & \text{OC} \; \text{CO} \\
\diagdown / \quad \diagup \text{OMe} & \xrightarrow{\;\;BX_3\;\;} & \diagdown / & \xrightarrow[-\,L]{+\,X^-} & \diagdown / \\
L - M = C & & L - M \equiv C\text{-}R & & X - M \equiv C\text{-}R \\
\diagup \diagdown \quad \diagdown R & & \diagup \diagdown & & \diagup \diagdown \\
\text{OC} \; \text{CO} & & \text{OC} \; \text{CO} & & \text{OC} \; \text{CO}
\end{array}
$$

Scheme 64.

Strong support for this proposed reaction mechanism can be seen in the
isolation of an adduct formed by the reaction of *cis*-bromo(hydroxymethyl-
carbene)pentacarbonylmanganese with boron tribromide (154) (Scheme 65).

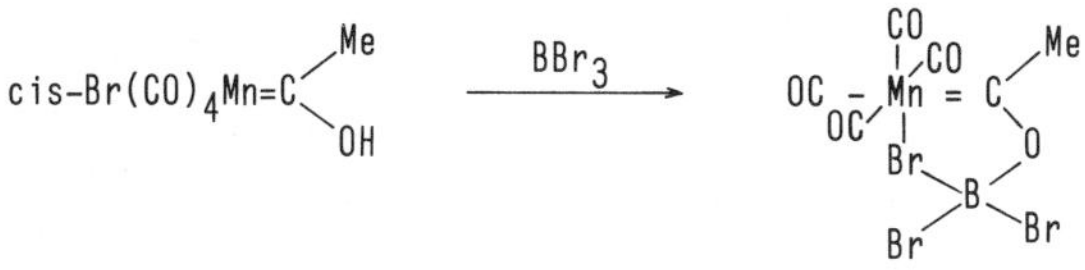

$$
\text{cis-Br(CO)}_4\text{Mn=C} \diagup^{\text{Me}}_{\text{OH}} \quad \xrightarrow{\;\;BBr_3\;\;} \quad
\begin{array}{c}
\text{CO} \\
\text{OC} - \overset{|}{\underset{|}{\text{Mn}}} = \text{C} \diagup^{\text{Me}} \\
\text{OC} \quad \text{Br} \diagdown_{O} \\
\text{Br} \diagup \text{B} \diagdown \text{Br}
\end{array}
$$

Scheme 65.

6.3 Rearrangement of Carbene to Carbyne Complexes

Pentacarbonyl carbene complexes of chromium with particular combina-
tions of substituents at the carbene carbon rearrange spontaneously by loss of
the *trans* carbonyl ligand to form carbyne complexes. The facility of this rearrangement depends on the nature of the group X and is, as yet, restricted to X
equal to Cl, Br (155), I (156), $SnPh_3$ (157), SeC_6H_4R-(4) (158) $PbPh_3$ (159) and
TePh (160) (Scheme 66).

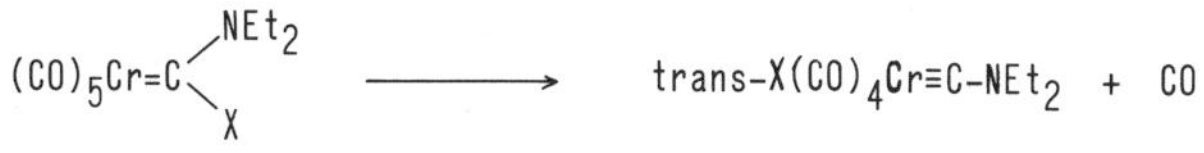

$$
\text{(CO)}_5\text{Cr=C} \diagup^{\text{NEt}_2}_{\text{X}} \qquad \longrightarrow \qquad \text{trans-X(CO)}_4\text{Cr}\equiv\text{C-NEt}_2 \; + \; \text{CO}
$$

Scheme 66. X = Cl, Br, I, $SnPh_3$, $PbPh_3$, SeC_6H_4R-(4), TePh; R = H, F, Br, CF_3, CH_3, OCH_3.

Runs with X equal to F, CN, SCN or $SiPh_3$ have not proved successful. The
rearrangement follows a first order rate law and free carbon monoxide does
not influence the reaction (155, 161).

A probable rearrangement of an intermediate carbene complex takes place
during the reaction of metal acylate with dibromotriphenylphosphine (162)
(Scheme 67).

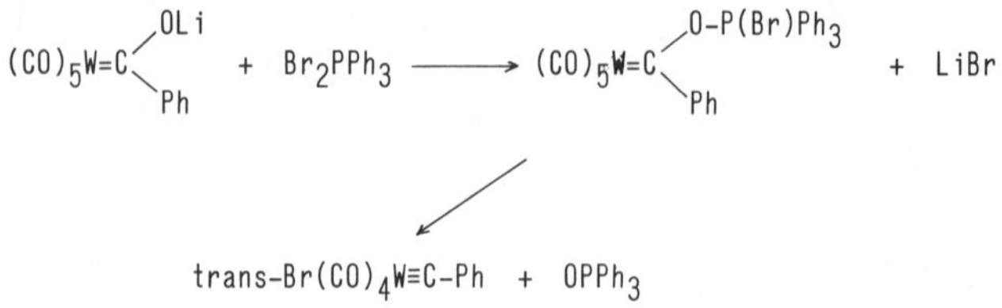

$$(CO)_5W=C\begin{smallmatrix}OLi\\Ph\end{smallmatrix} + Br_2PPh_3 \longrightarrow (CO)_5W=C\begin{smallmatrix}O-P(Br)Ph_3\\Ph\end{smallmatrix} + LiBr$$

$$trans\text{-}Br(CO)_4W\equiv C\text{-}Ph + OPPh_3$$

Scheme 67.

The reaction of hydroxycarbene complex anhydrides with tetraalkyl-ammonium halide probably proceeds by cleavage of one carbene carbon-oxygen bond to give a halocarbene complex which may then rearrange to give a *trans*-halogeno(tetracarbonyl)carbene complex by loss of carbon monoxide (89). The reaction will depend on both, the electronic properties of the substituent R and the kind of halogen. The cleavage is possible with bromide and iodide, where R is $C_6H_4CF_3$-(4); with R = phenyl only iodide could be used. No reaction occurred in the case of the combination chloride and $C_6H_4CF_3$-(4) (Scheme 68).

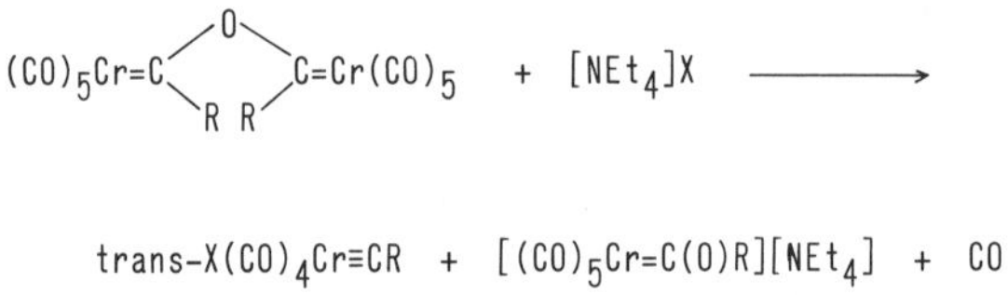

$$(CO)_5Cr=C\begin{smallmatrix}O\\R\ R\end{smallmatrix}C=Cr(CO)_5 + [NEt_4]X \longrightarrow$$

$$trans\text{-}X(CO)_4Cr\equiv CR + [(CO)_5Cr=C(O)R][NEt_4] + CO$$

Scheme 68. R = aryl.

6.4 Miscellaneous

Aluminium bromide can be shown to convert *trans*-pentacarbonylrhenio-(methoxyphenylcarbene)tetracarbonylrhenium into a species with a methylidyne bridge (163) (Scheme 69).

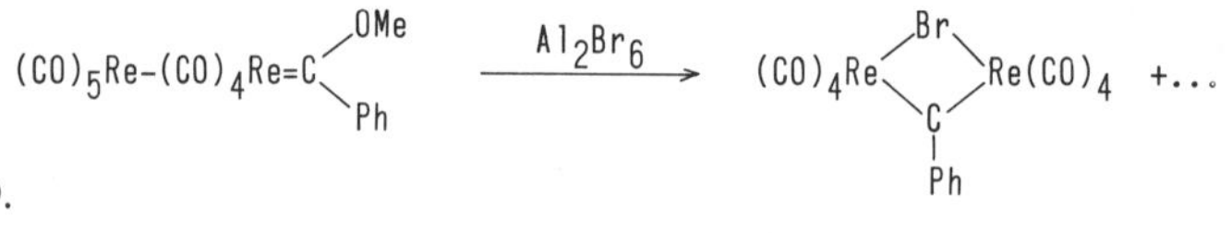

$$(CO)_5Re\text{-}(CO)_4Re=C\begin{smallmatrix}OMe\\Ph\end{smallmatrix} \xrightarrow{Al_2Br_6} (CO)_4Re\begin{smallmatrix}Br\\C\\|\\Ph\end{smallmatrix}Re(CO)_4 + \dots$$

Scheme 69.

Ethoxy(phenyldimethylaminovinyl)carbene complexes of chromium and tungsten show an enhanced basic character in the alkoxygroup. Upon reacting them with boron trihalides the corresponding adduct can be isolated (Scheme 70).

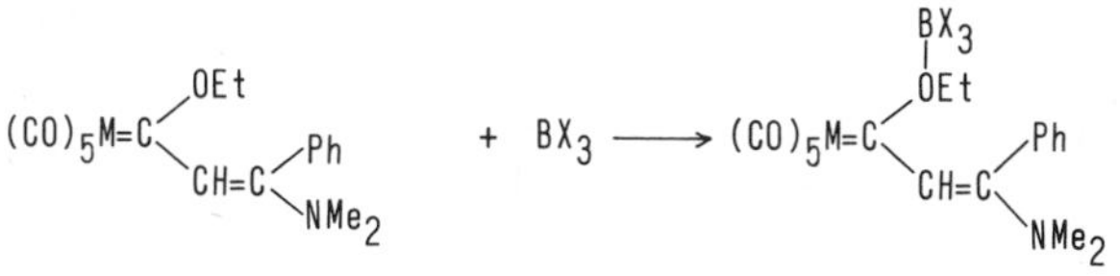

Scheme 70. M = Cr, X = F; M = W, X = Br.

In the presence of silica gel, the tungsten compound gives a carbyne complex, whereas the chromium adduct can be converted to an allenylidene complex by tetrabutylammonium fluoride (164) (Scheme 71).

Scheme 71.

Allenylidene complexes of chromium and tungsten may also be obtained by treating the corresponding carbene precursors with aluminium alkyls or boron trichloride, though now without the formation of a stable intermediate (165) (Scheme 72).

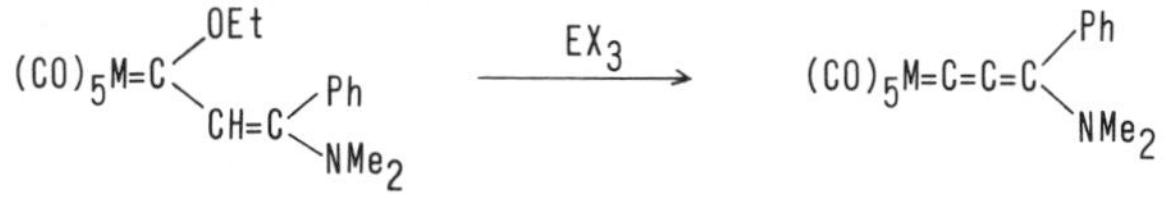

Scheme 72. M = Cr, EX_3 = $AlEt_3$, BCl_3; M = W, EX_3 = $AlEt_3$.

7 Oxidation and Reduction Reactions

Oxidation and reduction reactions can be carried out on both, the central metal and the metal-carbene bond.

Paramagnetic cationic carbene complexes can be obtained by one-electron oxidation of the corresponding neutral compounds using silver(I) salts as oxidizing agents (166, 167) (Scheme 73).

$$(CO)_3Fe[\overline{CN(CH_3)(CH_2)_2NCH_3}]_2 \qquad \xrightarrow{Ag[BF_4]}$$

$$[(CO)_3Fe[=\overline{CN(CH_3)(CH_2)_2NCH_3}]_2]^+[BF_4]^-$$

Scheme 73.

The electron density depends in both cases on the nature of the metal. For iron, e.s.r. studies indicate localisation of the odd electron mainly on the metal, whereas for the chromium compound a delocalisation of the unpaired electron is favoured.

Cyclic voltammetry has been attempted for molybdenum and tungsten carbene complexes with cyclic diaminocarbene ligands (168, 169).

Oxidative addition can also be observed for cyclic diaminocarbene complexes (170) as well as for cyclic osmium (171) and platinum carbene complexes (18) (Scheme 74).

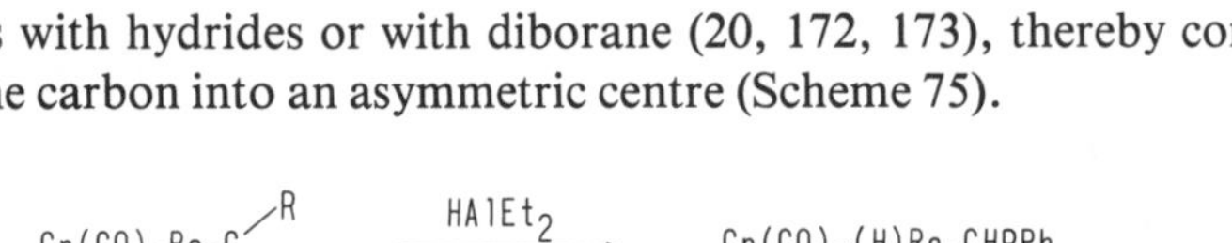

Scheme 74.

Reduction of the metal-carbene bond occurs on treating different carbene complexes with hydrides or with diborane (20, 172, 173), thereby converting the carbene carbon into an asymmetric centre (Scheme 75).

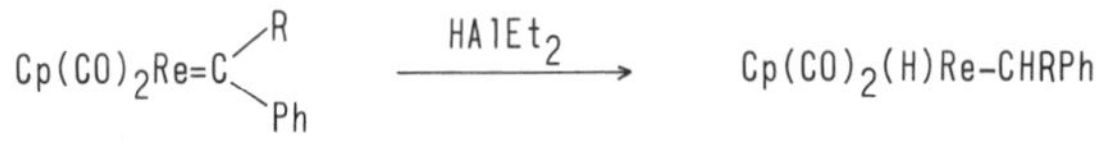

$$Cp(CO)_2Re=C\begin{smallmatrix} R \\ \\ Ph \end{smallmatrix} \qquad \xrightarrow{HAlEt_2} \qquad Cp(CO)_2(H)Re-CHRPh$$

Scheme 75.

With dihydrogen a reversible oxidation-reduction has been observed using an iridium carbene complex (174) (Scheme 76).

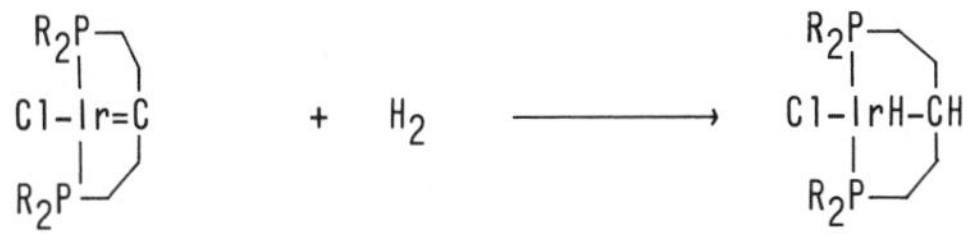

$$\underset{R_2P}{\overset{R_2P}{Cl-Ir=C}} \quad + \quad H_2 \quad \longrightarrow \quad \underset{R_2P}{\overset{R_2P}{Cl-IrH-CH}}$$

Scheme 76.

8 References

(1) E. O. Fischer, A. Maasböl, Angew. Chem. Int. Ed. Engl. **3** (1964) 580 – 581.
(2) D. J. Cardin, B. Cetinkaya, M. F. Lappert, Chem. Rev. **72** (1972) 545 – 574.
(3) F. A. Cotton, C. M. Lukehart, Prog. Inorg. Chem. **16** (1972) 487 – 614.
(4) F. J. Brown, Prog. Inorg. Chem. **27** (1980) 1 – 122.
(5) H. Fischer, The Chemistry of Metal Carbon Bond, in S. Patai (Ed.), 181 – 231, Wiley and Sons, New York 1982.
(6) W. B. Perry, T. F. Schaaf, W. L. Jolly, L. J. Todd, D. L. Cronin, Inorg. Chem. **13** (1974) 2038 – 2039.
(7) T. F. Block, R. F. Fenske, C. P. Casey, J. Am. Chem. Soc. **98** (1976) 441 – 444.
(8) T. F. Block, R. F. Fenske, J. Organomet. Chem. **139** (1967) 235 – 269.
(9) U. Klabunde, E. O. Fischer, J. Am. Chem. Soc. **89** (1967) 7141 – 7142.
(10) J. A. Connor, E. O. Fischer, J. Chem. Soc. A. 1969, 578 – 584.
(11) E. Moser, E. O. Fischer, J. Organomet. Chem. **16** (1969) 275 – 282.
(12) E. O. Fischer, H.-J. Kollmeier, Chem. Ber. **104** (1971) 1339 – 1346.
(13) E. O. Fischer, M. Leupold, Chem. Ber. **105** (1972) 599 – 608.
(14) E. O. Fischer, B. Heckl, H. Werner, J. Organomet. Chem. **28** (1971) 359 – 365.
(15) P. E. Baikie, E. O. Fischer, O. S. Mills, J. Chem. Soc. Chem. Commun. **1967**, 1199 – 1200.
(16) E. Moser, E. O. Fischer, Naturwissensch. **54** (1967) 615 – 616.
(17) C. G. Kreiter, Habilitationsschrift, Technische Universität München, 1971.
(18) M. H. Chrisholm, H. C. Clark, W. S. Johns, J. E. H. Ward, K. Yasufuku, Inorg. Chem. **14** (1975) 900 – 905.
(19) H. Brunner, E. O. Fischer, M. Lappus, Angew. Chem. Int. Ed. Engl. **10** (1971) 924 – 925.
(20) A. Davison, D. L. Reger, J. Am. Chem. Soc. **94** (1972) 9237 – 9238.
(21) K. Weiß, E. O. Fischer, Chem. Ber. **109** (1976) 1868 – 1886.
(22) E. O. Fischer, P. Rustemeyer, J. Organomet. Chem. **225** (1982) 265 – 277.
(23) B. Heckl, H. Werner, E. O. Fischer, Angew. Chem. Int. Ed. Engl. **7** (1968) 817 – 818.

(24) H. Werner, E. O. Fischer, B. Heckl, C. G. Kreiter, J. Organomet. Chem. **28** (1971) 367 – 389.

(25) E. O. Fischer, F. R. Kreissl, J. Organomet. Chem. **35** (1972) C47 – C51.

(26) E. O. Fischer, H. J. Kalder, J. Organomet. Chem. **131** (1977) 57 – 64.

(27) E. O. Fischer, H. Hollfelder, P. Friedrich, F. R. Kreissl, G. Huttner, Chem. Ber. **110** (1977) 3467 – 3480.

(28) E. O. Fischer, T. Selmayr, F. R. Kreissl, U. Schubert, Chem. Ber. **110** (1977) 2574 – 2583.

(29) E. O. Fischer, M. Leupold, C. G. Kreiter, J. Müller, Chem. Ber. **105** (1972) 150 – 162.

(30) C. T. Lam, C. V. Senoff, J. E. H. Ward, J. Organomet. Chem. **70** (1974) 273 – 281.

(31) E. O. Fischer, G. Kreis, F. R. Kreissl, C. G. Kreiter, J. Müller, Chem. Ber. **106** (1973) 3910 – 3919.

(32) C. P. Casey, W. R. Brunsvold, Inorg. Chem. **16** (1977) 391 – 396.

(33) R. A. Bell, M. H. Chrisholm, D. A. Couch, L. A. Rankel, Inorg. Chem. **16** (1977) 677 – 686.

(34) C. G. Kreiter, Angew. Chem. Int. Ed. Engl. **7** (1968) 390 – 391.

(35) F. R. Kreissl, C. G. Kreiter, E. O. Fischer, Angew. Chem. Int. Ed. Engl. **11** (1972) 643.

(36) F. R. Kreissl, E. O. Fischer, C. G. Kreiter, H. Fischer, Chem. Ber. **106** (1973) 1262 – 1276.

(37) F. R. Kreissl, T. Selmayr, unpubl. results.

(38) F. R. Kreissl, W. Held, Chem. Ber. **110** (1977) 799 – 804.

(39) F. R. Kreissl, P. Stückler, E. W. Meineke, Chem. Ber. **110** (1977) 3040 – 3045.

(40) H. Fischer, E. O. Fischer, C. G. Kreiter, H. Werner, Chem. Ber. **107** (1974) 2459 – 2467.

(41) C. P. Casey, S. W. Polichnowski, J. Am. Chem. Soc. **99** (1977) 6097 – 6098.

(42) F. R. Kreissl, K. Eberl, P. Stückler, Angew. Chem. Int. Ed. Engl. **16** (1977) 654 – 655.

(43) W. Uedelhoven, K. Eberl, W. Sieber, F. R. Kreissl, J. Organomet. Chem. **236** (1982) 301 – 307.

(44) F. R. Kreissl, E. O. Fischer, C. G. Kreiter, K. Weiss, Angew. Chem. Int. Ed. Engl. **12** (1973) 563.

(45) F. R. Kreissl, E. O. Fischer, Chem. Ber. **107** (1974) 183 – 188.

(46) W. K. Wong, W. Tam, J. A. Gladysz, J. Amer. Chem. Soc. **101** (1979) 5440 – 5442.

(47) C. P. Casey, S. W. Polichnowski, A. J. Shusterman, C. R. Jones, J. Amer. Chem. Soc. **101** (1979) 7282 – 7292.

(48) A. Sanders, L. Cohen, W. P. Giering, D. Kenedy, C. V. Magatti, J. Amer. Chem. Soc. **95** (1973) 5430 – 5431.

(49) C. G. Kreiter, V. Formacek, Angew. Chem. Int. Ed. Engl. **11** (1972) 141 – 142.

(50) E. O. Fischer, S. Riedmüller, Chem. Ber. **109** (1976) 3358 – 3361.

(51) E. O. Fischer, W. Held, F. R. Kreissl, Chem. Ber. **110** (1977) 3842 – 3848.

(52) T. J. Burkhardt, C. P. Casey, J. Amer. Chem. Soc. **95** (1973) 5833 – 5834.

(53) C. P. Casey, S. W. Polichnowski, R. L. Anderson, J. Amer. Chem. Soc. **97** (1975) 7375 – 7376.

(54) E. O. Fischer, W. Held, F. R. Kreissl, A. Frank, G. Huttner, Chem. Ber. **110** (1977) 656 – 666.

(55) C. P. Casey, T. J. Burkhardt, C. A. Bunnell, J. C. Calabrese, J. Amer. Chem. Soc. **99** (1977) 2127 – 2134.

(56) C. P. Casey, W. R. Brunsvold, J. Organomet. Chem. **77** (1974) 345 – 352.

(57) E. O. Fischer, W. Held, J. Organomet. Chem. **112** (1976) C59 – C62..

(58) A. J. Harthorn, M. F. Lappert, J. Chem. Soc. Chem. Commun. 1976, 761 – 762.

(59) E. O. Fischer, V. Kiener, Angew. Chem. Int. Ed. Engl. **6** (1967) 961.

(60) R. A. Pickering, R. J. Angelici, J. Organomet. Chem. **225** (1982) 253.

(61) E. O. Fischer, S. Fontana, U. Schubert, J. Organomet. Chem. **91** (1975) C7 – C8.

(62) E. O. Fischer, G. Kreis, Chem. Ber. **106** (1973) 2310 – 2314.

(63) E. O. Fischer, K. R. Schmid, W. Kalbfus, C. G. Kreiter, Chem. Ber. **106** (1973) 3893 – 3909.

(64) E. Lindner, H. Behrens, Spectrochim. Acta 23 (1967) 3025 – 3033.

(65) E. O. Fischer, S. Walz, G. Kreis, F. R. Kreissl, Chem. Ber. **110** (1977) 1615 – 1658.

(66) E. O. Fischer, R. Aumann, Chem. Ber. **101** (1968) 963 – 968.

(67) H. G. Raubenheimer, S. Lotz, G. Kruger, G. Gafner, J. Organomet. Chem. **173** (1979) C1 – C5.

(68) E. O. Fischer, L. Knauss, Chem. Ber. **103** (1970) 1262 – 1272.

(69) E. O. Fischer, L. Knauss, Chem. Ber. **103** (1970) 3744 – 3751.

(70) J. A. Connor, E. M. Jones, J. Chem. Soc. (A) 1971, 3368 – 3372.

(71) R. Aumann, E. O. Fischer, Chem. Ber. **101** (1968) 954 – 962.

(72) C. G. Kreiter, R. Aumann, Chem. Ber. **111** (1978) 1223 – 1227.

(73) H. Fischer, J. Organomet. Chem. **219** (1981) C34 – C36.

(74) H. Fischer, R. Märkl, Chem. Ber. **115** (1982) 1349 – 1354.

(75) F. R. Kreissl, E. O. Fischer, C. G. Kreiter, J. Organomet. Chem. **57** (1973) C9 – C11.

(76) K. H. Dötz, Chem. Ber. **110** (1977) 78 – 85.

(77) K. H. Dötz, I. Pruskil, Chem. Ber. **111** (1978) 2059 – 2063.

(78) H. Fischer, K. H. Dötz, Chem. Ber. **113** (1980) 193 – 302.

(79) K. H. Dötz, B. Fügen-Köster, D. Neugebauer, J. Organomet. Chem. **182** (1979) 489 – 498.

(80) K. H. Dötz, Chem. Ber. **113** (1980) 3597 – 3604.

(81) J. Levisalles, H. Rudler, D. Villemin, J. Organomet. Chem. **146** (1978) 259 – 265.

(82) J. Levisalles, H. Rudler, D. Villemin, J. Daran, Y. Jeannin, L. Martin, J. Organomet. Chem. **155** (1978) C1 – C4.

(83) K. H. Dötz, I. Pruskil, Chem. Ber. **114** (1981) 1980 – 1982.

(84) H. Fischer, U. Schubert, Angew. Chem. Int. Ed. Engl. **20** (1981) 461 – 463.

(85) K. H. Dötz, C. G. Kreiter, Chem. Ber. **109** (1976) 2026 – 2032.

(86) K. H. Dötz, D. Neugebauer, Angew. Chem. Int. Ed. Engl. **17** (1978) 851 – 852.

(87) E. O. Fischer, K. Weiss, J. Müller, Chem. Ber. **107** (1974) 3548 – 3553.

(88) E. O. Fischer, K. Weiss, C. G. Kreiter, Chem. Ber. **107** (1974) 3554 – 3561.

(89) K. Weiß, E. O. Fischer, Chem. Ber. **109** (1976) 1120 – 1127.

(90) E. O. Fischer, K. Weiss, Chem. Ber. **109** (1976) 1128 – 1139.

(91) K. H. Dötz, Angew. Chem. Int. Ed. Engl. **14** (1975) 644 – 645.

(92) K. H. Dötz, R. Dietz, A. von Imhof, H. Lorenz, G. Huttner, Chem. Ber. **109** (1976) 2033 – 2038.

(93) K. H. Dötz, R. Dietz, Chem. Ber. **111** (1978) 2517 – 2526.

(94) K. H. Dötz, B. Fügen-Köster, Chem. Ber. **113** (1980) 1449 – 1457.

(95) W. A. Herrmann, J. Plank, Angew. Chem. Int. Ed. Engl. **17** (1978) 525 – 526.

(96) Ch. Rüchardt, G. N. Schrauzer, Chem. Ber. **93** (1960) 1840 – 1848.

(97) G. Henrici-Olive, S. Olive, Angew. Chem. Int. Ed. Engl. **15** (1975) 136 – 141.

(98) T. A. Mitsudo, T. Sasaki, Y. Watanabe, Y. Takegami, J. C. S. Chem. Commun. 1978, 252 – 253.

(99) H. Werner, H. Rascher, Inorg. Chim. Acta **2** (1968) 181 – 185.

(100) E. O. Fischer, H. Fischer, Chem. Ber. **107** (1974) 657-672.

(101) E. O. Fischer, H. Fischer, H. Werner, Angew. Chem. Int. Ed. Engl. **11** (1972) 644 – 645.

(102) E. O. Fischer, K. Richter, Chem. Ber. **109** (1976) 1140 – 1157.

(103) H. Werner, H. Rascher, Helv. Chim. Acta 51 (1968) 1765 – 1775.

(104) E. O. Fischer, G. Besl, J. Organomet. Chem. **157** (1978) C33 – C34.

(105) E. O. Fischer, E. Louis, W. Bathelt, E. Moser, J. Müller, J. Organomet. Chem. **14** (1968) P9 – P12.

(106) E. O. Fischer, L. Knauss, Chem. Ber. **102** (1969) 223 – 229.

(107) E. O. Fischer, W. Plabst, Chem. Ber. **107** (1974) 3326 – 3331.

(108) E. O. Fischer, A. Maasböl, J. Organomet. Chem. **12** (1968) P15 – P17.

(109) C. P. Casey, R. L. Anderson, J. Chem. Soc. Chem. Commun. 1975, 895 – 896.

(110) H. G. Raubenheimer, C. J. A. Boeyens, S. Lotz, J. Organomet. Chem. **91** (1975) C23 – C26.

(111) E. O. Fischer, H.-J. Beck, Angew. Chem. Int. Ed. Engl. **9** (1970) 72 – 73.

(112) E. O. Fischer, H.-J. Beck, Chem. Ber. **104** (1971) 3101 – 3107.

(113) E. O. Fischer, H.-J. Beck, C. G. Kreiter, J. Lynch, J. Müller, E. Winkler, Chem. Ber. **105** (1972) 162 – 172.

(114) K. Öfele, M. Herberhold, Angew. Chem. Int. Ed. Engl. **9** (1970) 739 – 740.

(115) C. G. Kreiter, K. Öfele, G. W. Wiesner, Chem. Ber. **109** (1976) 1749 – 1758.

(116) K. Öfele, E. Roos, M. Herberhold, Z. Naturforsch. **31B** (1976) 1070 – 1077.

(117) T. V. Ashworth, M. Berry, J. A. K. Howard, M. Laguna, F. G. A. Stone, J. Chem. Soc. Chem. Comm. 1979, 43 – 45.

(118) T. V. Ashworth, J. A. K. Howard, M. Laguna, F. G. A. Stone, J. Chem. Soc. Dalton 1980, 1593 – 1600.

(119) M. Berry, J. A. K. Howard, F. G. A. Stone, J. Chem. Soc. Dalton 1980, 1601 – 1608.

(120) T. V. Ashworth, M. Berry, J. A. K. Howard, M. Laguna, F. G. A. Stone, J. Chem. Soc. Dalton 1980, 1615 – 1624.

(121) M. Berry, J. Martin-Gil, J. A. K. Howard, F. G. A. Stone, J. Chem. Soc. Dalton 1980, 1625 – 1629.

✶ (122) E. O. Fischer, G. Kreis, C. G. Kreiter, J. Müller, G. Huttner, H. Lorenz, Angew. Chem. Int. Ed. Engl. **12** (1973) 564.

(123) E. O. Fischer, U. Schubert, J. Organomet. Chem. **100** (1975) 59 – 81.

(124) E. O. Fischer, Angew. Chem. **86** (1974) 651 – 663.
Adv. Organomet. Chem. **14** (1976) 1 – 32.

(125) E. O. Fischer, U. Schubert, H. Fischer, Pure Appl. Chem. **50** (1978) 857 – 870.

(126) R. R. Schrock, Acc. Chem. Res. **12** (1979) 98 – 104.

(127) E. O. Fischer, G. Kreis, Chem. Ber. **109** (1976) 1673 – 1683.

(128) S. Fontana, O. Omara, E. O. Fischer, U. Schubert, F. R. Kreissl, J. Organomet. Chem. **149** (1978) C57 – C62.

(129) E. O. Fischer, A. Schwanzer, H. Fischer, D. Neugebauer, G. Huttner, Chem. Ber. **110** (1977) 53 – 66.

(130) E. O. Fischer, H. Hollfelder, F. R. Kreissl, Chem. Ber. **112** (1979) 2177 – 2189.

(131) E. O. Fischer, W. Kleine, G. Kreis, F. R. Kreissl, Chem. Ber. **111** (1978) 3542 – 3551.

(132) E. O. Fischer, W. R. Wagner, F. R. Kreissl, D. Neugebauer, Chem. Ber. 112 (1979) 1320 – 1328.

(133) E. O. Fischer, H. J. Kalder, F. H. Köhler, J. Organomet. Chem. **81** (1974) C23 – C27.

(134) E. O. Fischer, M. Schluge, J. O. Besenhard, P. Friedrich, G. Huttner, F. R. Kreissl, Chem. Ber. **111** (1978) 3530 – 3541.

(135) E. O. Fischer, V. N. Postnov, F. R. Kreissl, J. Organomet. Chem. **127** (1977) C19 – C21.

(136) E. O. Fischer, F. J. Gammel, J. O. Besenhard, A. Frank, D. Neugebauer, J. Organomet. Chem. **191** (1980) 261 – 282.

(137) E. O. Fischer, F. J. Gammel, D. Neugebauer, Chem. Ber. **113** (1980) 1010 – 1019.

(138) K. Weiss, E. O. Fischer, Chem. Ber. **109** (1976) 1868-1886.

(139) A. Schwanzer, Dissertation, Technische Universität München, 1976.

(140) E. O. Fischer, T. Selmayr, F. R. Kreissl, U. Schubert, Chem. Ber. **110** (1977) 2574 – 2583.

(141) E. O. Fischer, T. Selmayr, Z. Naturforsch. **32B** (1977) 105 – 107.

(142) E. O. Fischer, K. Richter, Chem. Ber. **109** (1976) 2547 – 2557.

(143) K. Richter, E. O. Fischer, C. G. Kreiter, J. Organomet. Chem. **122** (1976) 187 – 196.

(144) E. O. Fischer, F. J. Gammel, Z. Naturforsch. **34B** (1979) 1183 – 1185.

(145) E. O. Fischer, S. Walz, A. Ruhs, F. R. Kreissl, Chem. Ber. **111** (1978) 2765 – 2773.

(146) E. O. Fischer, K. Richter, Chem. Ber. **109** (1976) 3079 – 3088.

(147) E. O. Fischer, P. Stückler, H.-J. Beck, F. R. Kreissl. Chem. Ber. **109** (1976) 3089 – 3098.

(148) E. O. Fischer, E. W. Meineke, F. R. Kreissl, Chem. Ber. **110** (1977) 1140 – 1147.

(149) E. O. Fischer, P. Rustemeyer, D. Neugebauer, Z. Naturforsch. **35B** (1980) 1083 – 1087.

(150) E. O. Fischer, V. N. Postnov, F. R. Kreissl, J. Organomet. Chem. **231** (1982) C73 – C77.

(151) A. J. Harthorn, M. F. Lappert, J. Chem. Soc. Chem. Commun. 1976, 761 – 762.

(152) D. Wittmann, Dissertation, Technische Universität München, 1982.

(153) F. R. Kreissl, W. Uedelhoven, unpubl. result.

(154) E. W. Meineke, Disseratition, Technische Universität München, 1975.

(155) H. Fischer, A. Motsch, W. Kleine, Angew. Chem. Int. Ed. Engl. **17** (1978) 842 – 843.

(156) E. O. Fischer, W. Kleine, F. R. Kreissl, H. Fischer, P. Friedrich, G. Huttner, J. Organomet. Chem. **128** (1977) C49 – C53.

(157) E. O. Fischer, H. Fischer, U. Schubert, R. B. A. Pardy, Angew. Chem. Int. Ed. Engl. **18** (1979) 871.

(158) E. O. Fischer, D. Himmelreich, R. Cai, H. Fischer, U. Schubert, B. Zimmer-Gasser, Chem. Ber. **114** (1981) 3209 – 3219.

(159) H. Fischer, E. O. Fischer, R. Cai, Chem. Ber. **115** (1982) 2707 – 2713.

(160) H. Fischer, E. O. Fischer, R. Cai, D. Himmelreich, Chem. Ber., in press.

(161) H. Fischer, J. Organomet. Chem. **195** (1980) 55 – 61.

(162) H. Fischer, E. O. Fischer, J. Organomet. Chem. **69** (1974) C1 – C3.

(163) E. O. Fischer, G. Huttner, T. L. Lindner, A. Frank, F. R. Kreissl, Angew. Chem. Int. Ed. Engl. **15** (1976) 231 – 232.

(164) H. J. Kalder, Dissertation, Technische Universität München, 1976.

(165) E. O. Fischer, H. J. Kalder, A. Franl, F. H. Köhler, G. Huttner, Angew. Chem. Int. Ed. Engl. **15** (1976) 623 – 624.

(166) M. F. Lappert, R. W. McCabe, J. J. MacQuitty, P. L. Pye, P. I. Riley, J. Chem. Soc. Dalton 1980, 90 – 98.

(167) M. F. Lappert, J. J. MacQuitty, P. L. Pye, J. Chem. Soc. Chem. Commun. 1977, 411 – 412.

(168) R. D. Rieke, H. Kojima, K. Öfele, J. Am. Chem. Soc. **98** (1976) 6735 – 6737.

(169) R. D. Rieke, H. Kojima, K. Öfele, Angew. Chem. Int. Ed. Engl. **19** (1980) 538 – 540.

(170) M. F. Lappert, P. L. Pye, J. Chem. Soc. Dalton 1977, 1283 – 1291.

(171) M. Green, F. G. A. Stone, M. Underhill, J. Chem. Soc. Dalton 1975, 939 – 943.

(172) E. O. Fischer, A. Frank, Chem. Ber. **111** (1978) 3740 – 3744.

(173) M. L. H. Green, L. C. Mitchard, M. G. Sandwick, J. Chem. Soc. (A) 1971, 794.
(174) H. D. Empsall, E. M. Hyde, R. Markham, W. S. McDonald, M. C. Norton, B. L. Shaw, B. Weeks, J. Chem. Soc. Chem. Commun. 1977, 589 – 591.

Carbene Complexes in Organic Synthesis

By Karl Heinz Dötz

1 Introduction

Carbonyl complexes of transition metals are widely employed as powerful reagents in carbonylation reactions (1, 2). In a similar way a metal-coordinated carbene moiety can be used in reactions leading to carbon-carbon bond formation. This topic has been addressed in part in recent review articles (3, 4).

The "Fischer-type" carbene complexes **1** are characterized by an electrophilic carbene carbon atom. In this respect, their chemical behaviour is in total contrast to that of "Schrock-type" alkylidene complexes, *e.g.* **2**. In this class of compounds the metal-coordinated sp^2-carbon atom is nucleophilic in character and displays an ylide-like reactivity. The chemistry of these complexes has already been reviewed (5) and will not be discussed here in detail.

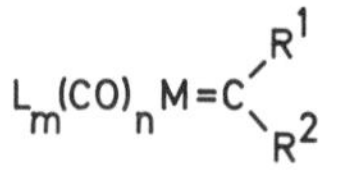

$$L_m(CO)_n M = C \overset{R^1}{\underset{R^2}{\diagdown}}$$

M = Mn, Re, Cr, Mo, W, Fe, Ni

$$(\underline{1})$$

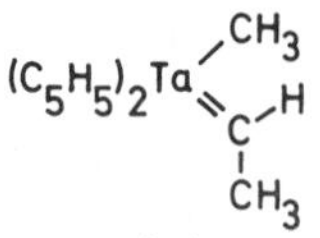

$$(C_5H_5)_2 Ta \overset{CH_3}{\diagdown} \underset{CH_3}{\overset{|}{C}} \diagup H$$

$$(\underline{2})$$

The chemical properties of carbonyl(carbene) complexes of type **1**, which form the topic of this chapter, are outlined in Figure 1.

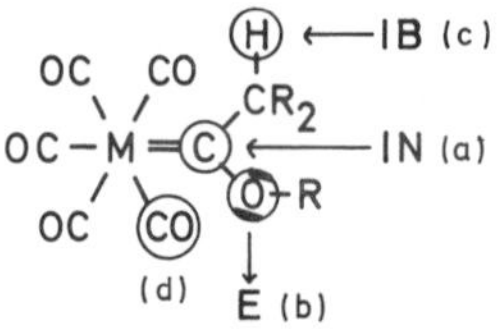

Figure 1. Reactivity of carbonyl(carbene) complexes

Nucleophilic agents, N, attack the electrophilic carbene carbon atom [via route (a)]. Electrophiles, E, e.g. Lewis acids, are coordinated to the alkoxy substituent [via route (b)]. This provides a simple route to metal-coordinated carbynes. Owing to the acidity of α-CH groups, alkylcarbene complexes are deprotonated by bases B leading to metal carbene anions [via route (c)]. Finally, a carbonyl ligand can be replaced by other ligands as is known in the case of binary metal carbonyl complexes [via route (d)]. In this scheme, routes (a), (c) and (d) have proved their value in the formation of new carbon-carbon bonds.

2 Carbon-Carbon Bond Formation via Metal Carbene Anions

The strategy of using metal complexes in organic synthesis is to establish a metal to carbon bond, to modify it subsequently by further reaction and finally to cleave the new ligand from the metal.

Hydrogen atoms attached to the α-carbon in the carbene side chain show remarkable acidity (6). The methoxy(methyl)carbene complex **3** is among the most acidic neutral carboxylic acids and its thermodynamic acidity has been estimated to be comparable to that of *p*-cyanophenol (7). Accordingly, alkyl-carbene complexes are deprotonated when treated with bases such as alkoxide or organolithium reagents (Scheme 1). The resulting carbene anions can be isolated as air-stable bis(triphenylphosphino)imminium salts. Spectroscopic studies indicate that the conjugate carbene anion is probably better regarded as a vinylchromium anion **3a** than a carbanion **3b**.

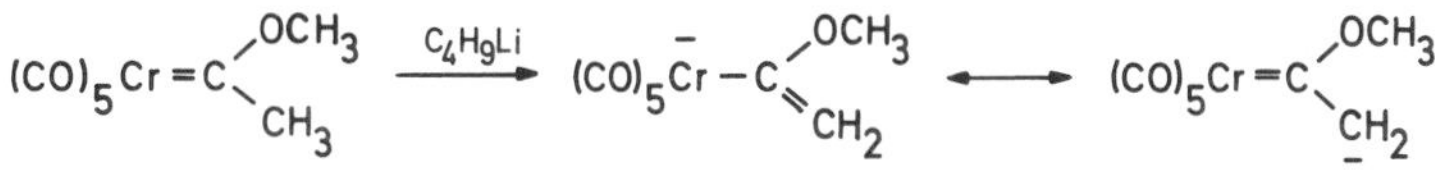

$$(CO)_5Cr=C\begin{smallmatrix}OCH_3\\CH_3\end{smallmatrix} \xrightarrow{C_4H_9Li} (CO)_5\overset{-}{Cr}-C\begin{smallmatrix}OCH_3\\CH_2\end{smallmatrix} \longleftrightarrow (CO)_5Cr=C\begin{smallmatrix}OCH_3\\\underset{-}{C}H_2\end{smallmatrix}$$

$$\underline{(3)} \qquad\qquad \underline{\underline{(3a)}} \qquad\qquad \underline{\underline{(3b)}}$$

Scheme 1.

The facile preparation and high thermodynamic stability of the carbene anions favor the introduction of additional functionality into the carbene side chain via reaction with electrophiles. This route is especially useful for the synthesis of carbene ligands containing carbonyl groups since these compounds are not directly accessible by the original procedure via organolithium reagents. The carbene anions – conveniently generated by the addition of stoichiometric amounts of base at low temperatures – react readily with alkylating reagents, allyl and benzyl halides, aldehydes or α-bromoesters at room temperature (8 – 10). Less active reagents, such as primary alkyl iodides, require moderate heating and afford only low yields (Schemes 2 and 3).

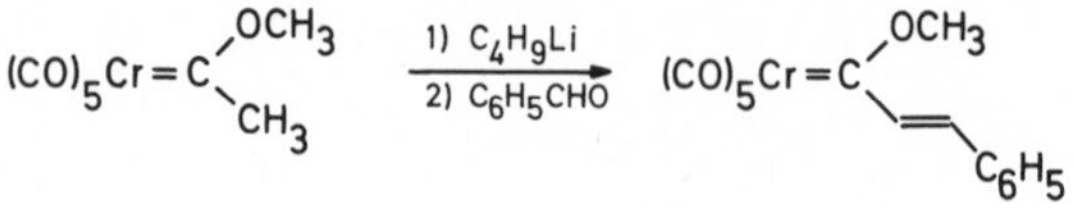

Scheme 2.

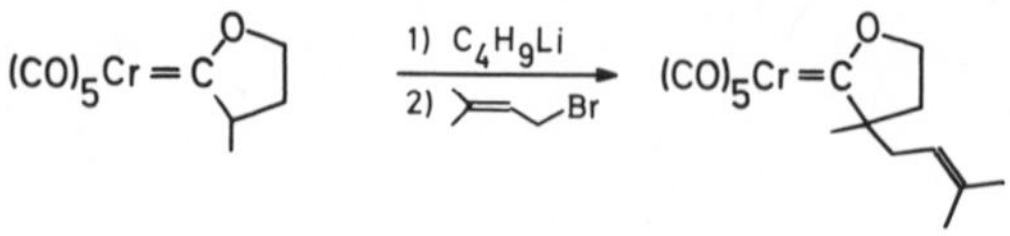

Scheme 3.

Carbene anions can effect a nucleophilic ring opening in epoxides. In the intermediate adduct **4**, methoxide elimination occurs to yield a five-membered cyclic carbene ligand. Substituted epoxides are attacked by the carbene anion at the less hindered carbon atom (Scheme 4).

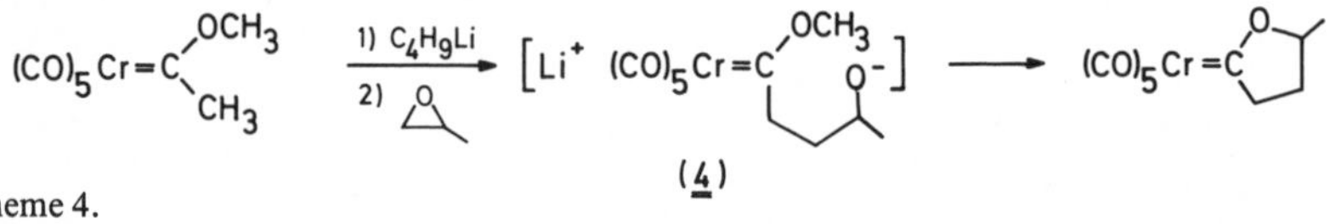

Scheme 4.

In the alkylation of carbene anions, dialkylation may become an important side reaction (Scheme 5).

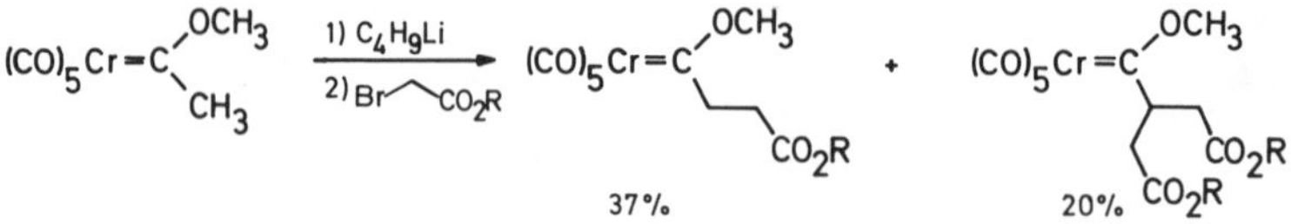

Scheme 5.

In the 2-oxacyclopentylidenechromium series, the secondary anion **5** and the methylated tertiary anion **6** (Scheme 6) were found to be of comparable basicity. The more substituted species, however, was shown to react some 2 to 5 times faster with allyl and benzyl bromides and thus gives rise to substantial amounts of dialkylation products.

Scheme 6.

The conjugate addition of the carbene anion **5** to α,β-unsaturated carbonyl compounds can be controlled to result in either mono- or dialkylation, depending on whether stoichiometric or catalytic amounts of base are used (11). Tertiary carbene anions such as **6** fail to undergo Michael addition. However, upon reaction with benzyl halides, alkylation products are obtained (Scheme 7).

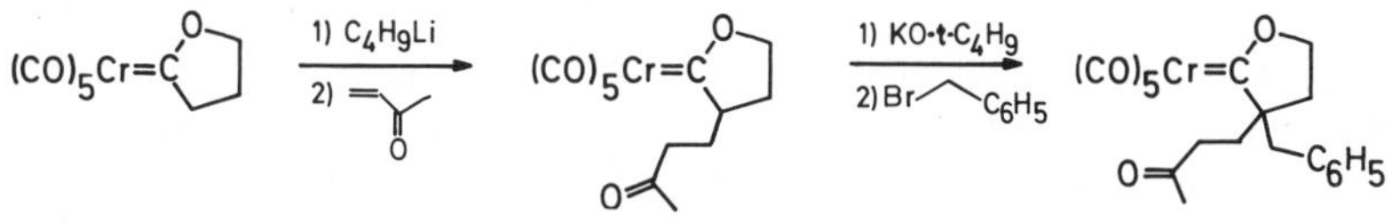

Scheme 7.

In a similar way acid chlorides react with carbene anions. If an enolizable hydrogen atom is present in the acylation product, the corresponding enol ester will be isolated (Scheme 8).

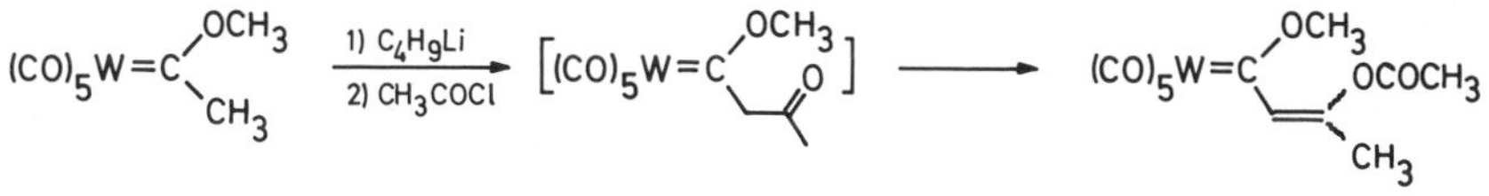

Scheme 8.

The reaction of carbene anions with enol ethers leads to addition products which can be converted into alkenylcarbene complexes by treatment with alumina (12) (Scheme 9).

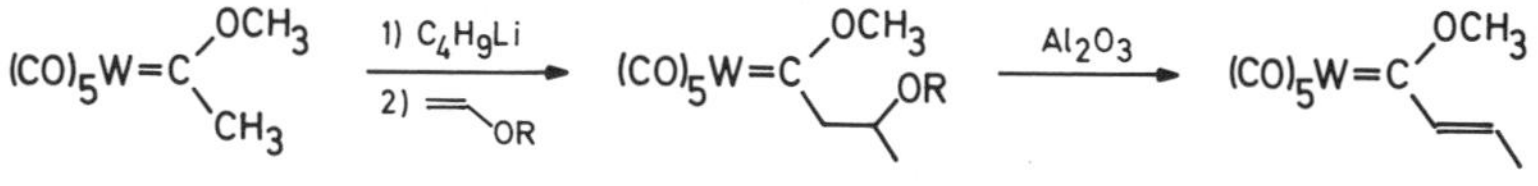

Scheme 9.

3 Cleavage of the Metal-Carbene Bond

For some time it remained an open question whether carbene complexes might be regarded as sources of free carbenes. Several procedures have been developed to release the carbene ligand from the metal, but in no case has the intermediacy of free carbenes been established.

Thermal decomposition of carbene complexes leads to carbene dimers. In the methoxy(phenyl)carbene series of the VIa metals, the *E/Z* ratio of the olefinic products is dependent on the metal. This observation indicates the influence of the metal on the dimerisation (13) (Scheme 10). Similar results have been obtained from the thermolysis of a 2-oxacyclopentylidenechromium complex (14). Cyclobutanone, which is known to be a characteristic product in the stabilization of the un-coordinated cyclic carbene, could not be detected in the reaction mixture (Scheme 11).

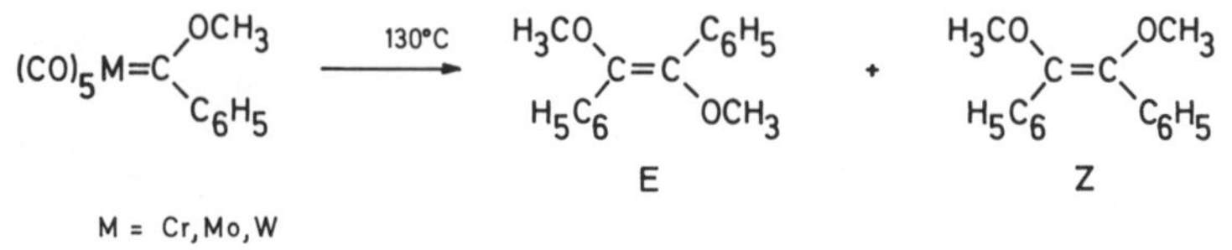

Scheme 10.

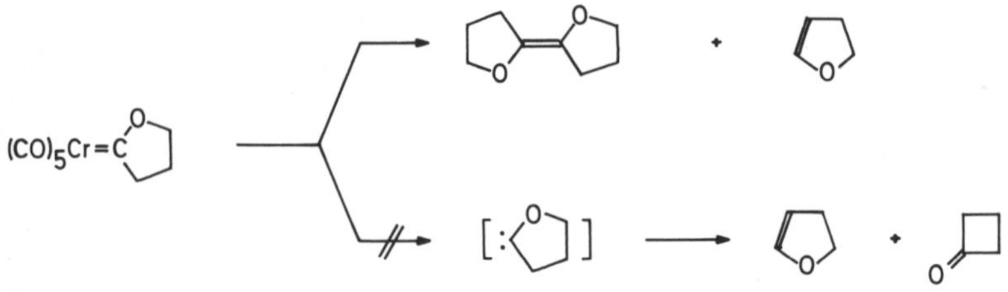

Scheme 11.

A less drastic route for the cleavage of the metal carbene bond involves replacement of the carbene ligand by carbon monoxide, tertiary phosphines or amines. Using amines, a base-induced 1,2-hydrogen shift was observed with carbene ligands containing α-hydrogen atoms (15, 16). Starting with alkoxy- or aminocarbene complexes, enol ethers or imines are obtained (Scheme 12).

Scheme 12.

From a synthetic point of view, the oxidative cleavage is the most convenient means of releasing the carbene ligand. Owing to the analogy of the
metal-carbene and the carbon-oxygen double bond, the metal is replaced by
oxygen and carbonyl compounds are obtained. Among a variety of oxidizing
agents, including pyridine *N*-oxide, dimethylsulfoxide, ceric compounds and
elemental oxygen, ceric compounds have proved to be the reagents of choice
(8, 17, 18). The oxidative cleavage is a clean, high yield process which has been
employed in the characterization of carbene ligands as well as for synthetic
purposes (Schemes 13 and 14).

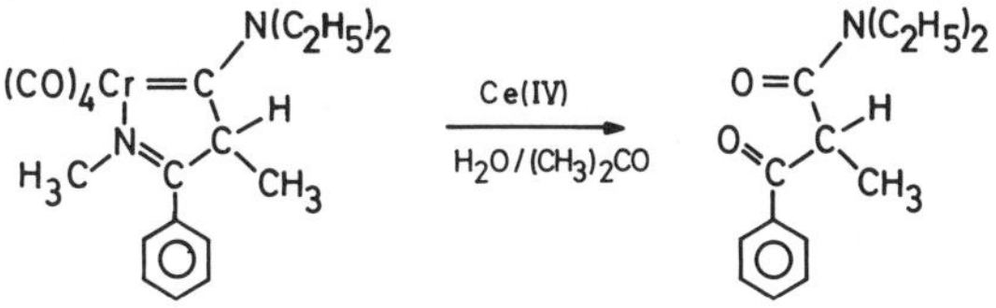

Scheme 13.

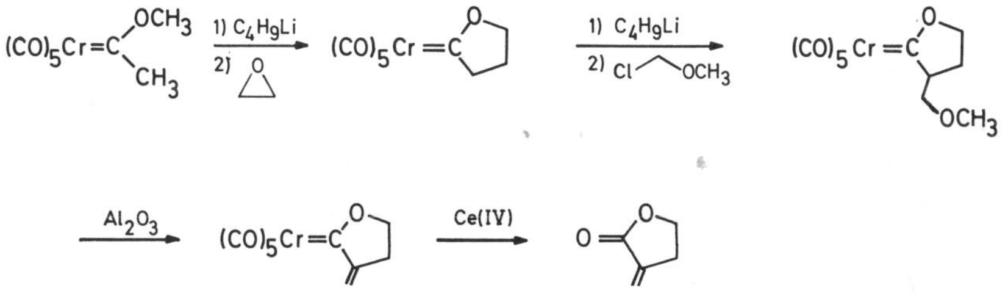

Scheme 14.

By contrast, reductive cleavage by hydrogenolysis leading to saturated
hydrocarbons requires much more drastic conditions (19) (Scheme 15).

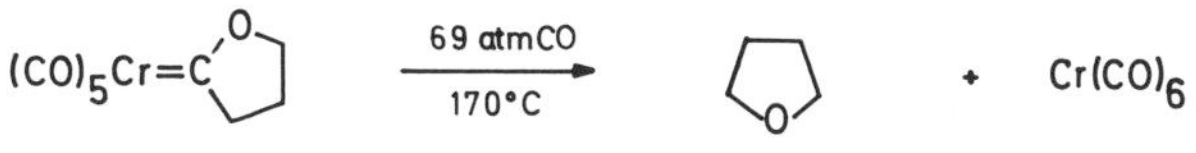

Scheme 15.

4 Addition of Nucleophiles

Numerous experimental results and supportive molecular orbital calculations may be considered as evidence for the fact that the carbene carbon atom in carbene complexes is highly susceptible to nucleophilic attack. Thus, ylide complexes have been produced from the reaction with phosphines and sterically rigid amines (20 – 22). Ylide formation may be used to trap carbene complexes which are unstable under the reaction conditions (23). A similar nucleophilic attack at the carbene carbon atom has been shown to occur in the aminolysis of alkoxycarbene complexes, indicating a mechanistic relationship to the aminolysis of carboxylic esters (24) (Scheme 16).

$$(CO)_5Cr=C\begin{smallmatrix}OCH_3\\C_6H_5\end{smallmatrix} \;+\; R-NH_2 \;\longrightarrow\; (CO)_5\,Cr=C\begin{smallmatrix}NHR\\C_6H_5\end{smallmatrix} \;+\; CH_3OH$$

Scheme 16.

The addition of carbon nucleophiles to the coordinated carbene carbon atom leads to the formation of a new carbon-carbon bond. The reaction with vinyl ethers and enamines has prompted further studies directed to the metal-catalyzed metathesis of olefins (25, 26). Owing to the thermal instability of the addition products, fragmentation occurs leading to alkene formation (Scheme 17 and 18).

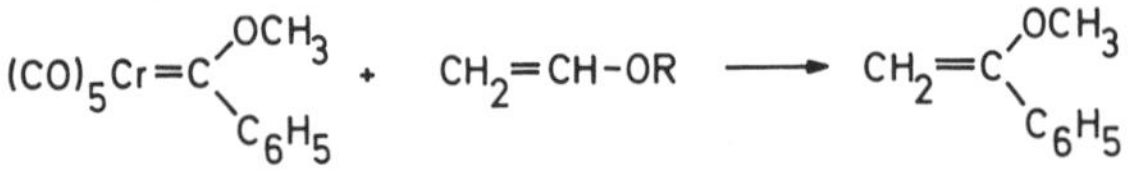

$$(CO)_5Cr=C\begin{smallmatrix}OCH_3\\C_6H_5\end{smallmatrix} \;+\; CH_2=CH-OR \;\longrightarrow\; CH_2=C\begin{smallmatrix}OCH_3\\C_6H_5\end{smallmatrix}$$

Scheme 17.

$$(CO)_5Cr=C\begin{smallmatrix}OCH_3\\C_6H_5\end{smallmatrix} \;+\; \begin{smallmatrix}R^1\\R^2\end{smallmatrix}C=C\begin{smallmatrix}H\\N\end{smallmatrix} \;\longrightarrow\; \begin{smallmatrix}R^1\\R^2\end{smallmatrix}C=C\begin{smallmatrix}OCH_3\\C_6H_5\end{smallmatrix}$$

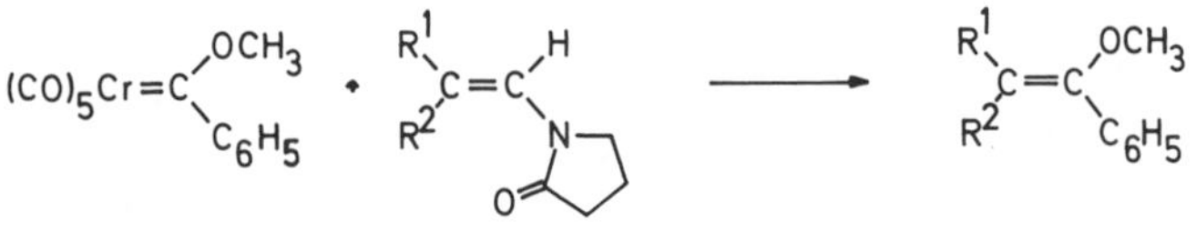

Scheme 18.

The reaction is regarded as involving a four-membered metallacycle intermediate **7** which is able to undergo a fragmentation process to yield the alkene and a modified carbene complex. This mechanism is supported by the reaction with 1-pyrrolidino-1-cyclopentene. Ring opening and insertion of the C_5-unit (arising from the enamine) into the metal-carbene bond are observed to occur

(27) (Scheme 19). The cleavage of the $C = C$ bond has attracted much attention in modelling olefin metathesis, a topic discussed in another chaper of this
volume.

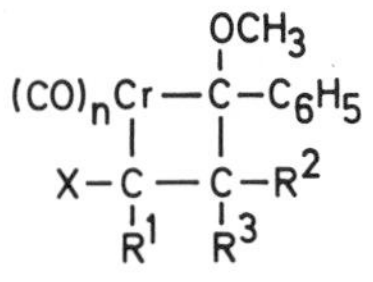

$$(CO)_nCr-\underset{\underset{R^1}{|}}{\overset{\overset{OCH_3}{|}}{C}}-C_6H_5$$

$$X-\underset{\underset{R^1}{|}}{C}-\underset{\underset{R^3}{|}}{C}-R^2$$

$$X = OR, NR_2$$

$$(\underline{7})$$

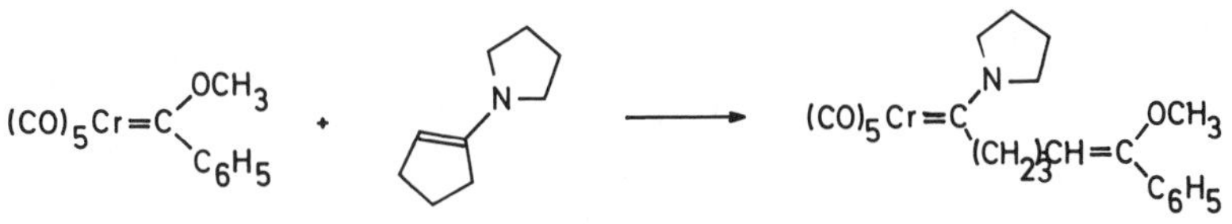

Scheme 19.

Phosphorus ylides have been found to convert alkoxy(aryl)carbene ligands
into enol ethers with high yields (Scheme 20). However, the scope of the reaction is restricted by both the steric requirements and the basicity of the Wittig
reagents. For instance, no C-C coupling has been observed with (isopropyl-
idene)triphenylphosphorane, and alkylcarbene complexes undergo a competing α-hydrogen abstraction to produce the conjugate base (28).

Scheme 20.

Unlike the phosphorus ylides, a carbonylnitrogen ylide is reported to give
an isolatable unique carbene complex. This product is believed to result from
nucleophilic attack of the ylide at the metal-coordinated cyclopropenylidene
carbon atom followed by ring expansion (29) (Scheme 21).

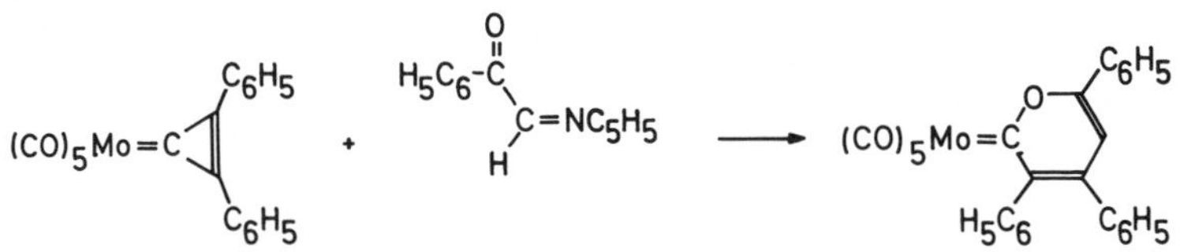

Scheme 21.

The synthesis of enol ethers from alkoxy(alkyl)carbene complexes can be achieved by use of the less basic diazoalkanes (Scheme 22). In some cases it has proved advantageous to add ligands such as pyridine to facilitate a clean, high yield reaction (30).

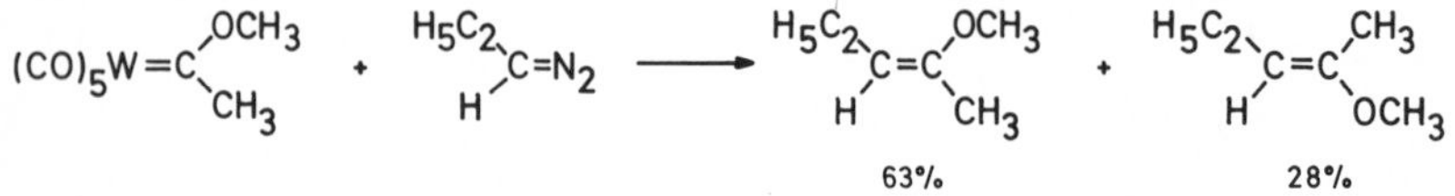

$$(CO)_5W=C\overset{OCH_3}{\underset{CH_3}{\diagup}} \;+\; \overset{H_5C_2}{\underset{H}{\diagup}}C=N_2 \longrightarrow \overset{H_5C_2}{\underset{H}{\diagup}}C=C\overset{OCH_3}{\underset{CH_3}{\diagdown}} \;+\; \overset{H_5C_2}{\underset{H}{\diagup}}C=C\overset{CH_3}{\underset{OCH_3}{\diagdown}}$$

$$63\% \qquad\qquad 28\%$$

Scheme 22.

Organolithium reagents can also be added to the carbene carbon. Using low temperature techniques, the metallate(1-) adducts **8** can be isolated as bis-(triphenylphosphine)imminum salts (31, 32). At ambient temperature, decomposition products characteristic of free radicals are observed (33). Thus, a persubstituted ethane is isolated from the reaction of the methoxy-(phenyl)carbene complex of chromium and phenyllithium (Scheme 23).

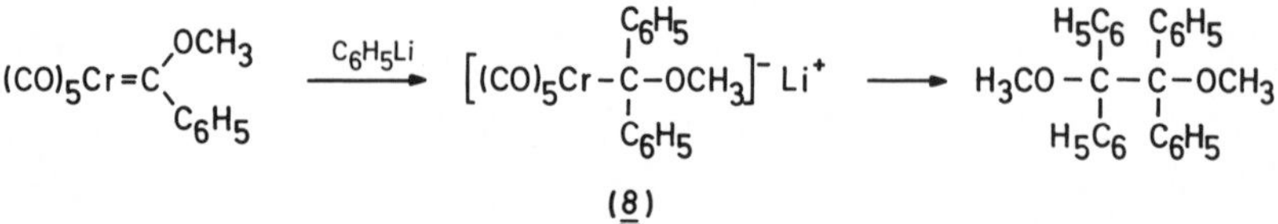

$$(CO)_5Cr=C\overset{OCH_3}{\underset{C_6H_5}{\diagup}} \xrightarrow{C_6H_5Li} \left[(CO)_5Cr-\overset{C_6H_5}{\underset{C_6H_5}{\overset{|}{\underset{|}{C}}}}-OCH_3 \right]^- Li^+ \longrightarrow H_3CO-\overset{H_5C_6}{\underset{H_5C_6}{\overset{|}{\underset{|}{C}}}}-\overset{C_6H_5}{\underset{C_6H_5}{\overset{|}{\underset{|}{C}}}}-OCH_3$$

$$(\underline{8})$$

Scheme 23.

Metallates(-1) such as **8** are used as intermediates in the synthesis of non-heteroatom-stabilized carbene complexes, which are most simply obtained by methoxide elimination over silica gel (34). Attempts to prepare alkyl(aryl)-carbene compounds by the addition of alkyllithium reagents were unsuccessful (35, 36). Instead, alkene products resulting from rearrangement or decomposition were isolated (Scheme 24).

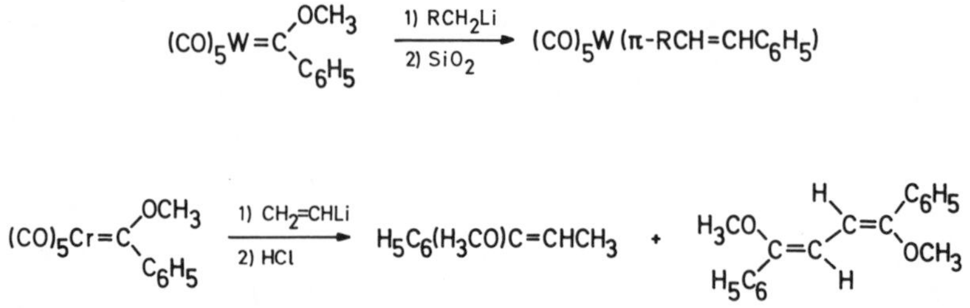

$$(CO)_5W=C\overset{OCH_3}{\underset{C_6H_5}{\diagup}} \xrightarrow[\text{2) SiO}_2]{\text{1) RCH}_2\text{Li}} (CO)_5W\,(\pi\text{-RCH}=CHC_6H_5)$$

$$(CO)_5Cr=C\overset{OCH_3}{\underset{C_6H_5}{\diagup}} \xrightarrow[\text{2) HCl}]{\text{1) CH}_2=\text{CHLi}} H_5C_6(H_3CO)C=CHCH_3 \;+\; \overset{H_3CO}{\underset{H_5C_6}{\diagup}}C=C\overset{H}{\diagdown}\overset{C_6H_5}{\underset{OCH_3}{\diagup}}C=C\overset{C_6H_5}{\underset{H}{\diagdown}}$$

Scheme 24.

Competition between a direct attack on the carbene carbon atom and conjugate addition at the carbon carbon double bond is observed with alkenylcarbene complexes (36, 37). Work-up under acidic conditions affords a mixture of carbene complex and enol ether. The relative yields depend on whether organolithium or organocopper reagents are used (Scheme 25).

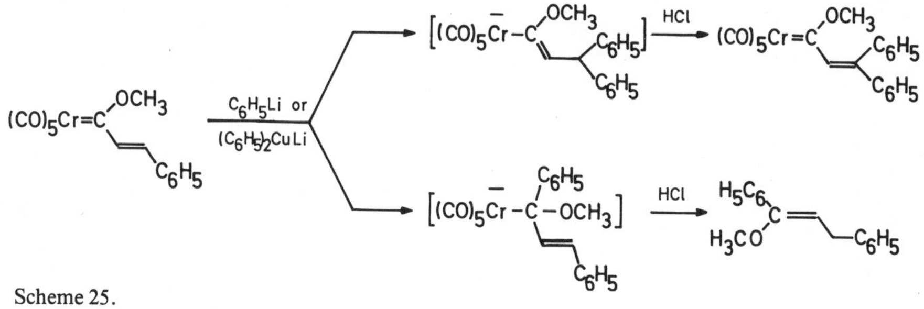

Scheme 25.

Steric requirements are invoked for the competing addition of enolates: Bulky nucleophiles, such as the enolate of isobutyrophenone, add to the more accessible remote vinylic position. In contrast, attack at the carbene carbon atom occurs with the less bulky enolate of acetone (Scheme 26).

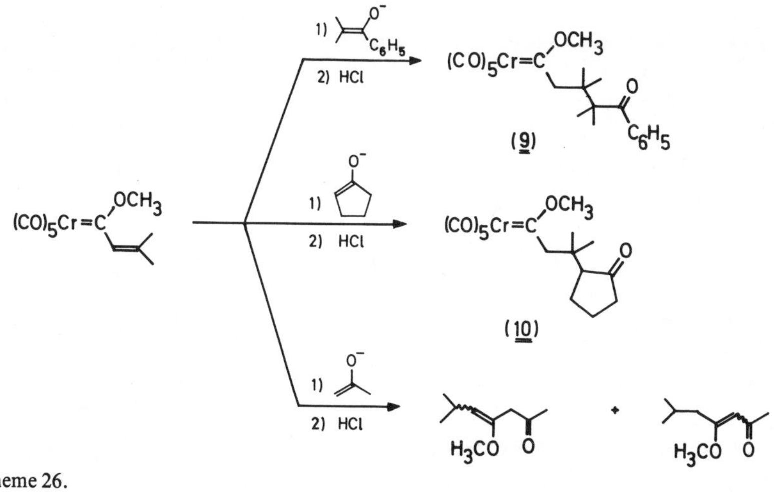

Scheme 26.

Various organic structures can be obtained depending on the method of subsequent cleavage of the metal-carbene bond (Scheme 27).

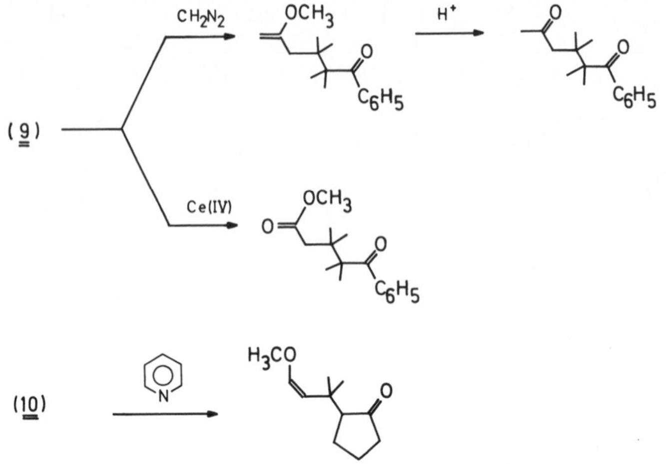

Scheme 27.

Related conjugate addition reactions have been observed with alkinyl-carbene complexes (38). The addition of diazomethane across the carbon-carbon triple bond leads to the formation of N-coordinated pyrazole complexes (39). Kinetic studies reveal that nucleophilic attack on the carbene atom is the first step in the formal insertion of 1-aminoalkynes into the metal-carbene bond (40) (Scheme 28). The C=C bond formation is highly stereoselective with the *E* configuration predominating (41 – 44). The direction of ring-opening of a four-membered metallacycle intermediate may influence the regioselectivity of the reaction. The insertion of multiple bond systems into the metal-carbene bond has proved to be a widely applicable type of reaction.

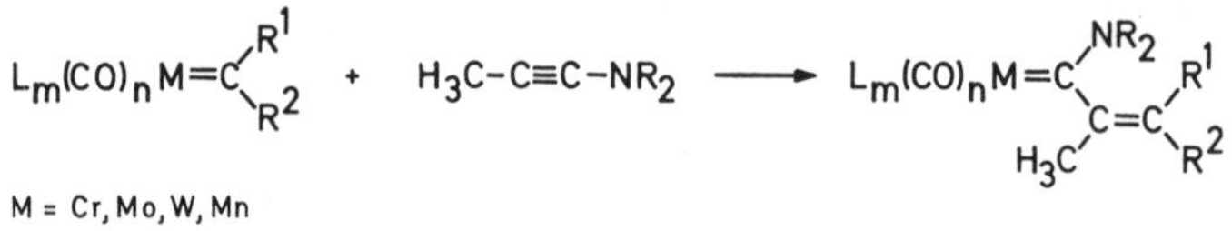

M = Cr, Mo, W, Mn

Scheme 28.

E-Styrylcarbene complexes prepared by this route undergo a cyclization when heated in an inert solvent to form coordinated indanone and indenone derivatives (Scheme 29). The *Z* isomer fails to react (44).

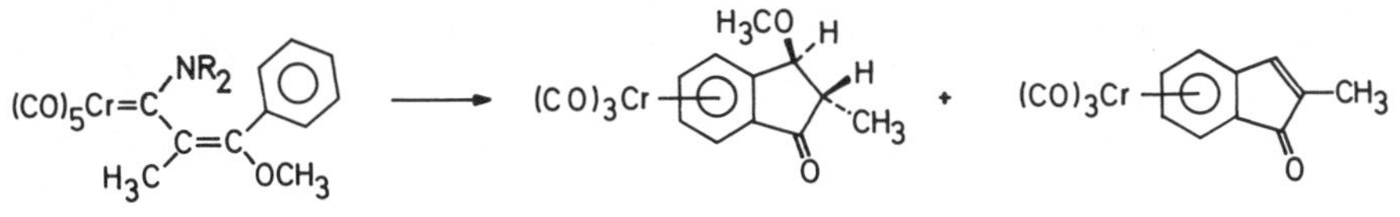

Scheme 29.

The role of the metal in these cyclization reactions is demonstrated in the 1,3-addition of bis(diethylamino)acetylene to the phenylcarbene ligand (45, 46). The first step in the temperature-controlled reaction sequence carried out at 14° C involves insertion of the yndiamine into the metal-carbene bond (Scheme 30). Warming up to 70° C induces an intramolecular substition reaction which affords a chelating carbene ligand. By the coordination of the enamine nitrogen atom to the metal the carbene ligand maintains a conformation favoring cyclization to the indene system upon further heating. The homologous methoxy(phenyl)carbene complex of tungsten yields the corresponding aminocarbene complexes, but fails to undergo the final cyclization step. This reaction sequence provides a prime example of the template role of the metal.

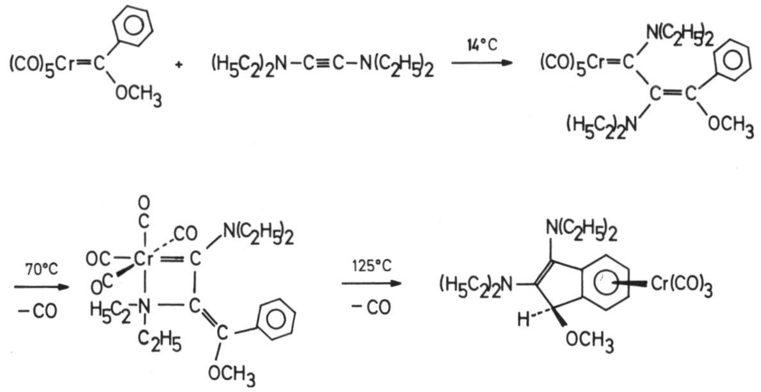

Scheme 30.

5 Carbene Transfer Reactions

Much chemical evidence is now available to suggest that no free carbenes are involved in the transfer reactions of carbene ligands. This applies both to intermetallic transfer reactions and to carbene transfer to organic substrates. The latter topic will be specially addressed in this chapter.

Whereas insertion into carbon-hydrogen bonds is a characteristic feature of free carbenes, no such reaction is known for metal-derived carbenes. However, alkoxy- and aminocarbene ligands bonded to chromium undergo insertion into the more reactive silicon, germanium and tin to hydrogen bonds (47, 48) (Scheme 31). Using triorganyl hydrides of these elements, the reaction rate was found to increase according to the sequence $Si < Ge \ll Sn$, which is believed to reflect the increasing likelihood of five-coordination (49, 50) occurring.

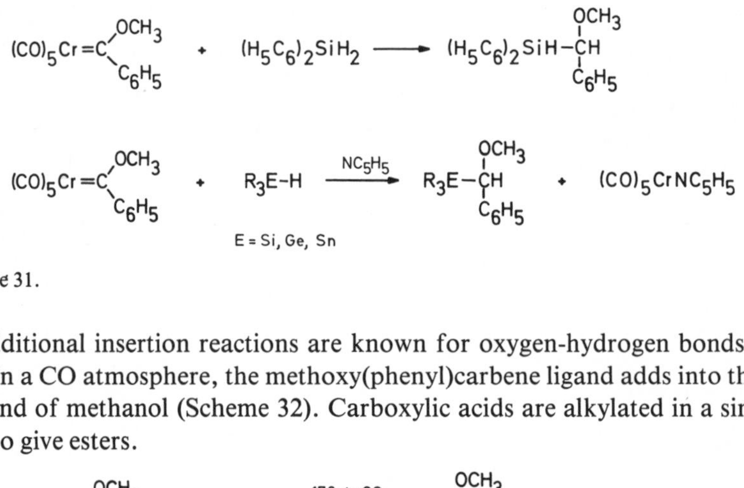

$$(CO)_5Cr=C \begin{smallmatrix} OCH_3 \\ \\ C_6H_5 \end{smallmatrix} + (H_5C_6)_2SiH_2 \longrightarrow (H_5C_6)_2SiH-\overset{OCH_3}{\underset{C_6H_5}{CH}}$$

$$(CO)_5Cr=C \begin{smallmatrix} OCH_3 \\ \\ C_6H_5 \end{smallmatrix} + R_3E-H \xrightarrow{NC_5H_5} R_3E-\overset{OCH_3}{\underset{C_6H_5}{CH}} + (CO)_5CrNC_5H_5$$

$$E = Si, Ge, Sn$$

Scheme 31.

Additional insertion reactions are known for oxygen-hydrogen bonds (51, 52). In a CO atmosphere, the methoxy(phenyl)carbene ligand adds into the O-H bond of methanol (Scheme 32). Carboxylic acids are alkylated in a similar way to give esters.

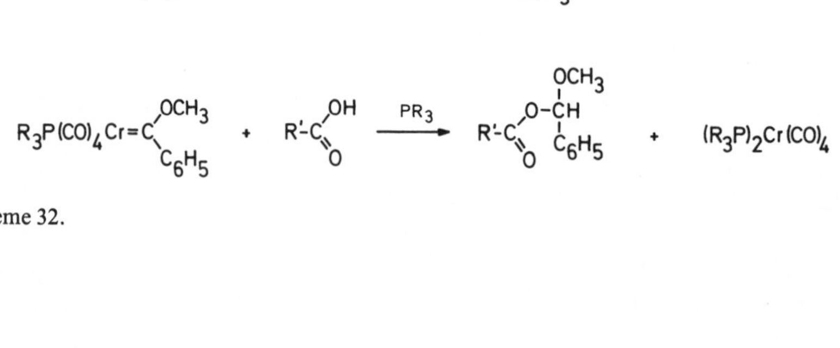

$$(CO)_5Cr=C \begin{smallmatrix} OCH_3 \\ \\ C_6H_5 \end{smallmatrix} + CH_3OH \xrightarrow{170\,atm\,CO} H_5C_6-\overset{OCH_3}{\underset{OCH_3}{CH}}$$

$$R_3P(CO)_4Cr=C \begin{smallmatrix} OCH_3 \\ \\ C_6H_5 \end{smallmatrix} + R'-C \begin{smallmatrix} OH \\ \\ O \end{smallmatrix} \xrightarrow{PR_3} R'-C \begin{smallmatrix} O-\overset{OCH_3}{\underset{C_6H_5}{CH}} \\ \\ O \end{smallmatrix} + (R_3P)_2Cr(CO)_4$$

Scheme 32.

Cycloaddition reactions of free carbenes normally fail with electron-deficient alkenes. By contrast, metal-derived carbenes were shown to add both to electron-rich and electron-deficient alkenes, depending on the reaction conditions. Starting from the methoxy(phenyl)carbene complexes of chromium, molybdenum and tungsten and α,β-unsaturated carbonyl compounds *syn-anti* cyclopropane isomers are obtained in a stereospecific reaction (53, 54) (Scheme 33).

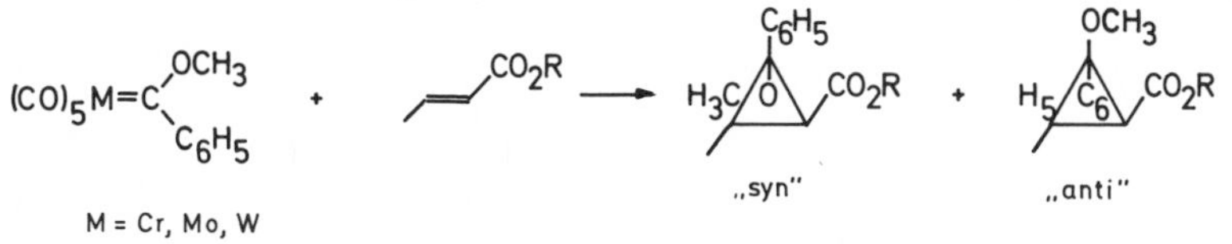

Scheme 33.

The *syn/anti* ratios are dependent on the metal, indicating that the metal is involved in the product-determining step. Further evidence for the active role of the metal is provided by the optical induction observed with a carbene complex bearing a chiral phosphine ligand (55) (Scheme 34).

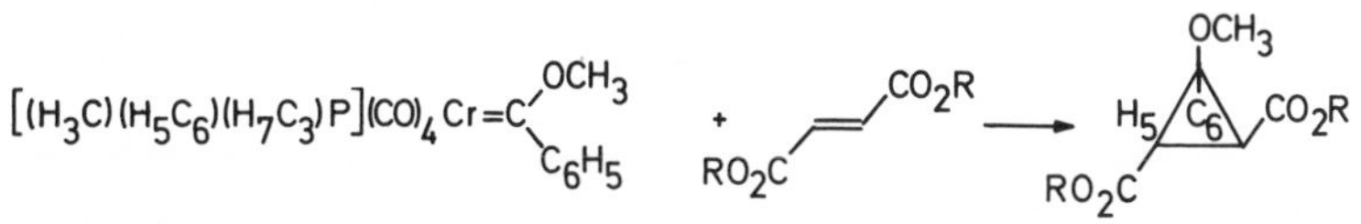

Scheme 34.

The cyclopropanation of α,β-unsaturated esters proceeds under conditions which are known to effect CO exchange in carbonyl(carbene) complexes. Thus, a reasonable mechanism has to account for CO elimination in the first step followed by coordination of the alkene. The alkene(carbene) complex may rearrange to a metallacyclobutane, from which reductive elimination might be expected to yield the cyclopropane below (Fig. 2).

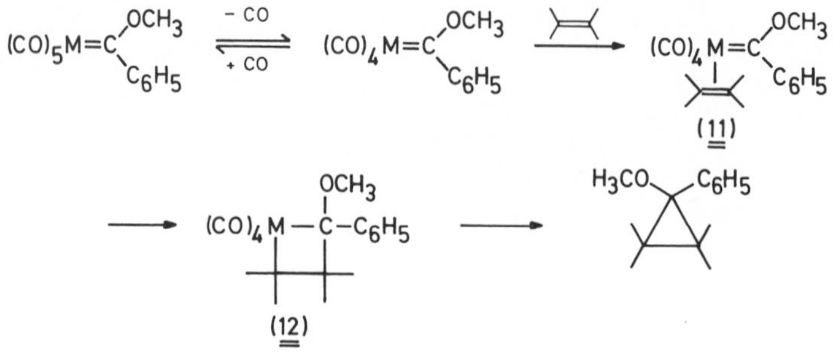

$$(CO)_5M=C\underset{C_6H_5}{\overset{OCH_3}{\big\langle}} \underset{+\ CO}{\overset{-\ CO}{\rightleftarrows}} (CO)_4M=C\underset{C_6H_5}{\overset{OCH_3}{\big\langle}} \longrightarrow (CO)_4M=C\underset{C_6H_5}{\overset{OCH_3}{\big\langle}}$$

(11)

$$\longrightarrow (CO)_4M-\overset{\overset{\displaystyle OCH_3}{|}}{C}-C_6H_5 \longrightarrow$$

(12)

$$H_3CO \diagdown \diagup C_6H_5$$

Figure 2: Cyclopropanation of alkenes via carbene complexes

Complexees of type **11** and **12** are thought to be involved in the non-pairwise exchange of alkylidene moieties in olefin metathesis.

The reaction of alkoxycarbene complexes and electron-rich alkenes, such as enol ethers, is strongly dependent on the reaction conditions. In an inert gas at atmospheric pressure alkene scission products will predominate (*c.f.*4). A carbon monoxide atmosphere, however, leads to the formation of cyclo-propanes in good yields (25). Again, the ratio of the resulting *syn* and *anti* isomers varies with the metal present in the carbene complex (Scheme 35).

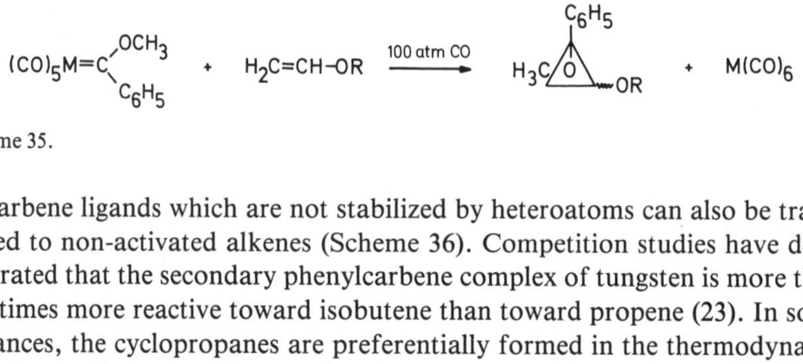

$$(CO)_5M=C\underset{C_6H_5}{\overset{OCH_3}{\big\langle}} + H_2C=CH-OR \xrightarrow{100\ atm\ CO} H_3C\diagdown O \overset{C_6H_5}{\triangle} {}_{\cdots OR} + M(CO)_6$$

Scheme 35.

Carbene ligands which are not stabilized by heteroatoms can also be trans-ferred to non-activated alkenes (Scheme 36). Competition studies have dem-onstrated that the secondary phenylcarbene complex of tungsten is more than 300 times more reactive toward isobutene than toward propene (23). In some instances, the cyclopropanes are preferentially formed in the thermodynami-cally less stable *syn* configuration, a fact which is not entirely understood.

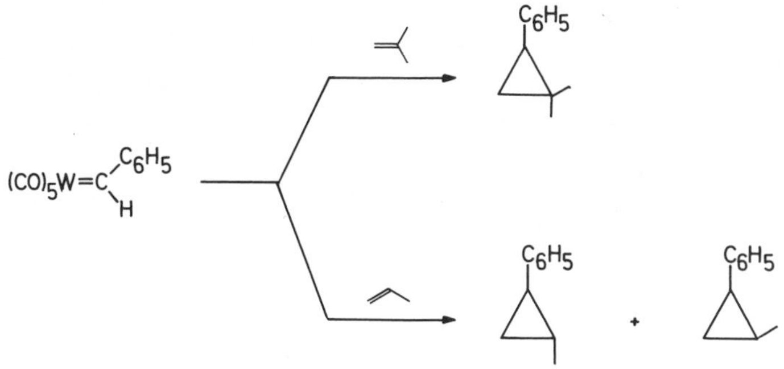

Scheme 36.

Cycloaddition reactions of carbene ligands have been extended to include the synthesis of heterocycles (56). Alkoxy- and aminocarbene ligands add to *N*-acylimines to give 2-oxazolines (Scheme 37). Thiocarbene complexes fail to yield [4 + 1]cycloaddition products. Instead, cleavage of the sulfur-carbene bond occurs and the thioalkyl group is added to the imine carbon atom.

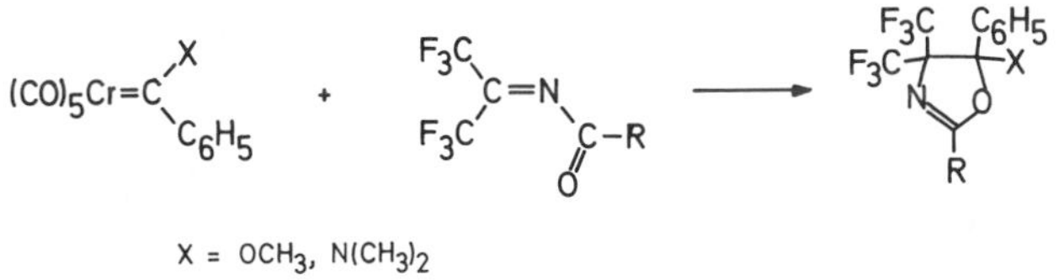

X = OCH₃, N(CH₃)₂

Scheme 37.

A novel 1,3-addition of an alkyne to the phenylcarbene ligand occurs in the reaction of bis(phosphino)acetylene with methoxy(phenyl)carbene complexes of chromium and tungsten (57). A CO substitution by the phosphine leads initially to binuclear biscarbene complexes having a bis(phosphino)alkyne bridge. In contrast to the tungsten compound, the chromium complex undergoes annelation of the carbene ligand to give a 1,2-bis(phosphino)indene which is coordinated to the metal through both phosphorous atoms (Scheme 38).

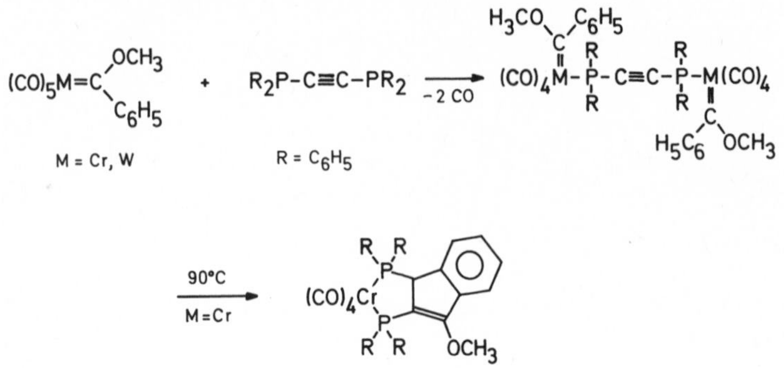

Scheme 38.

6 Reactions with Alkynes

The role of metal carbonyls in carbonylation reactions prompted attempts to make use of both the carbene and the carbonyl ligand in carbon-carbon bond formation. The bifunctionality of carbonyl carbene complexes is demonstrated by the reactions undergone with alkynes. Pentacarbonyl[methoxy-(phenyl)carbene]chromium reacts under mild conditions with a variety of non-heteroatom-substituted acetylenes to yield 4-methoxy-1-naphthols coordinated to a tricarbonylchromium fragment (58 – 62). The unsubstituted naphthol ring and the ring carbon atom with the methoxy group represent the former carbene ligand, while the naphthol functionality is formed by a carbonyl ligand. 1-Alkynes are incorporated regiospecifically into the positions 2 and 3 to give 2-alkylnaphthols. The regioselectivity decreases from 1-alkynes to 2-alkynes and diarylacetylenes (62, 63). Under kinetic control, the substituted naphthol ring is bonded to the metal. Further heating induces the migration of the carbonyl chromium fragment to produce the thermodynamically more stable 5-10-η-isomers (Scheme 39).

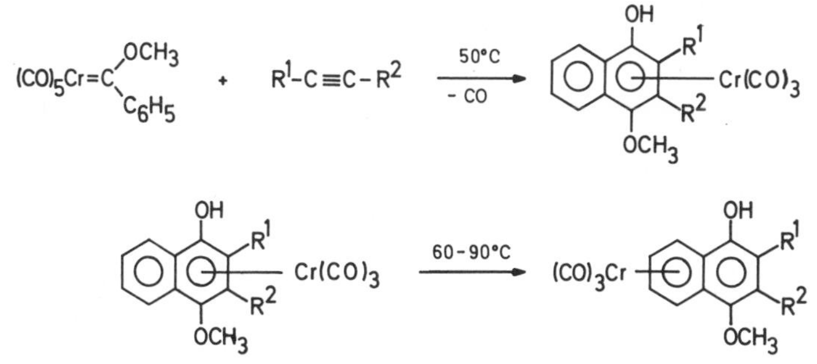

Scheme 39.

Naphthol formation requires effective acceptor ligands in the carbene complex. Thus, even under more vigorous conditions, the yields drastically decrease if one CO ligand is replaced by trialkylphosphine (60). Based on kinetic studies, a mechanism has been suggested which involves a primary CO elimination process (64). CO exchange experiments in carbonyl(carbene) complexes have indicated a strong preference for elimination of a *cis* CO ligand (65). The alkyne enters the vacant coordination site, thereby leading to formation of a *cis*-alkyne(carbene) complex. Interligand carbon-carbon bond formation between the alkyne and carbene ligands gives a chromacyclobutene expected to be in equilibrium with its ring-opening product. Carbonylation of the alkenylcarbene carbon atom by a CO ligand leads to a vinyl ketene system which is kept in *s-cis* conformation by a diene-like coordination to the metal. Experimental support for the involvement of vinyl ketenes is provided by the cyclization to cylcobutenones and by the isolation of *s-trans* vinyl ketenes starting from bulky alkynes (*vide infra*). In addition, trapping experiments in the presence of alcohols and amines have resulted in the formation of carboxylic acid derivatives (66). Cyclization of the vinyl ketene species, which should be regarded as the key intermediate in the reaction of carbonyl(carbene) complexes with alkynes, affords a metal-coordinated bicyclic cyclohexadienone which readily isomerizes to the $1-4:9, 10-\eta$-naphthol complex (Fig. 3).

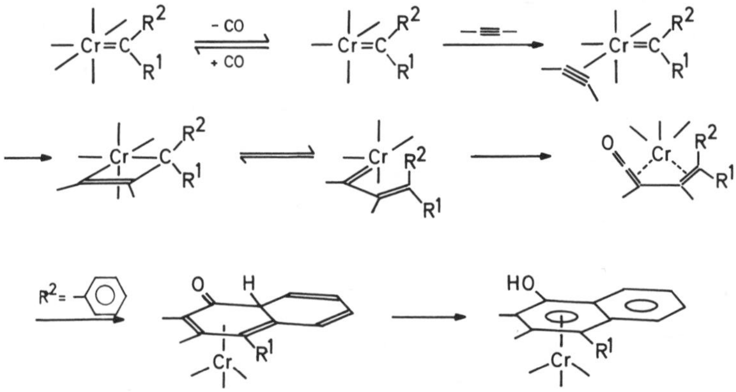

Figure 3: Proposed mechanism for the cocyclization of alkyne, carbene and carbonyl ligands

This mechanistic scheme suggests the role of the metal as that of a template holding the building blocks together in a position favorable for carbon-carbon bond formation. This idea is supported by formation of the thermodynamically less stable product under kinetic control as well as by the fact that chromium cannot be replaced by molybdenum or tungsten in these reactions.

Donor solvents such as THF or acyclic ethers are required to get a clean reaction with yields up to 90 %. The intermediacy of coordinatively unsaturated species may be held responsible for the lowered product selectivity when a non-coordinating solvent is used (67). Thus, from the reaction of the methoxy(phenyl)carbene complex and tolan performed in n-heptane, indene and furan derivatives are isolated in addition to moderate yields of naphthol compounds (Scheme 40). A similar formation of the furan system is observed with a ferrocenylcarbene complex (68) (Scheme 41).

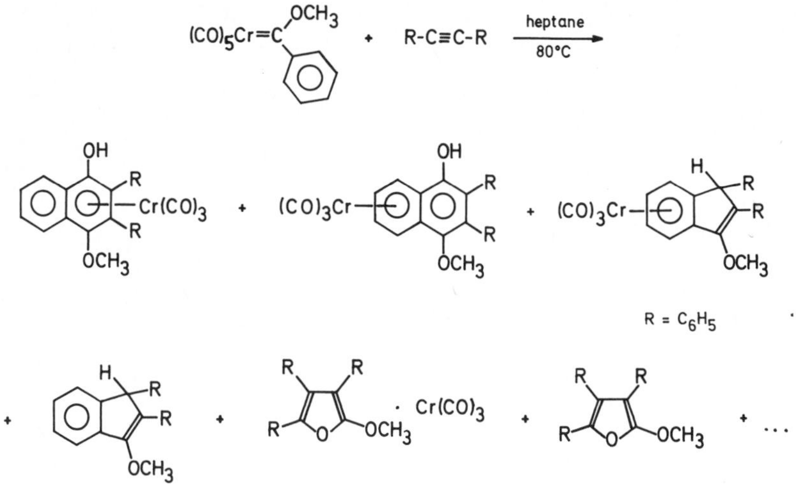

Scheme 40.

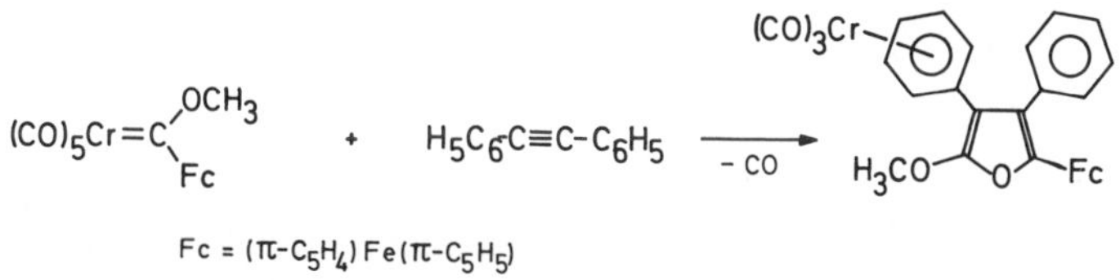

$Fc = (\pi\text{-}C_5H_4)\,Fe\,(\pi\text{-}C_5H_5)$

Scheme 41.

The annelation of a carbene substituent is not restricted to the phenyl group. Carbene ligands containing naphthyl groups, heterocyclic systems (such as furyl or thienyl), cycloalkenyl and alkenyl groups can react to give phenanthrene, benzofuran, benzothiophene, indan and benzene complexes (61, 69) (Scheme 42). The annelation occurs in a regiospecific manner. 2-Naphthyl-and 3-furylcarbene ligands which *a priori* offer two alternatives for cyclization lead to phenanthrene or benzofuran derivatives respectively.

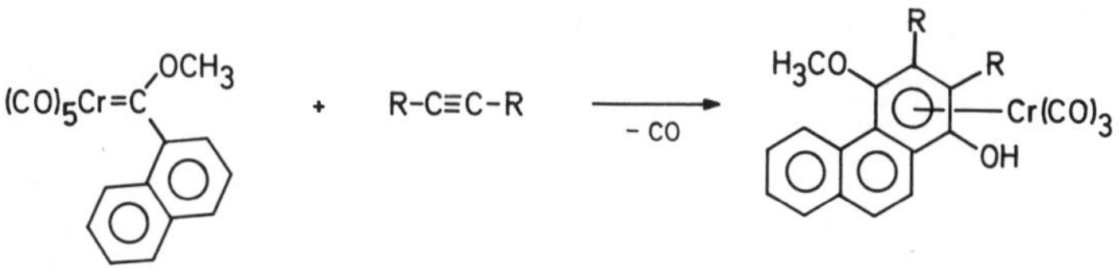

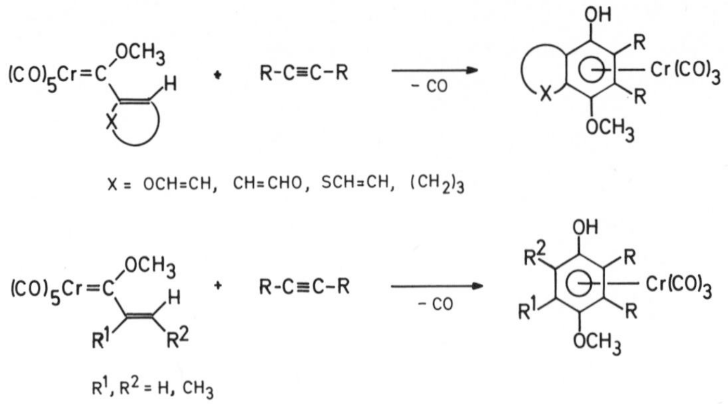

X = OCH=CH, CH=CHO, SCH=CH, (CH₂)₃

R¹, R² = H, CH₃

Scheme 42.

Competition experiments using diarylcarbene complexes have indicated preference for the annelation of a benzene ring rather than 2-naphthyl and 2-furyl systems (Scheme 43).In a similar way, starting with the *p*-(trifluor-methylphenyl)-*p*-tolylcarbene complex, the acceptor-substituted ring is preferentially annelated (Scheme 44).

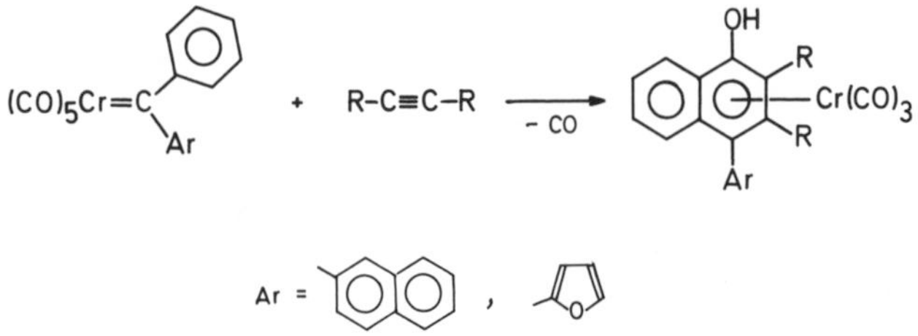

Ar =

Scheme 43.

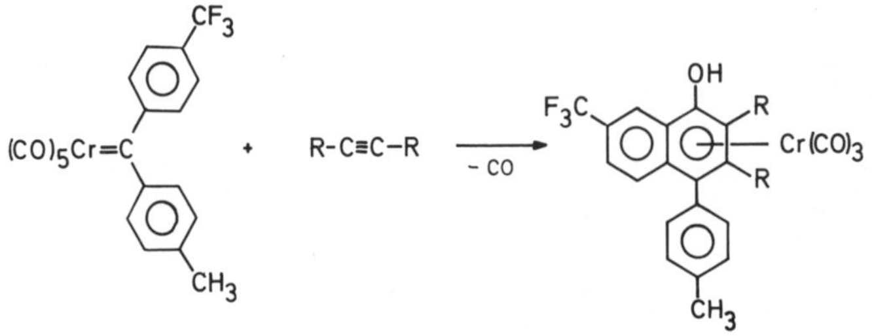

Scheme 44.

Very recently, an attempt has been made to improve the regioselectivity of incorporation of the dialkylalkynes using an intramolecular cyclization (70). Tricyclic products have been obtained from alkinyloxy(phenyl)carbene complexes (Scheme 45).

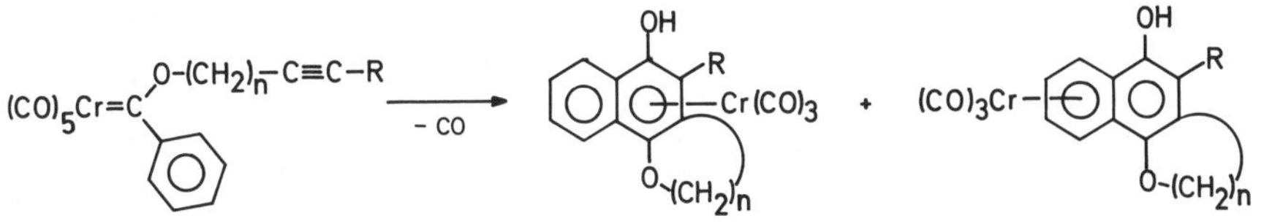

Scheme 45.

Because of the ease of CO elimination, the cyclization reaction is favored if the electron deficiency at the carbene carbon atom is increased. Thus, non-heteroatom-stabilized carbene complexes such as pentacarbonyl(diphenylcarbene)chromium can be reacted with acetylene or alkynoic acid derivatives even at room temperature (60).

An unusual 1:1:2 cycloaddition product is obtained from the methoxy-(methyl)carbene complex and an excess of phenylacetylene (71) (Scheme 46). The reaction may be explained in terms of double insertion of the alkyne into the metal-carbene bond, leading to a 1,3-dienylcarbene complex intermediate (see Fig. 3). It should be noted that the alkyne polymerization initiated by carbene complexes is discussed in terms of a multiple alkyne insertion.

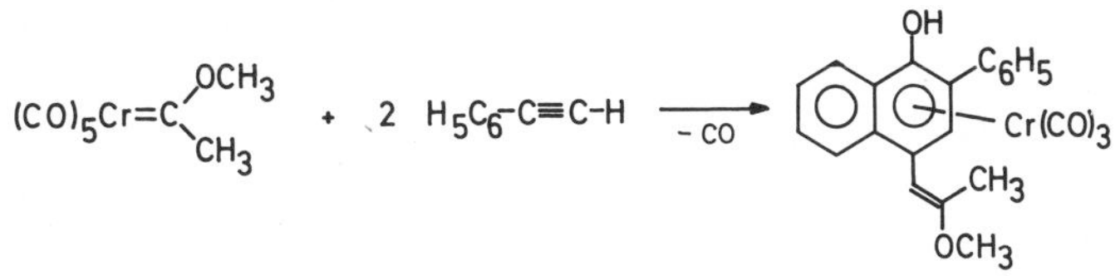

Scheme 46.

Aromatic ligands can be detached from the metal either by ligand substitution or by oxidation (67). By use of mild oxidants, the metal ring bond can be maintained in 5-10-η-naphthol complexes and thus naphthoquinone complexes become accessible (72, 73) (Scheme 47).

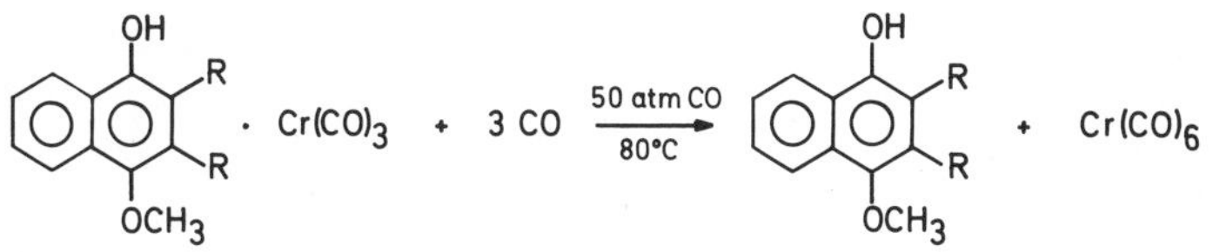

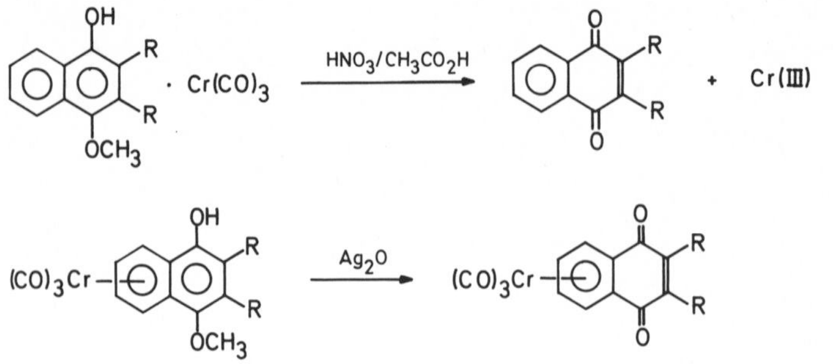

Scheme 47.

The vinyl ketene intermediate proposed in Figure 3 might be expected to lead to the formation of cyclobutenones by 1,4-cyclization as well. This proved to be the preferred reaction path, when no unsaturated carbene ligand is available (74) or when the *ortho* positions of an aromatic carbene substituent are blocked for annelation, *e.g.* in the 2,6-difluorophenyl(methoxy)-carbene complex (75) (Scheme 48). Blocking by 2,6-dimethyl substitution in the phenyl ring is not effective. Upon reaction with tolan, methyl migration occurs and an indene derivative is formed without incorporation of a carbonyl ligand (76).

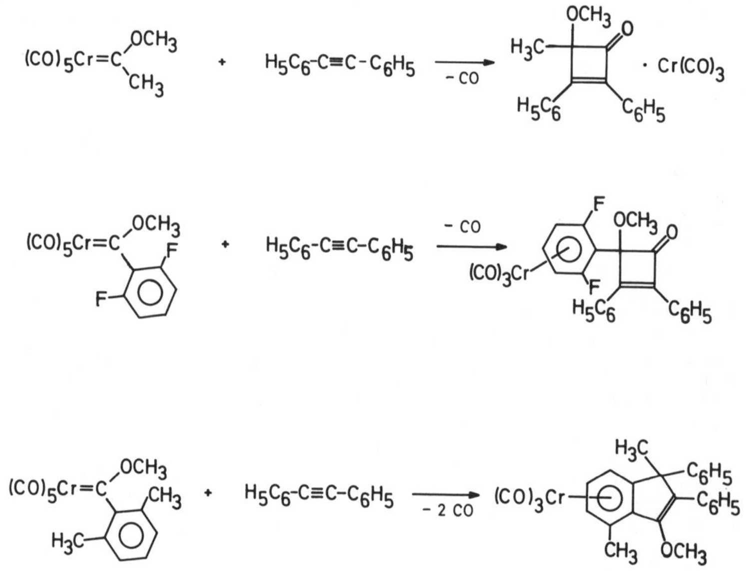

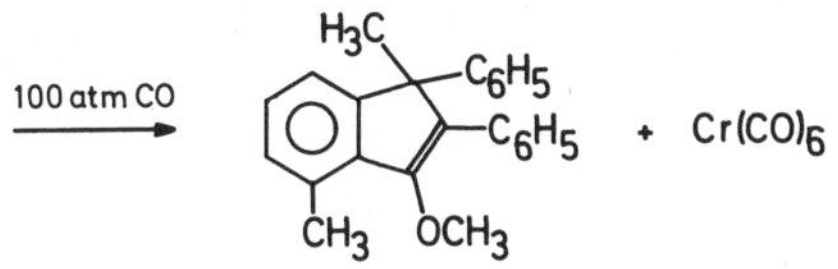

Scheme 48.

Vinyl ketenes can be isolated if bulky silylsubstituted alkynes are used (77, 78). In this case, the *s-cis* form cannot compete successfully with the *s-trans* conformation for steric reasons and cyclization to the aromatic system becomes impossible. Starting with bis(trimethylsilyl)acetylene, stable vinyl ketenes are accessible which have been shown by X-ray analysis to have the *s-trans* conformation in the solid state because of the large silyl substituents (Scheme 49).

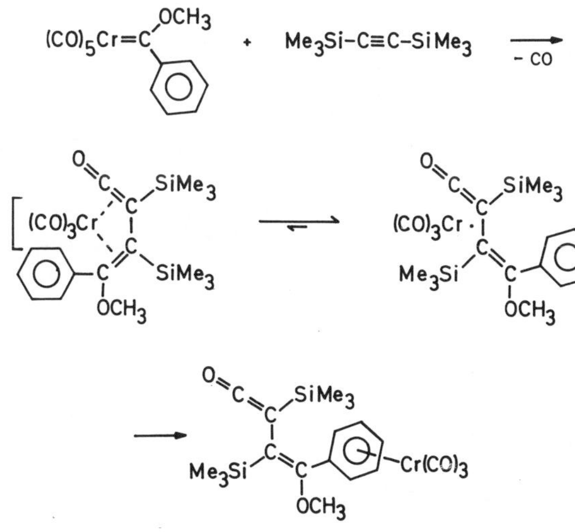

Scheme 49.

The vinyl ketene ligand is replaced by an excess of alkyne (79). The un-coordinated vinyl ketene is obtained along with a novel bis(alkyne)dicarbonylchromium complex originating from the carbonylmetal fragment (Scheme 50). In this compound the alkyne may be regarded as a four-electron donor ligand on the basis of ^{13}CNMR spectroscopy.

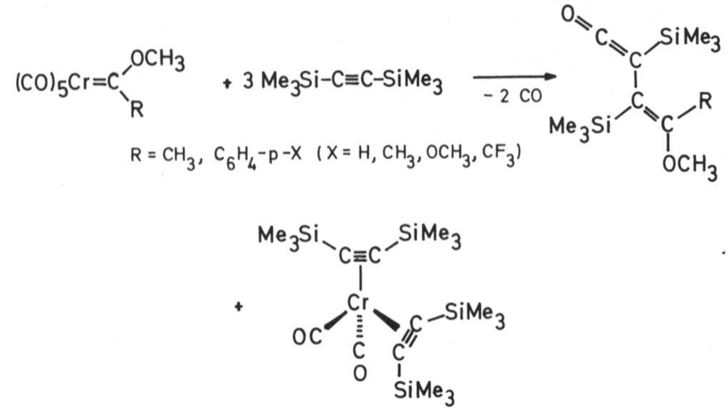

Scheme 50.

The competition which exists between the formation of vinyl ketenes and naphthols is governed by steric considerations (78). Using trimethylsilyl-acetylene in addition to a small amount of vinyl ketene, the naphthol complex is isolated as the major product (Scheme 51).

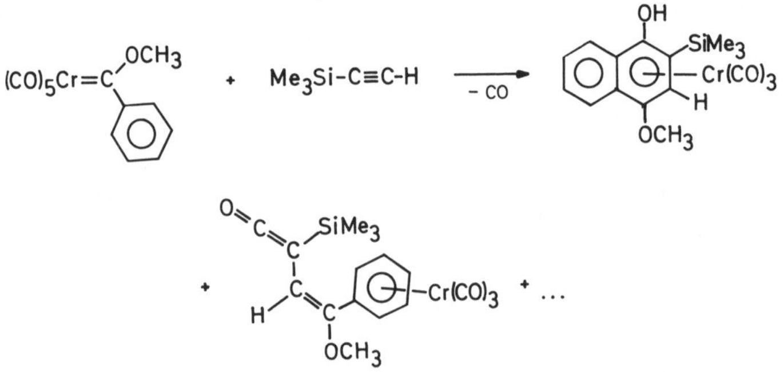

Scheme 51.

Vinyl ketenes are widely accepted to be intermediates in electrocyclic reactions. The synthesis of silylsubstituted derivatives — stable and readily tractable compounds — can be formally described as head-to-tail addition of the carbene and a carbonyl ligand to silylalkynes. Because of the presence of a ketene group, carboxylic acid derivatives are formed on treatment with alcohols and amines. Yet, surprisingly, no cycloaddition occurs with enol ethers and enamines (80). Strong nucleophiles, such as the ynamines, are required to achieve cyclization (81, 82). Bicyclo[3.1.0]hexenones with a predominating *endo*-aryl configuration are obtained which, owing to their silyl functionality, are susceptible to further synthetically useful transformation (Scheme 52).

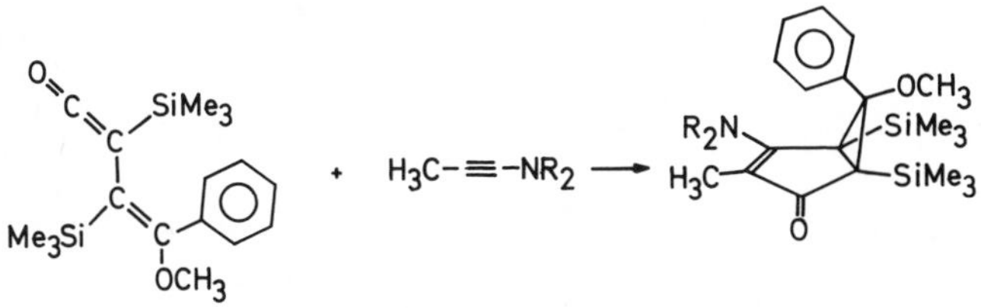

Scheme 52.

7 Synthesis of Natural Products

The facile and high yield aminolysis of alkoxycarbene complexes has promoted work on peptide synthesis using the metal carbenyl functionality as an amino-protecting group. Like primary and secondary amines, amino acid esters react with pentacarbonyl(alkoxycarbene) complexes of chromium and tungsten to give aminocarbene complexes (83, 84) (Scheme 53).

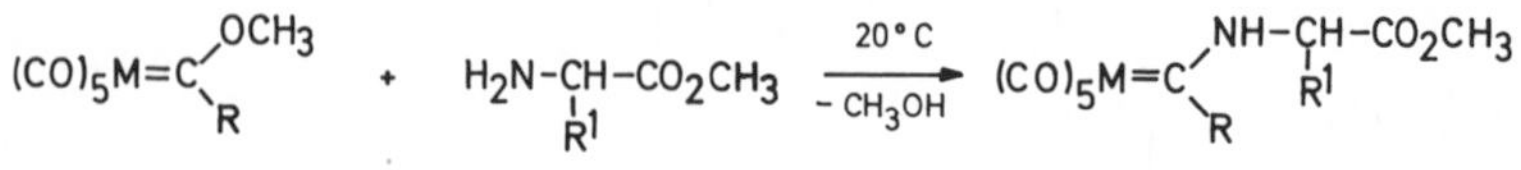

$$(CO)_5M=C\underset{R}{\overset{OCH_3}{}} \quad + \quad H_2N-\underset{R^1}{CH}-CO_2CH_3 \quad \xrightarrow[-CH_3OH]{20\,^\circ C} \quad (CO)_5M=C\underset{R}{\overset{NH-\underset{R^1}{CH}-CO_2CH_3}{}}$$

Scheme 53.

Owing to the nucleophilicity of carboxylate anions, protection by esterification is required for the aminolysis reaction. However, no further protection is necessary for heterocyclic nitrogen, hydroxy or mercapto groups as in histidine, serine or methionine esters. The aminolysis is not restricted to α-amino compounds. For example, both the α-and the ϵ-amino groups can be protected in lysine esters. The formation of the peptide bond is achieved by the conventional route of alkaline ester hydrolysis followed by coupling with a second amino ester using the dicyclohexylcarbodiimide/N-hydroxysuccinimide (DCCD/HOSU) method (Scheme 54).

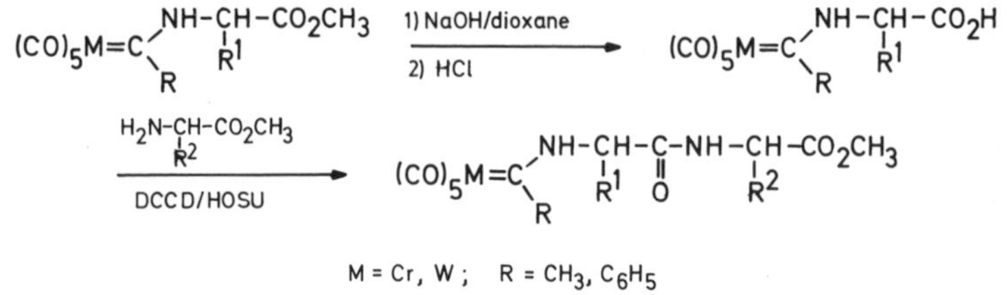

$$(CO)_5M=C\underset{R}{\overset{NH-\underset{R^1}{CH}-CO_2CH_3}{}} \quad \xrightarrow[\text{2) HCl}]{\text{1) NaOH/dioxane}} \quad (CO)_5M=C\underset{R}{\overset{NH-\underset{R^1}{CH}-CO_2H}{}}$$

$$\xrightarrow[\text{DCCD/HOSU}]{H_2N-\underset{R^2}{CH}-CO_2CH_3} \quad (CO)_5M=C\underset{R}{\overset{NH-\underset{R^1}{CH}-\underset{O}{\overset{\|}{C}}-NH-\underset{R^2}{CH}-CO_2CH_3}{}}$$

$$M = Cr, W; \quad R = CH_3, C_6H_5$$

Scheme 54.

The scope of this reaction sequence is demonstrated by the stepwise synthesis of the carbenyl, N-protected tetrapeptide gly-gly-pro-gly-OCH$_3$, which occurs as the sequence 14 − 17 in human proinsulin-C-peptide, with an overall yield of 14 %. The peptide can be conveniently released from the carbene complex by mild acid hydrolysis (Scheme 55). In general, chromium complexes give cleaner reactions than the homologous tungsten compounds.

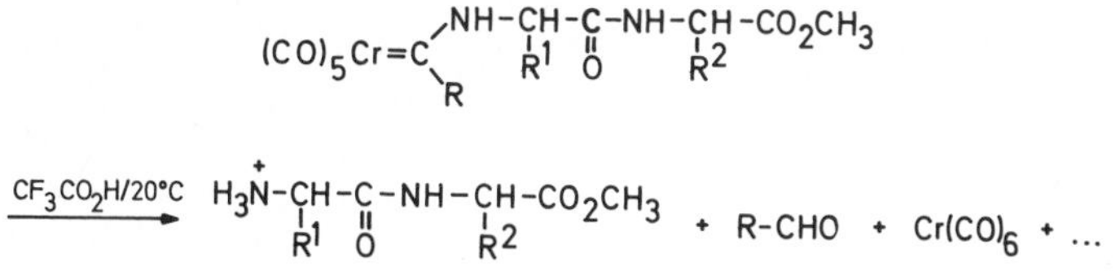

$$(CO)_5Cr=C\underset{R}{\overset{NH-CH-C-NH-CH-CO_2CH_3}{\big|}}$$

$$\xrightarrow{CF_3CO_2H/20°C} \overset{+}{H_3N}-\underset{R^1}{\underset{|}{CH}}-\underset{O}{\overset{||}{C}}-NH-\underset{R^2}{\underset{|}{CH}}-CO_2CH_3 \;+\; R-CHO \;+\; Cr(CO)_6 \;+\; ...$$

Scheme 55.

The carbenyl protection provides a method by which heavy metal atoms can be used for the labeling of amino groups.

As a rule, the thermal stability of aminocarbene complexes has prevented their wider application in organic synthesis. Nevertheless, the intramolecular hydrogen transfer leading to imines has been exploited in an indol alkaloid synthesis (16) (Scheme 56).

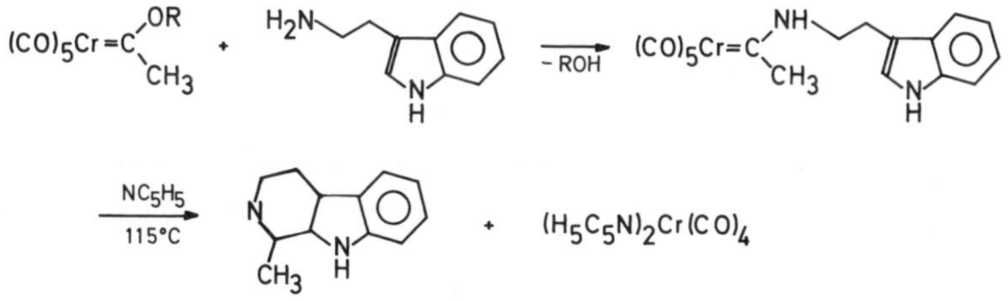

Scheme 56.

The clean and high yield cyclization of alkyne, phenylcarbene and carbonyl ligands offers an interesting route to naturally occurring quinones. Whereas alkenes are known to give cyclopropanes and/or metathesis-like products, a competitive study of non-conjugated enynes has established that the reaction with pentacarbonyl[methoxy(phenyl)carbene]chromium proceeds with complete regiospecific control to yield exclusively the six-membered aromatic system (72). By this route, metal-coordinated allyl-substituted 4-methoxy-1-naphthols become accessible from 1,4-enynes. Oxidation to 1,4-naphthoquinones is achieved by silver oxide (Scheme 57).

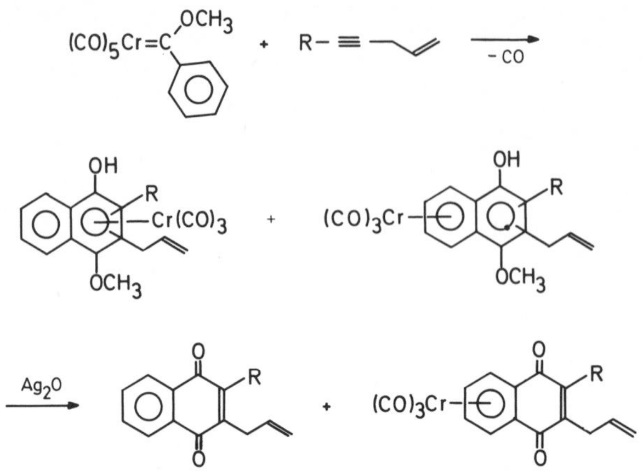

Scheme 57.

This route has been extended to the synthesis of vitamins in the K_1 and K_2 series (73, 85). The enyne components are readily accessible from commercially available isoprenoid alcohols by a two-step process (Scheme 58).

Scheme 58.

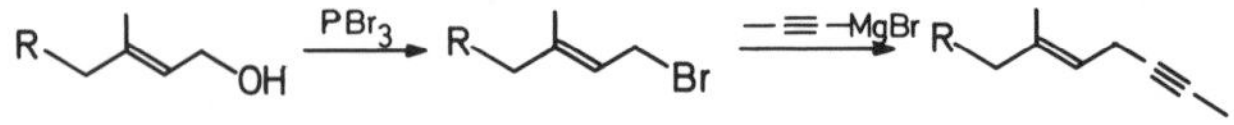

The cyclization to the naphthol system can be carried out under mild conditions using low boiling ethereal solvents such as tert.butyl methyl ether. By this procedure, subsequent migration of the metal to the unsubstituted ring is avoided and an approximately 2:1 mixture of 2- and 3-prenyl $1\text{-}4\text{:}9,10\text{-}\eta$-isomers is obtained in about 90 % yield. The cleavage of the chromium ring bond is achieved by oxidation with silver oxide to afford directly vitamin K. A more convenient alternative involves a two-step ligand substitution oxidation process (Scheme 59). The quantitative release of the naphthol under a CO atmosphere leads to hexacarbonylchromium which is recycled for the carbene synthesis. The subsequent naphthol oxidation is a well-known high yield process.

Unlike customary syntheses based on the condensation of the isoprenoid side chain to the ring system under acidic conditions, this route proceeds with complete retention of configuration at the allylic double bond. Thus, a stereospecific synthesis of the E isomer of vitamin K (which is the only biologically active one) is available.

The accessibility of benzohydroquinones from alkyne, alkenylcarbene and carbonyl ligands has prompted work on a carbene complex route to vitamin E

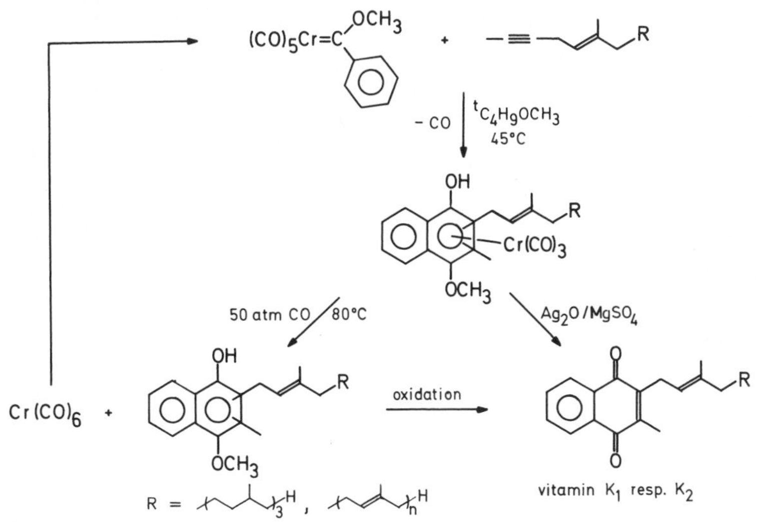

Scheme 59.

(Scheme 60). The methoxy(*E*-1-methyl-1-propenyl)carbene complex is obtained by stereospecific addition of *E*-2-lithio-2-butene to carbonylchromium. Cyclization with the C_{23} enyne produces the coordinated hydroquinone monoether as an approximately 2:1 mixture of isomers which are readily released from the metal under CO pressure. Boron tribromide induced ether cleavage is accompanied by HBr addition to the allylic double bond. Subsequent ring closure affords vitamin E (86).

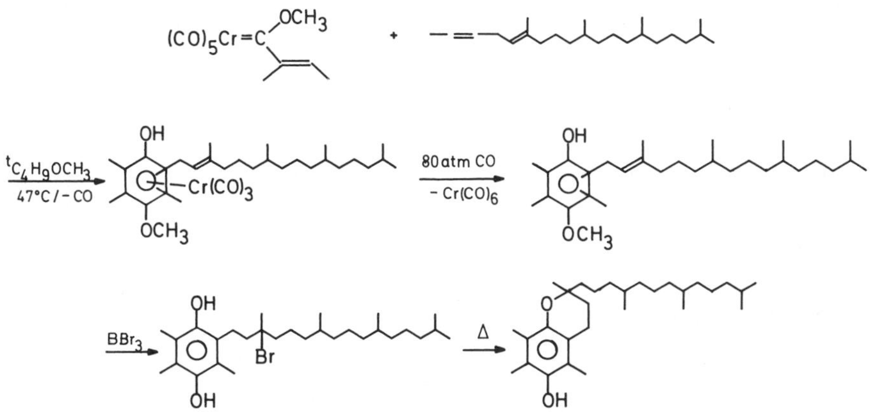

Scheme 60.

The cyclization reaction can be applied to the synthesis of naphthoquino-noid antibiotics as well. For instance, juglomycins A und B (87) and the tricyclic compounds nanaomycin A and deoxyfrenolicin (88) have been chosen as target molecules. The isochroman skeleton of the latter has been synthesized by means of an intramolecular variant of the cyclization reaction. The key step is based on the cyclization of an alkinyloxy(anisyl)carbene and a carbonyl ligand, followed by oxidation with 2,3-dichloro-5,6-dicyanoquinone (DDQ) as exemplified in the synthesis of deoxyfrenolicin (Scheme 61).

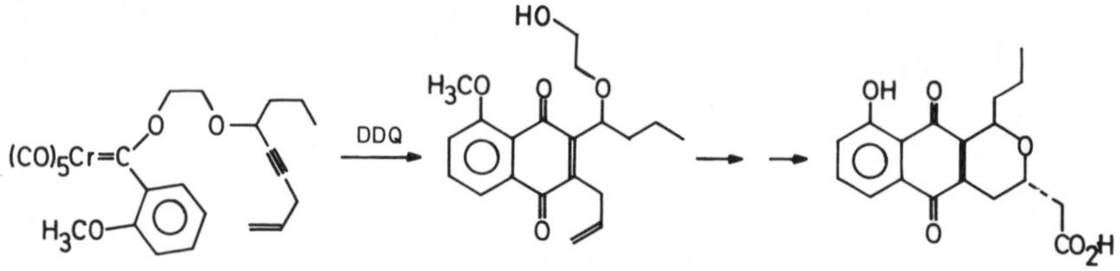

Scheme 61.

8 References

(1) I. Wender, P. Pino (Eds.): Organic Syntheses via Metal Carbonyls. Vol. 1, Interscience Publishers, New York 1968. Vol. 2, John Wiley, New York 1977.

(2) J. Falbe (Ed.): New Syntheses with Carbon Monoxide. Springer, Berlin 1980.

(3) C. P. Casey: Metal-Carbene Complexes, in M. Jones, Jr., R. A. Moss (Eds.): Reactive Intermediates. John Wiley, New York 1981, p. 135 – 174.

(4) K. H. Dötz: Carbene Complexes of Groups VIa, VIIa and VIIIa, in P. Braterman (Ed.): Reactions of Co-ordinated Ligands, Plenum Press, London, to be published.

(5) R. R. Schrock, Acc. Chem. Res. **12** (1979) 98 – 104.

(6) C. G. Kreiter, Angew. Chem. **80** (1968) 402; Angew. Chem. Int. Ed. Engl. **7** (1968) 390.

(7) C. P. Casey, R. L. Anderson, J. Am. Chem. Soc. **96** (1974) 1230 – 1231

(8) C. P. Casey, R. A. Boggs, R. L. Anderson, J. Am. Chem. Soc. **94** (1972) 8947 – 8949.

(9) C. P. Casey, R. L. Anderson, J. Organomet. Chem. **73** (1974) C28 – C30.

(10) C. P. Casey, W. R. Brunsvold, J. Organomet. Chem. **118** (1976) 309 – 323.

(11) C. P. Casey, W. R. Brunsvold, D. M. Scheck, Inorg. Chem. **16** (1977) 3059 – 3063.

(12) M. Rudler-Chauvin, H. Rudler, J. Organomet. Chem. **212** (1981) 203 – 210.

(13) K. H. Dötz, unpublished results 1972.

(14) C. P. Casey, R. L. Anderson, J. Chem. Soc., Chem. Commun. (1975) 895 – 896.

(15) E. O. Fischer, D. Plabst, Chem. Ber. **107** (1974) 3326 – 3331.

(16) J. A. Connor, P. D. Rose, J. Organomet. Chem. **46** (1972) 329 – 334.

(17) C. P. Casey, W. R. Brunsvold, J. Organomet. Chem. **102** (1975) 175 – 183.

(18) K. H. Dötz, B. Fügen-Köster, D. Neugebauer, J. Organomet. Chem. **182** (1979) 489 – 498.

(19) C. P. Casey, S. M. Neumann, J. Am. Chem. Soc. **99** (1977) 1651 – 1652.

(20) F. R. Kreissl, E. O. Fischer, C. G. Kreiter, H. Fischer, Chem. Ber. **106** (1973) 1262 – 1276.

(21) H. Fischer, E. O. Fischer, C. G. Kreiter, Chem. Ber. **107** (1974) 2459 – 2467.

(22) F. R. Kreissl, E. O. Fischer, Chem. Ber. **107** (1974) 183 – 188.

(23) C. P. Casey, S. W. Polichnowski, A. J. Shusterman, C. R. Jones, J. Am. Chem. Soc. **101** (1979) 7282 – 7292.

(24) H. Werner, E. O. Fischer, B. Heckl, C. G. Kreiter, J. Organomet. Chem. **28** (1971) 367 – 389.

(25) E. O. Fischer, K. H. Dötz, Chem. Ber. **105** (1972) 3966 – 3973.

(26) E. O. Fischer, B. Dorrer, Chem. Ber. **107** (1974) 1156 – 1161.

(27) K. H. Dötz, I. Pruskil, Chem. Ber. **114** (1981) 1980 – 1982.

(28) C. P. Casey, T. J. Burkhardt, J. Am. Chem. Soc. **94** (1972) 6543 – 6544.

(29) C. W. Rees, E. v. Angerer, J. Chem. Soc. Chem. Commun. **1972**, 420.

(30) C. P. Casey, S. H. Bertz, T. J. Burkhardt, Tetrahedron Lett. **1973**, 1421 – 1424.

(31) T. J. Burkhardt, C. P. Casey, J. Am. Chem. Soc. **95** (1973) 5833 – 5834.

(32) E. O. Fischer, W. Held, F. R. Kreissl, Chem. Ber. **110** (1977) 3842 – 3848.

(33) E. O. Fischer, S. Riedmüller, Chem. Ber. **109** (1976) 3358 – 3361.

(34) E. O. Fischer, W. Held, F. R. Kreissl, A. Frank, G. Huttner, Chem. Ber. **110** (1977) 656 – 666.

(35) E. O. Fischer, W. Held, J. Organomet. Chem. **112** (1976) C 59 – C 62.

(36) C. P. Casey, W. R. Brunsvold, J. Organomet. Chem. **77** (1974) 345 – 352.

(37) C. P. Casey, W. R. Brunsvold, Inorg. Chem. **16** (1977) 391 – 396.

(38) E. O. Fischer, F. R. Kreissl, J. Organomet. Chem. **35** (1972) C 47 – C 51.

(39) F. R. Kreissl, E. O. Fischer, C. G. Kreiter, J. Organomet. Chem. **57** (1973) C 9 – C 11.

(40) H. Fischer, K. H. Dötz, Chem. Ber. **113** (1980) 193 – 202.

(41) K. H. Dötz, C. G. Kreiter, J.Organomet. Chem. **99** (1975) 309 – 314.

(42) K. H. Dötz, I. Pruskil, J. Organomet. Chem. **132** (1977) 115 – 120.

(43) K. H. Dötz, Chem. Ber. **110** (1977) 78 – 85.

(44) K. H. Dötz, I. Pruskil, Chem. Ber. **111** (1978) 2059 – 2063.

(45) K. H. Dötz, C. G. Kreiter, Chem. Ber. **109** (1976) 2026 – 2032.

(46) K. H. Dötz, D. Neugebauer, Angew. Chem. **90** (1978) 898 – 899; Angew. Chem. Int. Ed. Engl. **17** (1978) 851 – 852.

(47) E. O. Fischer, K. H. Dötz, J. Organomet. Chem. **36** (1972) C 4 – C 6.

(48) J. A. Connor, P. D. Rose, R. M. Turner, J. Organomet. Chem. **55** (1973) 111 – 119.

(49) J. A. Connor, J. P. Day, R. M. Turner, J. Chem. Soc. Dalton **1976**, 108 – 112.

(50) J. A. Connor, J. P. Day, R. M. Turner, J. Chem. Soc. Dalton **1976**, 283 – 285.

(51) K. H. Dötz, Naturwissenschaften **62** (1975) 365 – 371.

(52) U. Schubert, E. O. Fischer, Chem. Ber. **106** (1973) 3882 – 3892.

(53) E. O. Fischer, K. H. Dötz, Chem. Ber. **103** (1970) 1273 – 1278.

(54) K. H. Dötz, E. O. Fischer, Chem. Ber. **105** (1972) 1356 – 1367.

(55) M. D. Cooke, E. O. Fischer, J. Organomet. Chem. **56** (1973) 279 – 284.

(56) E. O. Fischer, K. Weiss, K. Burger, Chem. Ber. **106** (1973) 1581 – 1588.

(57) K. H. Dötz, I. Pruskil, U. Schubert, K. Ackermann, Chem. Ber. **116** (1983) 2337 – 2343.

(58) K. H. Dötz, Angew. Chem. **87** (1975) 672 – 673; Angew. Chem. Int. Ed. Engl. **14** (1975) 644 – 645.

(59) K. H. Dötz, R. Dietz, A. v. Imhof, H. Lorenz, G. Huttner, Chem. Ber. **109** (1976) 2033 – 2038.

(60) K. H. Dötz, R. Dietz, Chem. Ber. **110** (1977) 1555 – 1563.

(61) K. H. Dötz, R. Dietz, Chem. Ber. **111** (1978) 2517 – 2526.

(62) K. H. Dötz, J. Mühlemeier, U. Schubert, O. Orama, J. Organomet. Chem. **247** (1983) 187 – 201.

(63) W. D. Wulff, P.-C. Tang, J. S. Mc Callum, J. Am. Chem. Soc. **103**(1981) 7677 – 7678.

(64) H. Fischer, J. Mühlemeier, R. Märkl, K. H. Dötz, Chem. Ber. **115** (1982) 1355 – 1362.

(65) C. P. Casey, M. C. Cesa, Organometallics **1** (1982) 87 – 94.

(66) A. Yamashita, T. A. Scahill, Tetrahedron Lett. **23** (1982) 3765 – 3768.

(67) K. H. Dötz, J. Organomet. Chem. **140** (1977) 177 – 186.

(68) K. H. Dötz, R. Dietz, D. Neugebauer, Chem. Ber. **112** (1979) 1486 – 1490.

(69) K. H. Dötz, W. Kuhn, J. Organomet. Chem. **252** (1983) C 78 – C 80.

(70) M. F. Semmelhack, J. J. Bozell, Tetrahedron Lett. **23** (1982) 2931 – 2934.

(71) R. Dietz, K. H. Dötz, D. Neugebauer, Nouv. J. Chim. **2** (1978) 59 – 61.

(72) K. H. Dötz, I. Pruskil, Chem. Ber. **113** (1980) 2876 – 2883.

(73) K. H. Dötz, I. Pruskil, J. Mühlemeier, Chem. Ber. **115** (1982) 1278 – 1285.

(74) K. H. Dötz, R. Dietz, J. Organomet. Chem. **157** (1978) C 55 – C 57.

(75) K. H. Dötz, G. Raudaschl, unpublished results.

(76) K. H. Dötz, R. Dietz, Ch. Kappenstein, D. Neugebauer, U. Schubert, Chem. Ber. **112** (1979) 3682 – 3690.

(77) K. H. Dötz, Angew. Chem. **91** (1979) 1021 – 1022; Angew. Chem. Int. Ed. Engl. **18** (1979) 954 – 955.

(78) K. H. Dötz, B. Fügen-Köster, Chem. Ber. **113** (1980) 1449 – 1457.

(79) K. H. Dötz, J. Mühlemeier, Angew. Chem. **94**(1982) 936 – 937; Angew. Chem. Int. Ed. Engl. **21** (1982) 929; Angew. Chem. Suppl. **1982**, 2023 – 2029.

(80) K. H. Dötz, B. Trenkle, unpublished results.

(81) K. H. Dötz, B. Trenkle, U. Schubert, Angew. Chem. **93** (1981) 296 – 297; Angew. Chem. Int. Ed. Engl. **20** (1981) 287.

(82) K. H. Dötz, B. Trenkle, J. Mühlemeier, unpublished results.

(83) K. Weiss, E. O. Fischer, Chem. Ber. **106** (1973) 1277 – 1284.

(84) K. Weiss, E. O. Fischer, Chem. Ber. **109** (1976) 1868 – 1886.

(85) K. H. Dötz, I. Pruskil, J. Organomet. Chem. **209** (1981) C 4 – C 6.

(86) K. H. Dötz, W. Kuhn, Angew. Chem., in the press.

(87) K. H. Dötz, W. Sturm, unpublished results.

(88) M. F. Semmelhack, J. J. Bozell, T. Sato, W. Wulff, E. Spiess, A. Zask, J. Am. Chem. Soc. **104** (1982) 5850 – 5852.

Carbene Complexes as Intermediates in Catalytic Reactions

By Karin Weiss

Currently, carbene complexes appear to be much in favor as intermediates in catalytic reactions. For virtually every catalytic process, a mechanism involving a carbene complex as an intermediate can now be postulated. However, it is only in a limited number of instances, such as in metathesis reactions, that genuine carbene complexes are able to initiate a catalytic reaction.

Intermediates in catalytic processes are usually scarcely detectable because of their short lifetime and low concentration. Accordingly, catalytic processes are commonly simulated using fairly stable model complexes. Some of the most frequently encountered catalytic reactions and their postulated mechanisms based on carbene complex intermediates will be discussed.

1 Synthesis of Cyclopropanes by Catalytic Reactions of Diazoalkanes and Olefins

Since their discovery by Curtius and Pechmann, aliphatic diazoalkanes have been used as source materials for carbenes in a variety of organic syntheses (1 – 4). However, they have not yet been used in industrial processes.

Stable carbene complexes have been synthesized from diazoalkanes by Herrmann (5) (Scheme 1). In such complexes the carbene ligands occupy not only terminal but also bridging positions. The reactions of diazoalkanes with alkenes to yield cyclopropanes is a well-known carbene reaction. Moreover, Fischer-type carbene complexes (with electrophilic $C_{carbene}$), though not Schrock-type carbene complexes (with nucleophilic $C_{carbene}$), are able to form cyclopropanes in stoichiometric reactions with alkenes (6 – 9).

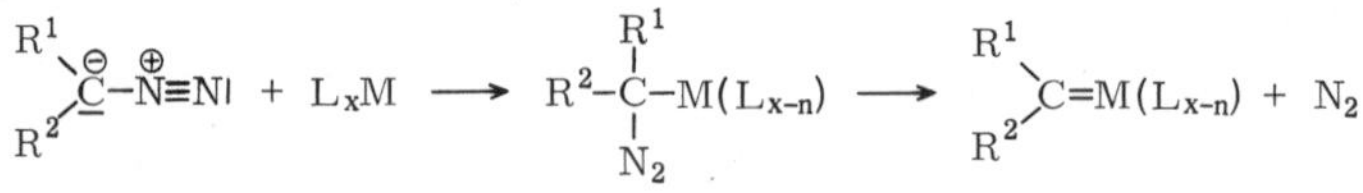

$$\underset{R^2}{\overset{R^1}{>}}\!\!\overset{\ominus}{C}\!\!-\!\!\overset{\oplus}{N}\!\!\equiv\!\!N| \; + \; L_xM \; \longrightarrow \; R^2\!\!-\!\!\underset{\underset{N_2}{|}}{\overset{\overset{R^1}{|}}{C}}\!\!-\!\!M(L_{x-n}) \; \longrightarrow \; \underset{R^2}{\overset{R^1}{>}}\!\!C\!\!=\!\!M(L_{x-n}) \; + \; N_2$$

Scheme 1.

In the mechanism, two intermediates are postulated. Initially there is formation of an olefin complex and this is followed by the formation of a metallacyclobutane (Scheme 2). Confirmation of the first step was documented by Casey with the synthesis of carbene alkene complexes (10, 11).

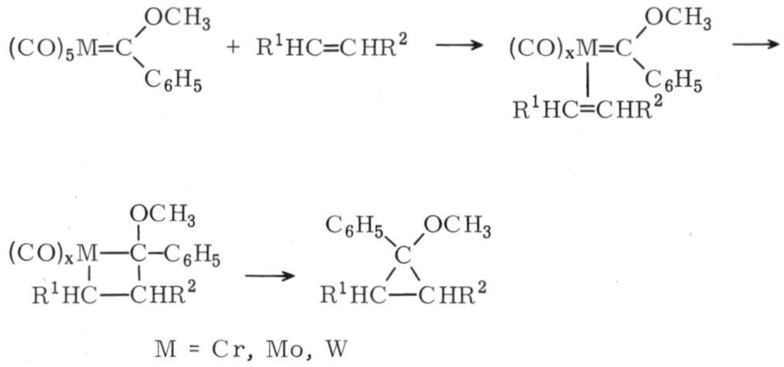

$$(CO)_5M=C\begin{subarray}{l} OCH_3 \\ \\ C_6H_5 \end{subarray} + R^1HC=CHR^2 \longrightarrow (CO)_xM=C\begin{subarray}{l} OCH_3 \\ \\ C_6H_5 \end{subarray} \longrightarrow$$
$$R^1HC=CHR^2$$

$$(CO)_xM-\overset{\overset{\displaystyle OCH_3}{|}}{\underset{\underset{\displaystyle R^1HC-CHR^2}{|}}{C}}-C_6H_5 \longrightarrow \begin{subarray}{c} C_6H_5 \quad OCH_3 \\ \diagdown \;\; C \;\; \diagup \\ \diagup \qquad \diagdown \\ R^1HC-CHR^2 \end{subarray}$$

$$M = Cr, Mo, W$$

Scheme 2.

The reactions of diazoalkanes with alkenes to yield cyclopropanes are catalyzed by many transition metals as well as their salts and complexes, including Cu, Ag, Pd, Fe, Rh, Ru, Re, Co and Mo compounds (12 – 14). Wulfman and Poling have postulated a reaction mechanism for the catalytic reactions of diazoalkanes which is very similar to that (see above) for the stoichiometric formation of cyclopropanes from Fischer-type carbene complexes (4) (Scheme 3).

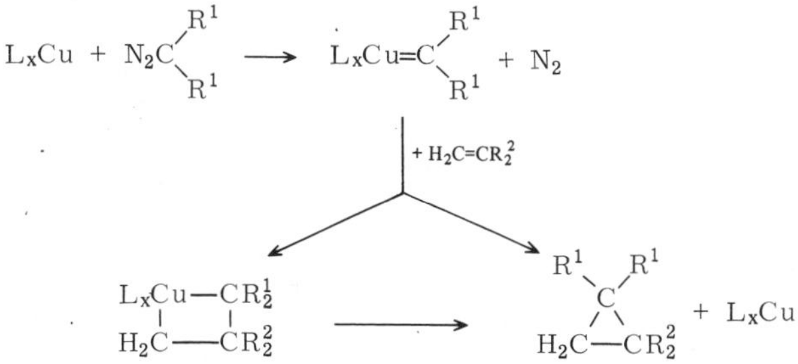

$$L_xCu + N_2C\begin{subarray}{l} R^1 \\ \\ R^1 \end{subarray} \longrightarrow L_xCu=C\begin{subarray}{l} R^1 \\ \\ R^1 \end{subarray} + N_2$$

$$+ H_2C=CR_2^2$$

$$L_xCu-\overset{\overset{\displaystyle }{|}}{\underset{\underset{\displaystyle H_2C}{|}}{C}}R_2^1 \qquad\qquad \begin{subarray}{c} R^1 \quad R^1 \\ \diagdown \;\; C \;\; \diagup \\ \diagup \qquad \diagdown \\ H_2C-CR_2^2 \end{subarray} + L_xCu$$

Scheme 3.

One proof for the cyclopropanation reaction in the coordination sphere of the metal atom is the optical induction brought about by catalysts containing chiral ligands. The optical yields can contain up to 80 % (15), and even 90 % (16), of optically active cyclopropanes. Such results − and the existence of comparable intermediates in the stoichiometric cyclopropanation reaction (where most of the compounds can be identified) − provide strong evidence for the postulated mechanism of the catalytic reaction.

2 Carbene Complexes as Intermediates in Fischer-Tropsch Syntheses

Fischer-Tropsch syntheses using cobalt or iron catalysts, doped with alkali metals, and synthesis gas yield mainly saturated and unsaturated aliphatic hydrocarbons, ranging from methane to waxes, and alcohols (17) (Scheme 4).

$$n\ CO\ +\ 2\ n\ H_2\ \longrightarrow\ -(CH_2)_n-\ +\ n\ H_2O$$

Scheme 4.

Elementary carbon (which may perhaps form carbides with the catalyst) could be obtained via the Boudouard equilibrium ($2\ CO \rightleftarrows C + CO_2$).

Apart from the most commonly employed heterogeneous alkali-doped Fe and Co catalyst, Ni, Rh, Ru and Cu catalysts are also of interest here. Many of the catalysts are not selective and therefore a wide variety of reaction products are obtained. Alkanes and 1-alkenes are usually the dominant reaction products and these have a Schulz-Flory type of molecular mass distribution (18). Reviews of the Fischer-Tropsch synthesis by Herrmann (19), Muetterties and Stein (20) and Masters (21) illustrate the complexities associated with finding a common mechanism for all the possible products. The application of homogeneous catalysis is thought to produce more selective products. Fischer and Tropsch announced in their first paper (22) that the polymerization of methylene groups was dependent upon surface carbide species. The carbides may be reduced to $(CH)_x$ (x = 1, 2, 3), which is then capable of producing a hydrocarbon chain. This notion is still in favor and is referred to as the carbide/methylene mechanism. Another reaction which forms surface methylene groups is the step by step hydrogenation of coordinated CO (23) (Scheme 5).

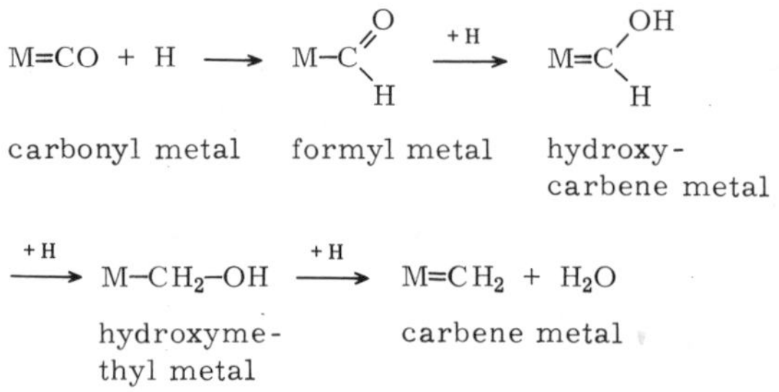

$$M=CO + H \longrightarrow M-C\overset{O}{\underset{H}{\diagup}}\ \overset{+H}{\longrightarrow}\ M=C\overset{OH}{\underset{H}{\diagup}}$$

carbonyl metal   formyl metal   hydroxy-
                                        carbene metal

$$\overset{+H}{\longrightarrow}\ M-CH_2-OH\ \overset{+H}{\longrightarrow}\ M=CH_2 + H_2O$$

hydroxyme-   carbene metal
thyl metal

Scheme 5.

The isolation of all essential intermediates was achieved in an osmium cluster model compound [starting from $M = Os_3(CO)_{11}$] (24).

An interesting experiment on the polymerization of methylene groups involved the use of H_2/CH_2N_2 instead of H_2/CO with Fischer-Tropsch catalysts (25, 26). The reaction products are very similar to those normally obtained, which suggests a common intermediate composed of CH_2 surface fragments. Chain growth is believed to proceed by the combination of these CH_2 surface fragments.

Another model reaction entails the stoichiometric reaction shown in Scheme 6 (27).

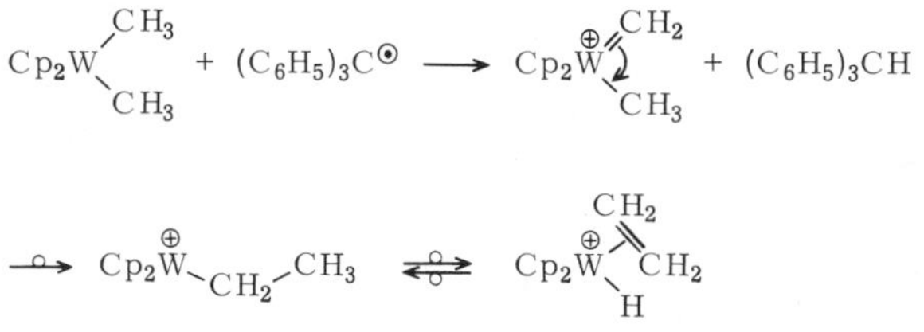

$$Cp_2W\begin{smallmatrix}CH_3\\CH_3\end{smallmatrix} + (C_6H_5)_3C^{\ominus} \longrightarrow Cp_2\overset{\oplus}{W}\begin{smallmatrix}CH_2\\CH_3\end{smallmatrix} + (C_6H_5)_3CH$$

$$\longrightarrow Cp_2\overset{\oplus}{W}{-}CH_2{-}CH_3 \rightleftharpoons Cp_2\overset{\oplus}{W}\overset{CH_2}{\underset{H}{\diagup\diagdown}}CH_2$$

Scheme 6.

This methylene carbene insertion into a metal-methyl bond may explain the chain growth whereas the β-H elimination provides a possible terminal step in the polymerization.

A second mechanism for Fischer-Tropsch reaction, first proposed by Storch (28), is known as the hydroxy carbene mechanism (29). Chain growth is thought to occur by water elimination from surface hydroxycarbenes (Scheme 7).

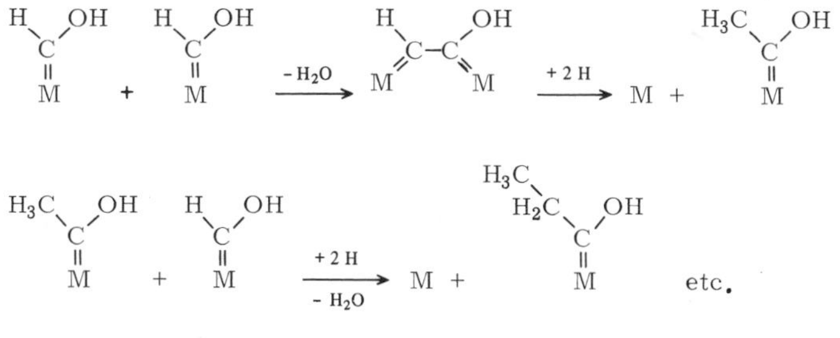

$$\underset{M}{\overset{H\diagdown\diagup OH}{C}}\;+\;\underset{M}{\overset{H\diagdown\diagup OH}{C}}\;\xrightarrow{-H_2O}\;\underset{M}{\overset{H\diagdown}{C}}{=}\underset{M}{\overset{\diagup OH}{C}}\;\xrightarrow{+2H}\;M\;+\;\underset{M}{\overset{H_3C\diagdown\diagup OH}{C}}$$

$$\underset{M}{\overset{H_3C\diagdown\diagup OH}{C}}\;+\;\underset{M}{\overset{H\diagdown\diagup OH}{C}}\;\xrightarrow[-H_2O]{+2H}\;M\;+\;\underset{M}{\overset{H_3C\diagdown\atop H_2C\diagdown\diagup OH}{C}}\quad \text{etc.}$$

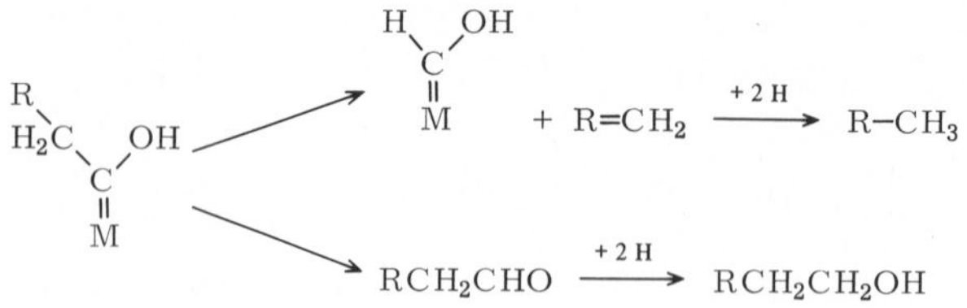

Scheme 7.

In contrast to the carbide/methylene mechanism, the hydroxycarbene route involves oxygen-containing groups and may therefore offer an explanation for the synthesis of alcohols, aldehydes, acids and esters.

Stable hydroxycarbene complexes may be models for the proposed mechanism. Hydroxycarbene complexes were obtained as intermediates in the first synthesis of methoxycarbene complexes by Fischer (30). Their reaction with diazomethane yields stable methoxycarbene complexes. Several years later isolation of these hydroxycarbene complexes was achieved (31). They react like acids and are not stable at room temperature.

Water elimination from these acidic hydroxycarbene complexes with dicyclohexylcarbodiimide (DCCD) results in a variety of reaction products (32 – 34) which depend on the metal atom and the substituent R. For R = CH_3 and M = Cr intramolecular water elimination is feasible (Scheme 8).

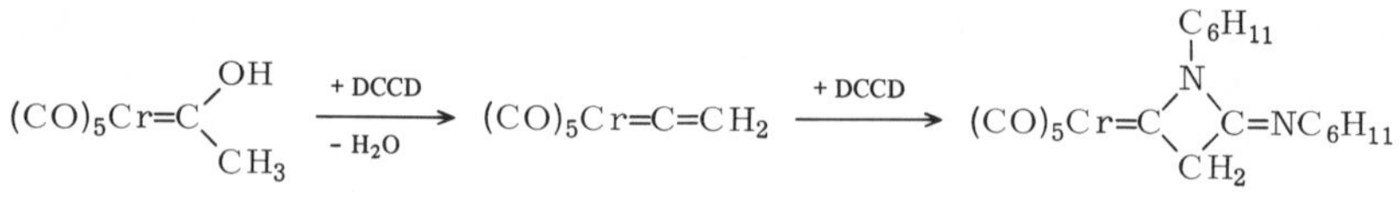

Scheme 8.

The methylene carbene complex cannot be isolated since it reacts immediately with DCCD to form a carbena-azetidine complex by (2 + 2)cycloaddition. Such intramolecular water elimination may represent a chain termination step in Fischer-Tropsch synthesis.

For R = C_6H_5 and M = Cr, intermolecular water elimination yields the expected hydroxycarbene anhydride complex (Scheme 9).

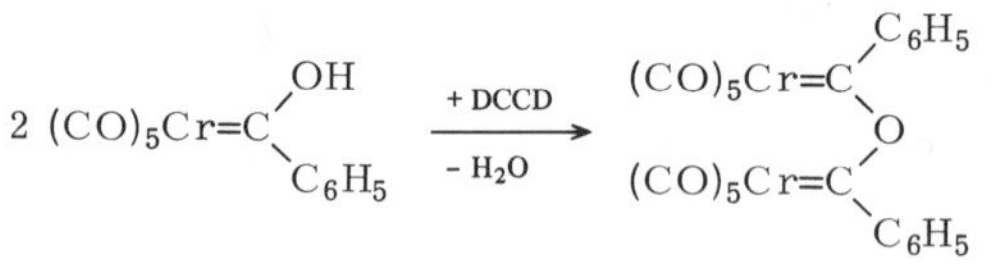

Scheme 9.

This well-known way for the intermolecular elimination of water cannot be neglected in Fischer-Tropsch syntheses.

The reaction of hydroxycarbene complexes (for R = CH_3 or C_6H_5 and M = W) with DCCD came as a great surprise. The reaction products isolated were carbene-carbyne complexes (Scheme 10).

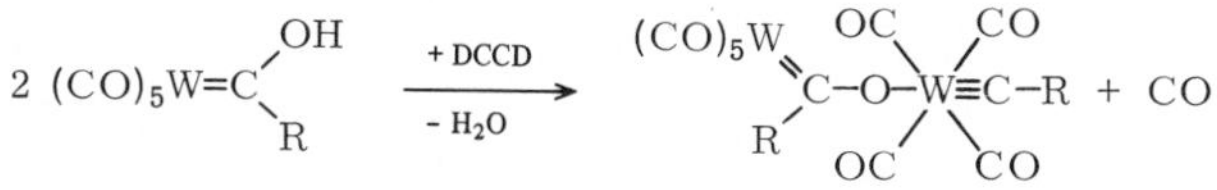

$$2 \ (CO)_5W{=}C{\overset{OH}{\underset{R}{}}} \quad \xrightarrow[- H_2O]{+ \ DCCD} \quad (CO)_5W{\underset{R}{\overset{}{}}}C{-}O{-}W{\equiv}C{-}R \ + \ CO$$

Scheme 10.

Carbyne complex intermediates formed from hydroxycarbenes may also be reactive intermediates in Fischer-Tropsch syntheses.

Such reactions of simple isolated hydroxycarbene complexes provide an appropriate insight into the great variety of compounds formed on hetero-geneous surface metal atoms.

3 Ziegler-Natta Polymerization of Olefins via Carbene Complex Intermediates

The polymerization of ethylene and α-olefins at low pressure can be brought about by means of catalysts formed during the reaction of transition element compounds, such as halides, alkoxides or alkyls with main group element alkyls or alkyl halides, e.g. aluminium alkyls. Most of the products have a stereoregular structure (35, 36). Generally speaking, catalysts are heterogeneous though some homogeneous ones are also known to exist. Whether stereoregulation and stereoselectivity result from heterogeneity is not clear at present. For recent reviews on Ziegler-Natta polymerization see (37 – 40).

Like many other catalytic reactions, the mechanism of Ziegler-Natta polymerization is poorly understood at the molecular level. Nevertheless, this polymerization is a major industrial organometallic process.

The Cossee-Arlman mechanism (41, 42) (Scheme 11) provides a widely accepted model which involves no carbene complex intermediate, although some of the reaction steps are similar to those in the carbene complex mechanism shown in Scheme 12. The interchange of unoccupied coordination sites is characteristic of the postulated mechanism.

A two-stage mechanism for catalysis is envisaged within the coordination sphere of the metal. First there is coordination of the monomer and second stereoregulated insertion of the activated monomer into a metal-carbon bond (37, 40).

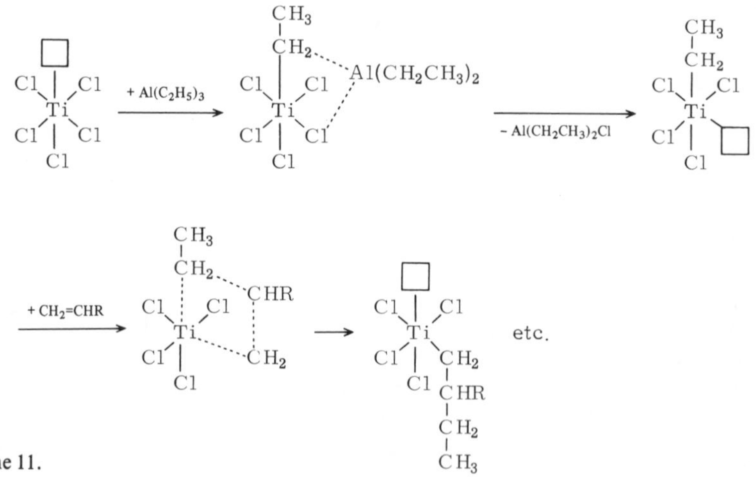

Scheme 11.

A new mechanism involving carbene and metallacyclobutane intermediates has been proposed by Green, Ivin, Rooney et al. (43 – 45) (Scheme 12).

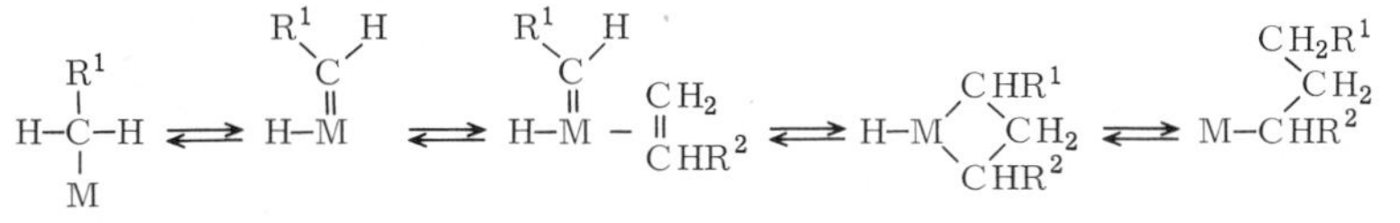

$$
\begin{array}{ccccccccc}
& & R^1 \quad H & & R^1 \quad H & & & & CH_2R^1 \\
R^1 & & \diagdown C \diagup & & \diagdown C \diagup & & \diagup CHR^1 & & \diagdown CH_2 \\
| & & \| & & \| & CH_2 & H-M & \diagdown CH_2 & M-CHR^2 \\
H-C-H & \rightleftarrows & H-M & \rightleftarrows & H-M - \| & & \diagup & \diagup & \\
| & & & & & CHR^2 & CHR^2 & & \\
M & & & & & & & &
\end{array}
$$

Scheme 12.

It may serve as a common mechanism for both olefin polymerization and metathesis reactions. When the key step in the reaction, namely the elimination of hydrogen, is slow or if hydrogen can be removed from the metal atom, metathesis may occur (45).

Turner and Schrock reported recently (46) that they had succeeded in isolating a well-characterized ethylene polymerization catalyst – a tantalum neopentylidene hydride complex (Scheme 13). With this particular catalyst the reaction proceeds slowly enough for polymer chain growth to be observed; a Lewis acid co-catalyst is not required.

$$
2 \ nx \ CH_2{=}CH_2 \ \xrightarrow{\text{Ta (CHCMe}_3)\text{ (H) (PMe}_3)_3\text{I}_2} \ (C_nH_{2n})_x \ + \ (C_nH_{2n+2})_x
$$

Scheme 13.

Most of the polymer chains are olefinic presumably as a result of chain transfer. There are two ways of viewing the reaction (Scheme 14).

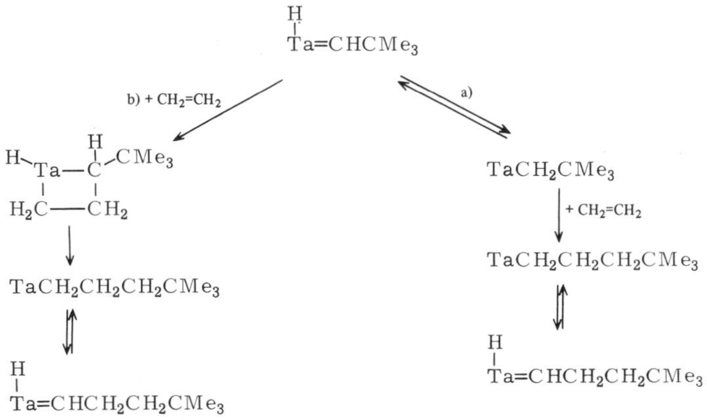

Scheme 14.

The first (a) incorporates some of the classical Cossee mechanism in which ethylene is inserted into a tantalum(III)-neopentyl bond and subsequently tantalum(III)-alkyl bonds are formed. The alkyl complexes are in equilibrium with tantalum(V) carbene hydride complexes.

The second (b) is analogous to that proposed by Green and Rooney where there is a metallacyclobutane intermediate. Turner and Schrock prefer the second alternative since alkylidene ligands in other tantalum(V) complexes react extremely rapidly with alkenes (47) and there are only a few examples of isolatable metal alkyl complexes that react rapidly with ethylene (48). In both of the proposed mechanisms, (a) and (b), the polymerization of ethylene is initiated by a Schrock-type carbene complex: $Ta(CHCMe_3)(H)(PMe_3)_3I_2$.

4 The Olefin Metathesis Reaction Induced by Metal Carbene Complexes

In 1964 Banks and Bailey discovered a new catalytic disproportionation reaction in which linear olefins were converted to homologs having shorter or longer carbon chains (49) (Scheme 15). The heterogeneous catalysts used for this reaction were prepared from molybdenum or tungsten hexacarbonyl and aluminia.

$$2 \ R^1HC=CHR^2 \longrightarrow R^1HC=CHR^1 + R^2HC=CHR^2$$

Scheme 15.

Calderon was the first who called this reaction, i.e. the interchange of alkylidene units between olefins, "metathesis" (50). Metathesis reactions are possible not only with linear alkenes but also with cycloalkenes to form polyalkenes by ring-opening polymerization (Scheme 16).

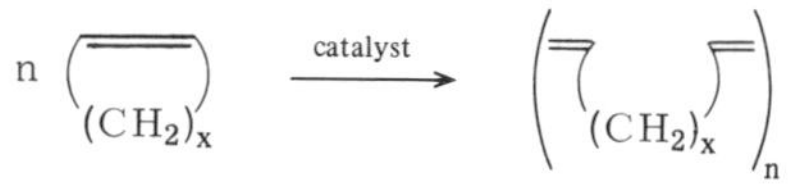

$$n \left(\overline{} \atop (CH_2)_x\right) \xrightarrow{\text{catalyst}} \left(\overline{} \atop (CH_2)_x\right)_n$$

Scheme 16.

The catalysts used in metathesis reactions resemble Ziegler-Natta catalysts. The most active ones are formed by reacting transition element compounds, such as halogenides, oxides, sulfides or carbonyls, with main group element alkyls, alkyl halides or oxides, for instance WCl_6/Et_3Al or Re_2O_7/Al_2O_3. Catalysts used in olefin metathesis reaction may be heterogeneous or homogeneous. Carbene complexes are considered as homogeneous catalysts in mechanistic studies.

The many reviews of metathesis reactions published over the past few years bear witness to the importance of this reaction and to the numerous mechanistic studies which have been carried out (51 – 58).

Bradshaw proposed the first mechanism for a metathesis reaction. He suggested an intermediate "quasi cyclobutane metal complex" (59) (Scheme 17).

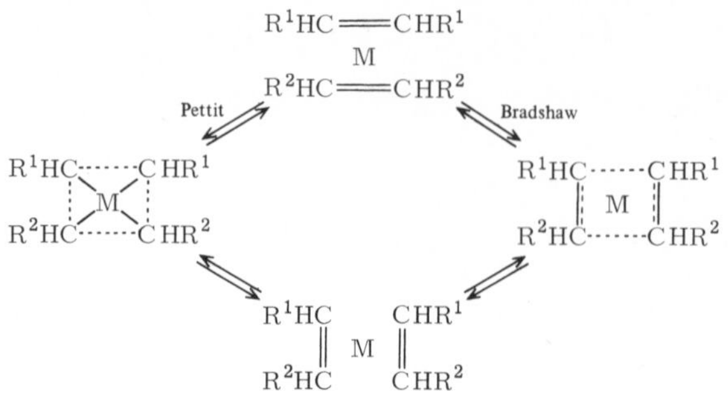

$$R^1HC = CHR^1$$
$$M$$
$$R^2HC = CHR^2$$

Pettit ⇗   Bradshaw ⇖

$$R^1HC\cdots\cdots CHR^1$$
$$M$$
$$R^2HC\cdots\cdots CHR^2$$

$$R^1HC\cdots\cdots CHR^1$$
$$M$$
$$R^2HC\cdots\cdots CHR^2$$

⇙   ⇗

$$R^1HC \qquad CHR^1$$
$$\| \ \ M \ \ \|$$
$$R^2HC \qquad CHR^2$$

Scheme 17.

This "pairwise" reaction is symmetry-forbidden for simple olefins, though the metal d electrons were supposed to remove the symmetry restrictions on such a reaction. Pettit suggested that the intermediate had little cyclobutane character because cyclobutanes are unreactive under the usual metathesis conditions (60). He postulated a tetracarbene complex intermediate which is symmetry-allowed on molecular orbital grounds (61) (Scheme 17).

Although a lot of reaction products can be explained in terms of the pairwise mechanism, no model cmpounds for the postulated intermediates could be found.

Several years later Herisson and Chauvin proposed a "non-pairwise" mechanism which starts out from a carbene complex and has a metallacyclo-butane intermediate (62) (Scheme 18).

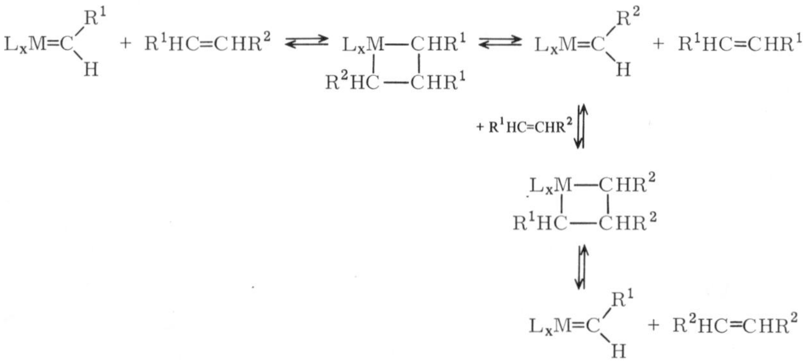

$$L_xM=C\begin{smallmatrix}R^1\\ \\H\end{smallmatrix} + R^1HC=CHR^2 \rightleftarrows \begin{smallmatrix}L_xM-CHR^1\\ | \qquad |\\ R^2HC-CHR^1\end{smallmatrix} \rightleftarrows L_xM=C\begin{smallmatrix}R^2\\ \\H\end{smallmatrix} + R^1HC=CHR^1$$

$$+ R^1HC=CHR^2 \big\Updownarrow$$

$$\begin{smallmatrix}L_xM-CHR^2\\ | \qquad |\\ R^1HC-CHR^2\end{smallmatrix}$$

$$\Updownarrow$$

$$L_xM=C\begin{smallmatrix}R^1\\ \\H\end{smallmatrix} + R^2HC=CHR^2$$

Scheme 18.

Carbene complexes may serve as compounds used to model this mechanism.

The first metathesis reaction of phenyl(methoxy)carbene complexes with alkenes was that with vinyl ethers to yield α-methoxystyrene (63, 64) (Scheme 19).

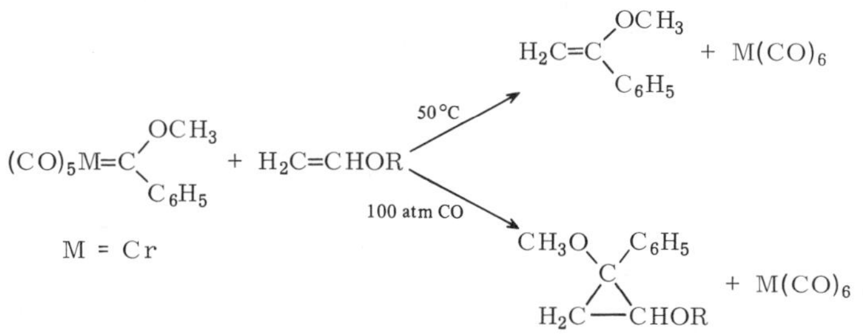

Scheme 19.

Under CO pressure (100 atm) cyclopropanes are formed. The Herisson-Chauvin mechanism can account for the reaction products (65) (Scheme 20).

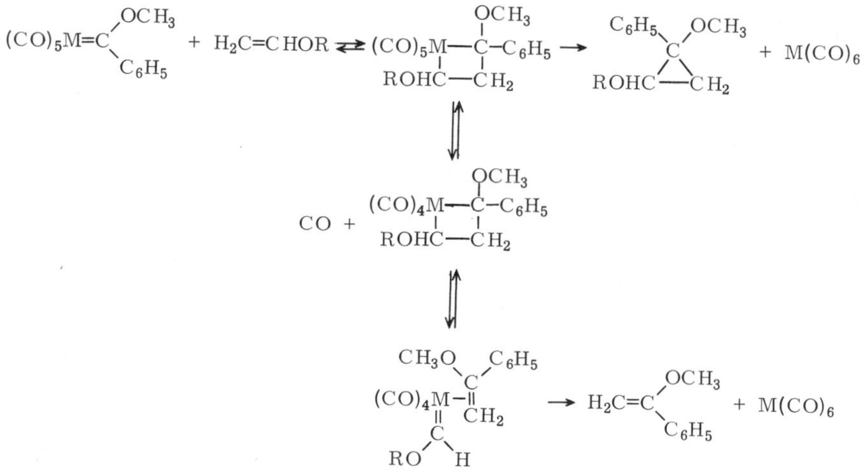

Scheme 20.

Formation of a seven-coordinated metal atom incorporated into a metallacyclobutane would be favored by high CO pressure. A subsequent reaction will yield cyclopropanes by means of reductive elimination. At low CO pressure, dissociation of one CO ligand will occur to produce a six-coordinated metallacyclobutane complex. A ring-opening reaction then generates the metathesis products: a new carbene alkene complex (which could not be isolated) and α-methoxystyrene.

The major breakthrough by Casey (66) was the synthesis of diphenylcarbe-nepentacarbonyltungsten (0) and the demonstration of its reactivity toward a variety of alkenes to yield metathesis products and cyclopropanes. Casey outlined a general reaction scheme involving a metallacyclobutane and a carbene alkene complex intermediate (68) (Scheme 21).

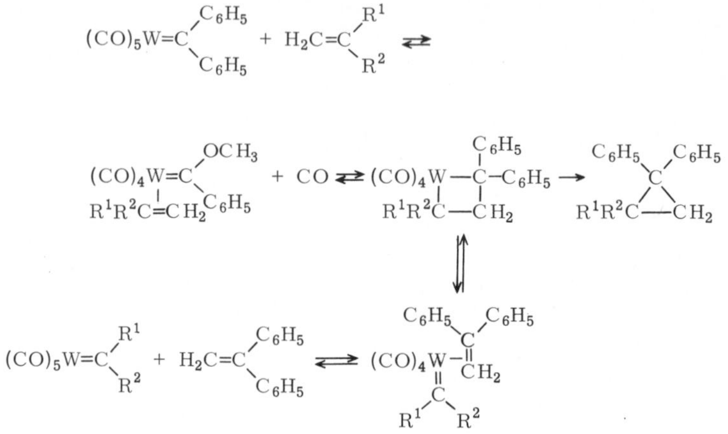

$$(CO)_5W=C\overset{C_6H_5}{\underset{C_6H_5}{}} + H_2C=C\overset{R^1}{\underset{R^2}{}} \rightleftharpoons$$

$$(CO)_4W=C\overset{OCH_3}{\underset{R^1R^2C=CH_2\overset{|}{C_6H_5}}{}} + CO \rightleftharpoons (CO)_4W\overset{C_6H_5}{\underset{R^1R^2C-CH_2}{\overset{|}{-}C-C_6H_5}} \rightarrow \overset{C_6H_5}{\underset{R^1R^2C-CH_2}{\overset{C}{\diagup}\diagdown\overset{C_6H_5}{}}}$$

$$\Updownarrow$$

$$(CO)_5W=C\overset{R^1}{\underset{R^2}{}} + H_2C=C\overset{C_6H_5}{\underset{C_6H_5}{}} \rightleftharpoons (CO)_4W\overset{\overset{C_6H_5}{\underset{C}{\diagup}\diagdown\overset{C_6H_5}{}}}{\underset{\overset{C}{\diagdown}\underset{R^1}{} \overset{}{} \underset{R^2}{}}{\overset{\|}{-}\overset{CH_2}{\|}}}$$

Scheme 21.

Recently, Casey isolated a stable model compound representing a key intermediate. This was a carbene alkene complex in which the carbene and alkene ligands are nearly perpendicular to each other (11). A metastable metal carbene alkene complex was observed by NMR spectroscopy (56) and was shown to decompose to yield cyclopropanes.

The reaction of a titanacyclobutane complex with olefins entails a degenerate olefin metathesis which may form the basis for metallacyclobutane reactions (68) (Scheme 22).

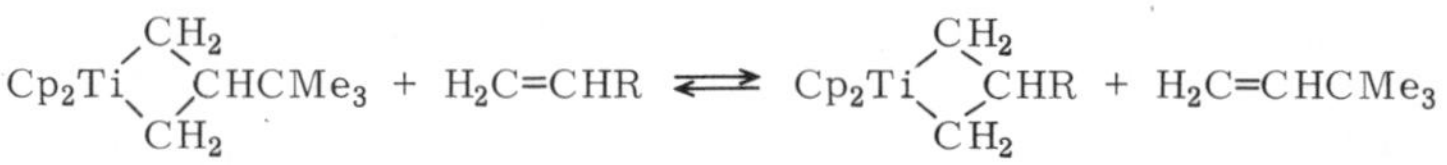

$$Cp_2Ti\overset{CH_2}{\underset{CH_2}{\diagup}\diagdown}CHCMe_3 + H_2C=CHR \rightleftharpoons Cp_2Ti\overset{CH_2}{\underset{CH_2}{\diagup}\diagdown}CHR + H_2C=CHCMe_3$$

Scheme 22.

Model compounds of this type provide strong evidence for the non-pairwise olefin metathesis mechanism of Herisson and Chauvin.

Poor yields are obtained in ring-opening polymerizations of cycloalkenes by metathesis reactions based on carbene complexes if no co-catalyst (Lewis acids such as $EtAlCl_2$) (69), no strained cycloalkene, such as bicyclo[4,2,0]oct-7-ene (70), or no cyclic vinylether like 2,3-dihydrofuran (71) are used.

Fischer-type carbene complexes are reactive catalysts for acetylene polymerization (72). In the postulated reaction mechanism a metallacyclobutene intermediate is put forward (Scheme 23).

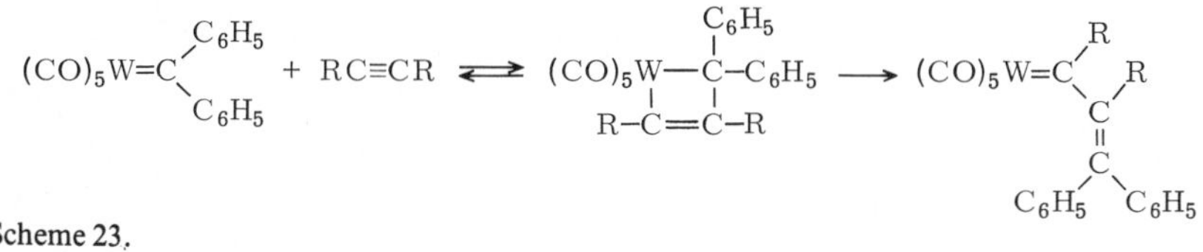

$$(CO)_5W{=}C\overset{C_6H_5}{\underset{C_6H_5}{\big<}} \;+\; RC{\equiv}CR \;\rightleftharpoons\; (CO)_5W{-}\overset{C_6H_5}{\underset{R-C{=}C-R}{\underset{|}{\overset{|}{C}}}}{-}C_6H_5 \;\longrightarrow\; (CO)_5W{=}C\overset{R}{\underset{\underset{C_6H_5\;\;C_6H_5}{C}}{\overset{\big<}{\underset{\|}{C}}}}R$$

Scheme 23.

Such intermediates catalyze the metathesis of cycloalkenes to yield polymers. An induction of cycloalkene metathesis by acetylene is thus possible (73).

Not only Fischer-type carbene complexes with electrophilic $C_{carbene}$, but also Schrock-type alkylidene (= carbene) complexes with nucleophilic $C_{carbene}$, may catalyze the metathesis of alkenes. Schrock isolated a five-coordinated tungsten-oxo-alkylidene complex (Scheme 24) which catalyzes olefin metathesis (74). He postulated that alkylidene complexes of tantalum and tungsten having "hard" ligands, such as oxo or alkoxo groups are metathesis catalysts whereas those having "soft" ligands, such as chloride, bromide or cyclopentadienyl, have no activity. If alkylidene complexes containing "soft" ligands react with alkenes, the initial metallacyclobutane complex rearranges rapidly into an olefin complex.

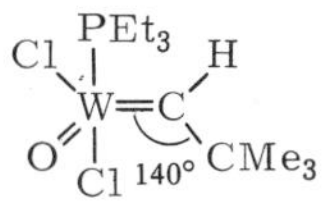

$$\overset{\overset{\textstyle PEt_3}{\textstyle |}}{\underset{\overset{\textstyle |}{\textstyle Cl}}{Cl{\diagdown}\underset{O{\diagup}}{W}{=}C}}\overset{\textstyle H}{\underset{\textstyle 140°\;CMe_3}{\diagup}}$$

Scheme 24.

Schrock's alkylidene ligands are usually formed by the intramolecular abstraction of an α-hydrogen atom from one alkyl ligand by another alkyl ligand (75) (Scheme 25).

$$Ta(CH_2CMe_3)_2X_3 \;+\; 2\,L \;\longrightarrow\; Ta(CHCMe_3)L_2X_3 \;+\; CMe_4$$

$$X = Cl,\; Br,\; L = PMe_3,\; PPhMe_2$$

Scheme 25.

It is very likely that the most reactive metathesis catalysts, e.g. $WCl_6/SnMe_4$, are able to form such alkylidene complexes under the reaction conditions, with the co-catalyst $SnMe_4$ acting as an alkylating agent (Scheme 26).

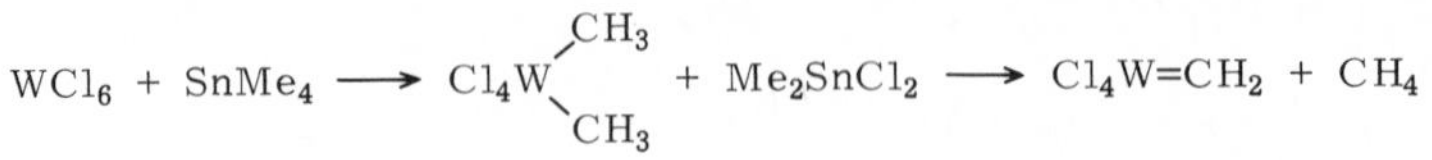

$$WCl_6 + SnMe_4 \longrightarrow Cl_4W\underset{CH_3}{\overset{CH_3}{<}} + Me_2SnCl_2 \longrightarrow Cl_4W{=}CH_2 + CH_4$$

Scheme 26.

These alkylidene complexes generated "in situ" could be the actual catalyst.

5 Conclusions

All the catalytic reactions discussed here can be interpreted by numerous mechanism involving one common intermediate: a carbene complex. Carbene complexes react with alkenes to form another common intermediate: a metal-lacyclobutane. The reaction products may depend on the decomposition of this metallacyclobutane species (Scheme 27).

A $(2+2)$ ring cleavage forms metathesis products, whereas a $(3+1)$ ring cleavage yields cyclopropanes. The chain growth in Ziegler-Natta polymerization could proceed by a 1,2 hydrogen shift of a metal hydride ligand and ring cleavage.

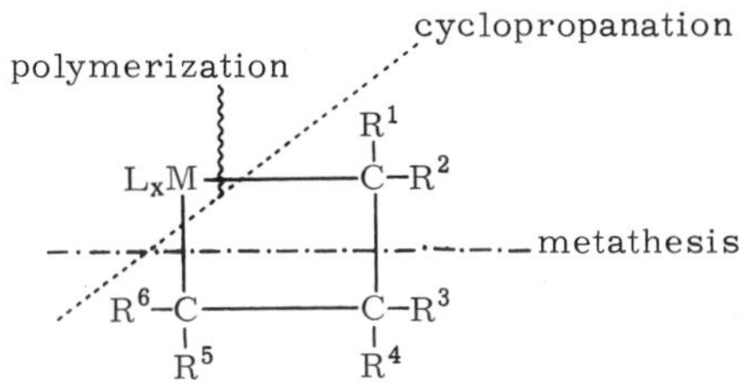

Scheme 27.

6 References

(1) S. Patai, The Chemistry of Diazonium and Diazo groups, Parts 1 and 2, Wiley, New York 1978.

(2) B. Eistert, M. Regitz, G. Heck and H. Schwall, Alipathische Diazoverbindungen in: Houben-Weyl Methoden der Organischen Chemie, Vol. X/4, Thieme, Stuttgart 1968.

(3) W. Kirmse, Carbene Chemistry, Academic Press, New York 1971.

(4) D.S. Wulfman and B. Poling, Metal Salt Catalyzed Carbenoids in Reactive Intermediates, Plenum Press, New York 1980, Vol. 1, p. 321.

(5) W.A Herrmann, Angew. Chem. **90**, (1978) 855; Angew. Chem. Int. Ed. Engl. **17** (1978) 800.

(6) K.H. Dötz and E.O. Fischer, Chem. Ber. **105** (1972) 1356.

(7) E.O. Fischer and K.H. Dötz, Chem. Ber. **103** (1970) 1273.

(8) C.P. Casey, E.H. Tuinstra and M.C. Saeman, J. Am. Chem. Soc. **98** (1976) 608.

(9) C.P. Casey and S.W. Polichnowski, J. Am. Chem. Soc. **99** (1977) 6097.

(10) C.P. Casey and A.J. Shustermann, J. Mol. Cat. **8** (1980) 1.

(11) C.P. Casey, A.J. Shusterman, N.W. Vollendorf and K.J. Haller, J. Am. Chem. Soc. **104** (1982) 2417.

(12) W.H. Taniblyn, S.R. Hoffmann and M.P. Doyle, J. Organomet. Chem. **216** (1981) C 64.

(13) W.R. Moser, J. Am. Chem. Soc. **91** (1969) 1135.

(14) W.R. Moser, J. Am. Chem. Soc. **91** (1969) 1141.

(15) A. Nakamura, A. Konishi, Y. Tatsuno and S. Otsuka, J. Am. Chem. Soc. **100** (1978) 3443.

(16) T. Aratani, Y. Yoneyoshi and N. Nagase, Tetrahedron Lett. (1975) 1707 and (1977) 2599.

(17) J. Tkatchenko, in M. Tsutsui, Fundamental Research in Homogeneous Catalysis, Plenum Press, New York 1979, Vol. 3, p. 119.

(18) G. Henrici-Olivé and S. Olivé, Angew. Chem. Int. Ed. Engl. **15** (1976) 136.

(19) W.A. Herrmann, Angew. Chem. **94** (1982) 118; Angew. Chem. Int. Ed. Engl. **21** (1982) 117.

(20) E.L. Muetterties and J. Stein, Chem. Reviews **79** (1979) 479.

(21) C. Masters, Adv. Organomet. Chem. **17** (1979) 61.

(22) F. Fischer and H. Tropsch, Brennst. Chem. **7** (1926) 97.

(23) R.R. Schrock, Acc. Chem. Res. **12** (1979) 98.

(24) G. Steinmetz and G.L. Geoffry, J. Am. Chem. Soc. **103** (1981) 1278.

(25) R.C. Brady III and R. Pettit, J. Am. Chem. Soc. **102** (1980) 6182.

(26) G. Henrici-Olivé and S. Olivé, J. Mol. Cat. **16** (1982) 111.

(27) J.C. Hayes, G.D.N. Pearson and N.J. Cooper, J. Am. Chem. Soc. **103** (1981) 4648.

(28) H.H. Storch, N. Golumbic and R.B. Anderson, The Fischer-Tropsch and Related Syntheses, Wiley, New York 1951.

(29) R.B. Anderson, L.J. Hofer and H.H. Storch, Chem. Ing. Tech. **30** (1958) 560.

(30) E.O. Fischer and A. Maasböl, Chem. Ber. **100** (1967) 2445.

(31) E.O. Fischer, G. Kreis and F.P. Kreissl, J. Organomet. Chem. **56** (1973) C 37.

(32) E.O. Fischer, K. Weiss and C.G. Kreiter, Chem. Ber. **107** (1974) 3554.

(33) E.O. Fischer and K. Weiss, Chem. Ber. **109** (1976) 1128.

(34) K. Weiss, E.O. Fischer and J. Müller, Chem. Ber. **107** (1974) 3548.

(35) K. Ziegler, E. Holzkamp, M. Martin and H. Beil, Angew. Chem. **67** (1955) 541.

(36) G. Natta, J. Polym. Sci. **16** (1955) 143.

(37) A.D. Count, Ziegler Polymerization in: Catalysis, The Chemical Society, London 1977, Vol. 1, p. 234.

(38) H. Sinn and W. Kaminsky, Adv. Organomet. Chem. **18** (1980) 99.

(39) P. Pino and R. Mülhaupt, Angew. Chem. **92** (1980) 869; Angew. Int. Ed. Engl. (1980) 857.

(40) J. Boor, Ziegler-Natta Catalysts and Polymerization, Academic Press, New York 1979.

(41) P. Cossee, Tetrahedron Lett. **17**, 12 (1960): J. Cat. **3** (1964) 80.

(42) E.J. Arlman, J. Cat. **3** (1964) 89.

(43) K.J. Ivin, I.I. Rooney, C.D. Stewart, M.L.H. Green and R. Mahtab, J. Chem. Soc. Chem. Com. (1978) 604.

(44) M.L.H. Green, Pure Appl. Chem. **50** (1978) 27.

(45) H.M. Colquhoun, I. Holton and M.V. Twigg, Ann. Rep. Prog. Chem. **11** (1977) 277.

(46) H.W. Turner and R.R. Schrock, J. Am. Chem. Soc. **104** (1982) 2331.

(47) J.D. Fellman, R.R. Schrock and G.A. Rupprecht, J. Am. Chem. Soc. **102** (1981) 5752.

(48) E.R. Evitt and R.G. Bergman, J. Am. Chem. Soc. **101** (1979) 3973.

(49) R.L. Banks and G.C. Bailey, Ind. and Eng. Chem. **3** (1964) 170.

(50) N. Calderon, H.Y. Chen and K.W. Scott, Tetrahedron Lett. (1967) 3327.

(51) N. Calderon, E.A. Ofstead and W.A. Indy, Angew. Chem. **88** (1976) 433; Angew. Chem. Int. Ed. Engl. **15** (1976) 401.

(52) T.J. Katz, Adv. Organomet. Chem. **16** (1977) 283.

(53) R.H. Grubbs, Prog. Inorg. Chem. **24** (1977) 1.

(54) I.I. Rooney and A. Stewart, Catalysis, The Chemical Society, London 1977, Vol. **1**, p. 277.

(55) N. Calderon, I.P. Lawrence and E.A. Ofstead, Adv. Organomet. Chem. **17** (1979) 449.

(56) C.P. Casey, D.M. Scheck and A.I. Shusterman in M. Tsutsui, Fundamental Research in Homogeneous Catalysis, Plenum Press, New York 1979, Vol. **3**, p. 141.

(57) E. Thorn-Cśanyi, Nachr. Chem. Techn. Lab. **29** (1981) 700.

(58) R.L. Banks in Catalysis, The Royal Society of Chemistry, London 1982, Vol. **4**, p. 100.

(59) C.P.C. Bradshaw, E.I. Howman and L. Turner, J. Catal. **7** (1967) 269.

(60) G.S. Lewandos and R. Pettit, J. Am. Chem. Soc. **93** (1971) 7087.

(61) F.S. Lewandos and R. Pettit, Tetrahedron Lett. (1977) 780.

(62) I.L. Herrisson and Y. Chauvin, Makromol. Chem. **141** (1970) 161.

(63) K.H. Dötz and E.O. Fischer, Chem. Ber. **105** (1972) 1356.

(64) E.O. Fischer and K.H. Dötz, Chem. Ber. **105** (1972) 3966.

(65) F.I. Brown, Progr. Inorg. Chem. **27** (1980) 1.

(66) C.P. Casey and T.I. Burkhardt, J. Am. Chem. Soc. **96** (1974) 7808.

(67) C.P. Casey, H.E. Tunistra and M.C. Soleman, J. Am. Chem. Soc. **98** (1976) 608.

(68) I.B. Lee, K.C. Ott and R.H. Grubbs, J. Am. Chem. Soc. **104** (1982) 7491.

(69) C. Larroche, I.P. Laval, A. Lattes, M. Leconte, F. Quinard, and I.M. Basset, J. Org. Chem. **47** (1982) 2019.

(70) Cuc Tran Thu, T. Bastelberger and H. Höcker, Makrom. Chem., Rapid Commun. **2** (1981) 7.

(71) Cuc Tran Thu, T. Bastelberger and H. Höcker, Makrom. Chem., Rapid Commun **2** (1981) 383.

(72) T.I. Katz and S.I. Lee, J. Am. Chem. Soc. **102** (1980) 422.

(73) T.I. Katz, S.I. Lee, M. Nair and E.B. Savage, J. Am. Chem. Soc. **102** (1980) 7940.

(74) I.A. Wengrovius, R.R. Schrock, M.R. Churchill, I.R. Missert and W.I. Youngs, J. Am. Chem. Soc. **102** (1980) 4515.

(75) G.A. Rupprecht, L.W. Messerle, I.D. Fellman and R.R. Schrock, J. Am. Chem. Soc. **102** (1980) 6236.

Mechanistic Aspects of Carbene Complex Reactions

By Helmut Fischer

1 Introduction

This chapter deals only with the mechanistic aspects of stoichiometric reactions of carbene complexes since the role of carbene complexes in metathesis is discussed in another chapter. Most detailed investigations of reactions of electrophilic carbene complexes in which the carbene ligand participates in some way have been performed with compounds of the type $(CO)_5M$-$[C(R^1)R^2]$ (M = Cr, W; R^1 = alkyl, R^2 = OR or R^1 = aryl; R^2 = alkyl, aryl, H, OR or R^1 = OR, SR; R^2 = OR, SR, SeR). Our discussion of the mechanisms will therefore focus on their reactions. The characteristics displayed can – with minor alterations – be transferred to carbene complexes of other metals. However, the carbene ligand in amino- and diaminocarbene complexes is – with few exceptions – far less reactive due to the effective electron transfer from the nitrogen to the carbene carbon atom. This has the effect of greatly reducing the electrophilicity of the carbene carbon atom.

Mulliken population analyses of $(CO)_5Cr[C(OMe)Me]$ (I) and other related carbene complexes indicate that the carbon atoms of the carbonyl ligands carry a greater positive charge than that on the carbene carbon atom. The lowest unoccupied molecular orbital (LUMO) is energetically isolated and spatially localized mainly on the carbene carbon atom (1). For such pentacarbonyl(carbene) complexes, the following possible initial reaction step(s) have to be taken into account (Scheme 1):

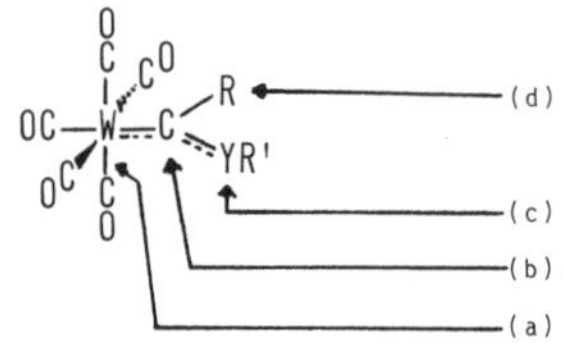

Scheme 1.

(a) Dissociation of a *cis* or *trans* carbonyl-metal bond with formation of a penta-coordinated (solvent-stabilized) intermediate.
(b) Addition of a nucleophile to the carbene carbon atom.
(c) Addition of an electrophile to the heteroatom (O, S or N) bonded to the carbene carbon atom.
(d) (For alkylcarbene complexes only) base-assisted dissociation of the hydrogen bonded to the carbon alpha to the carbene carbon atom.

Similar considerations apply to carbene(carbonyl)cyclopentadienyl complexes, whereas for nucleophilic carbene complexes — owing to the reversed polarization of the metal-alkylidene bond — electrophilic attack can be expected at the alkylidene carbon and nucleophilic attack at the central metal.

Other conceivable paths for the reaction of $(CO)_5M[C(R)YR']$, such as nucleophilic attack at a carbonyl carbon or electrophilic attack at a carbonyl oxygen atom or at the central metal, are less likely to occur and have not, in fact, been observed.

2 Reactions of Electrophilic Carbene Complexes

2.1 Thermolysis

The thermal decomposition of carbonyl(carbene) complexes in the solid state or in saturated hydrocarbon solvents is, in general, initiated by the dissociation of a M-CO bond (very likely a *cis* M-CO bond) forming a penta-coordinated intermediate. The rate of the dissociation depends on the metal, the substituents at the carbene carbon atom and on other substituents in the complex. At 40°C, the rate constants for CO dissociation in hexane or decane are:

For $(CO)_5Cr[C(OMe)Me]$: $0.647 \cdot 10^{-5}$ s^{-1}
For $(CO)_5Cr[C(OMe)Ph]$: 13.3 $\cdot 10^{-5}$ s^{-1} (both calculated from (2))
For $(CO)_5W[C(Ph)_2]$:   4.9 $\cdot 10^{-5}$ s^{-1} (3).

At 77°C CO dissociation in $(CO)_5Cr[C(OMe)Ph]$ is faster by a factor of approximately 500 than in the tungsten analogue (in toluene).

The rate of decomposition of these carbene complexes (in the absence of another potential donor ligand) is substantially slower than the rate of CO dissociation. Thus, re-addition of carbon monoxide to the coordinatively unsaturated intermediate must be faster than the further decomposition. The main organic products of thermolysis are generally the dimers of the former carbene ligands (4-6). The decomposition of pentacarbonyl(2-oxacyclopenty-lidene)chromium (II) was investigated in detail (7). From the results (second order in II, CO inhibits the decomposition, rate of ^{13}CO exchange with II exceeds decomposition rate) the mechanistic Scheme 2 was deduced.

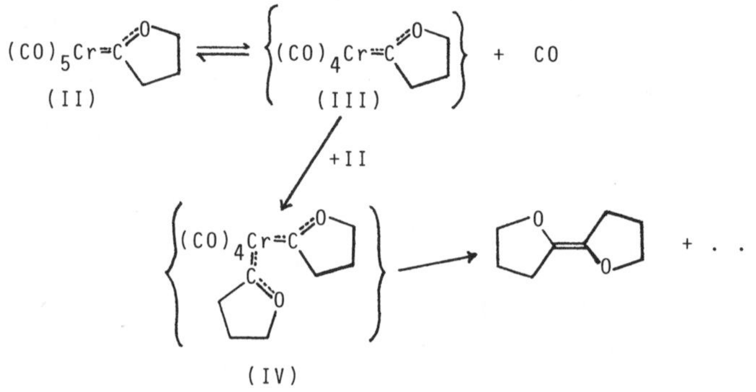

Scheme 2.

The slowest step in this sequence is formation of the biscarbene intermediate IV (III + II → IV + "(CO)$_5$Cr") which requires an intermolecular
transfer of the carbene ligand from one metal to another. Such transfer reactions are indeed possible, as has been demonstrated in several instances (see
Chapter 4.1 in "The Synthesis of Carbene Complexes" in this book.) This
reaction scheme probably also applies to the thermolysis of other alkyl- and
arylcarbene complexes.

The addition of pyridine or tertiary amines to solutions of alkoxy(alkyl)-
carbene or some amino(organyl)carbene complexes [but not to alkoxy(aryl)-
carbene complexes] (a) enhances the rate of complex decomposition, and (b)
gives differing organic products: vinyl ether derivatives in the case of alkoxy-
(alkyl)carbene complexes (*e. g.* 5, 7, 8) and imines in the case of carbene complexes containing a primary amino substituent at the carbene carbon atom
(*e. g.* 9). In both types of complex the α-hydrogen atom (alpha with respect to
the carbene carbon) has been shown to be acidic (10-12). The decomposition
of these compounds in the presence of nitrogen bases is thus probably initiated
by deprotonation of the carbene complexes (path d). The resulting carbene
anions are subsequently reprotonated at the carbene-metal bond. M-C bond
dissociation finally yields "(CO)$_5$M" and the organic product (Scheme 3).

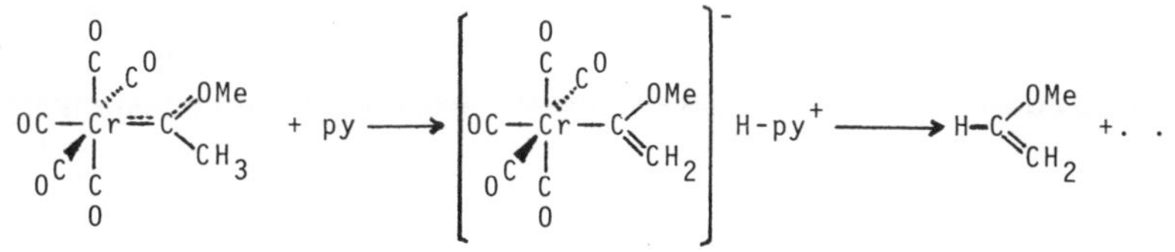

Scheme 3.

A third path obtains in the cases of (CO)$_5$W[C(Ph)CH$_2$R] (13) and Cp-
(CO)$_2$Fe[CMe$_2$]$^+$ (14): migration of the methyl hydrogen onto the carbene
carbon atom followed by rearrangement to an olefin complex. A similar
mechanism may also account for the formation of aldehydes in the thermolysis of hydroxycarbene complexes (15, 16).

An unusual type of thermolysis reaction is observed with some
(CO)$_5$Cr[C(NR$_2$)X] complexes (R = Me, Et; X = Cl, Br, I, SeR, TePh,
SnPh$_3$ or PbPh$_3$): intramolecular migration of X onto the central metal with
simultaneous extrusion of one CO ligand (17-23) (Scheme 4).

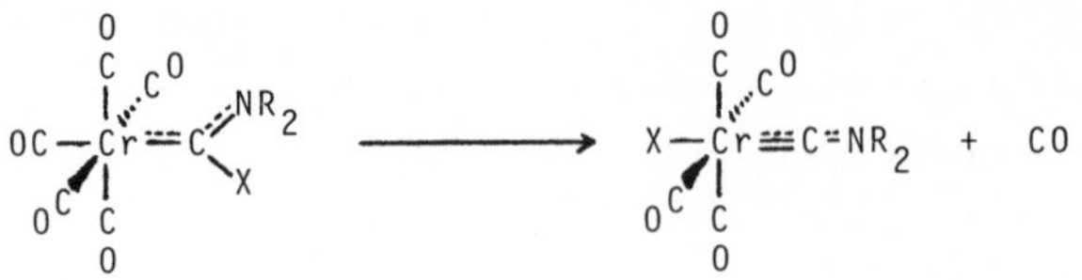

Scheme 4.

There is no conclusive evidence at present that in any of these carbene complex reactions a free carbene appears as an intermediate.

2.2 Reactions with Nucleophiles

In the reactions of carbene complexes with nucleophiles, the initial reaction steps a, b, and d (d only in the case of alkylcarbene complexes) can be observed and, under certain conditions, a and b simultaneously. The relative importance of the paths a and b can be influenced by the choice of carbene complex, altering the nucleophile or the reaction conditions.

Thus, at low temperatures alkyl- and aryl(alkoxy)carbene complexes of Cr and W add tertiary phosphines reversibly to the carbene carbon atom (24, 25). The equilibrium constants are dependent on the polarity of the solvent and on the steric and electronic properties of the carbene complex and phosphine, the steric properties representing the major factor (25, 26) (Scheme 5).

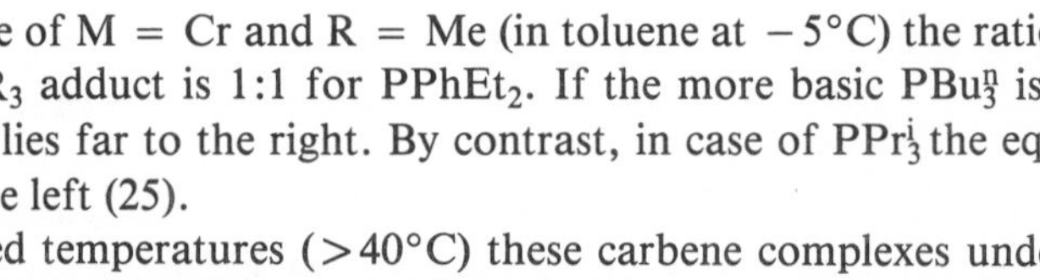

Scheme 5.

In the case of M = Cr and R = Me (in toluene at $-5°C$) the ratio carbene complex/PR_3 adduct is 1:1 for $PPhEt_2$. If the more basic PBu_3^n is used the equilibrium lies far to the right. By contrast, in case of PPr_3^i the equilibrium lies far to the left (25).

At elevated temperatures ($>40°C$) these carbene complexes undergo substitution of phosphine for one CO ligand giving a mixture of *cis*- and *trans*-carbene(tetracarbonyl)phosphine complexes (27, 28). Small amounts of $(CO)_5MPR_3$ and *trans*-$(CO)_4M(PR_3)_2$ are also formed. The reaction shown in

Scheme 6 was found to follow a generalized rate law (2) with two terms:

$$-d[I]/dt = k_1[I] + k_2[I][PR_3].$$

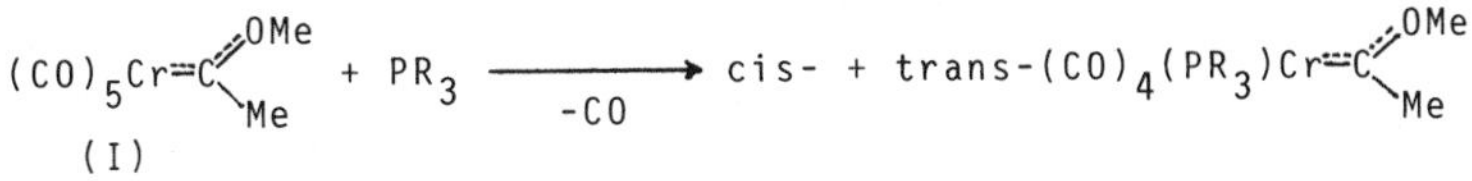

$$(CO)_5Cr{=}C\underset{Me}{\overset{OMe}{\diagup}} + PR_3 \xrightarrow[-CO]{} cis\text{-} + trans\text{-}(CO)_4(PR_3)Cr{=}C\underset{Me}{\overset{OMe}{\diagup}}$$

(I)

Scheme 6.

The relative importance of the second order term vis-à-vis the first order term depends on the type of PR_3. For phosphines of low basicity (PPh_3, $PEtPh_2$ or PEt_2Ph) or sterically very demanding phosphines ($P(C_6H_{11})_3$), the second order term is negligible. For highly basic phosphines having a small Tolman cone angle (PEt_3 or PBu_3), *i.e.* those having a marked tendency to carbene complex-PR_3 adduct formation (*vide supra*), this term is no longer negligible. The first-order term is associated with a rate limiting dissociation of CO from the carbene complex (path a). The exclusive formation of the *cis* isomer in the photochemical reaction of phosphines with carbene complexes at low temperatures (28) and the results of IR spectroscopic investigations and X-ray analyses of carbene complexes suggest that a *cis* M-CO bond must be broken. The penta-coordinated intermediate then adds PR_3 in a fast second step. The phosphine-dependent second order rate term can be ascribed to a second path (path b) in which CO (and probably also a *cis* CO) dissociates from the PR_3-carbene complex adduct followed either by migration of phosphine from the "carbene" carbon atom to the metal or capture of an external PR_3. Both paths very likely lead to the *cis* isomer as the kinetically controlled reaction product. However, at the temperatures required for thermal substitution, isomerization occurs and thus an equilibrium mixture of the *cis* and *trans* isomers is isolated (29). The same equilibrium mixture is obtained when solutions of either the pure *cis* or *trans* isomer are warmed. The equilibrium constant is dependent on the solvent, the steric requirements and the electronic properties of the carbene and phosphine ligand and the central metal. Again, the steric factors dominate (29). From kinetic study of the isomerization reaction, an intramolecular mechanism has been deduced (30).

By substituting tertiary amines for PR_3 in the reaction with carbene complexes, a similar addition of NR_3 to the carbene carbon atom can be observed. In the cases of some sterically hindered amines (quinuclidine and 1,4-diaza [2. 2. 2]bicyclooctane) the corresponding adducts were isolated (31). The substitution of NR_3 for CO − by analogy with the reaction of PR_3 with carbene complexes − was not observed. This is probably due to the low stability of amine(carbene) complexes which might be expected in the temperature range employed for their thermal preparation.

Whereas HPMe$_2$ at low temperatures also adds to the carbene carbon atom to give an isolatable adduct (path b) (24), the corresponding reaction of alkoxycarbene complexes with HNR$_2$ (or H$_2$NR or H$_3$N) proceeds further by exchanging an amino group for the alkoxy group to yield aminocarbene complexes ("aminolysis") (32) (Scheme 7).

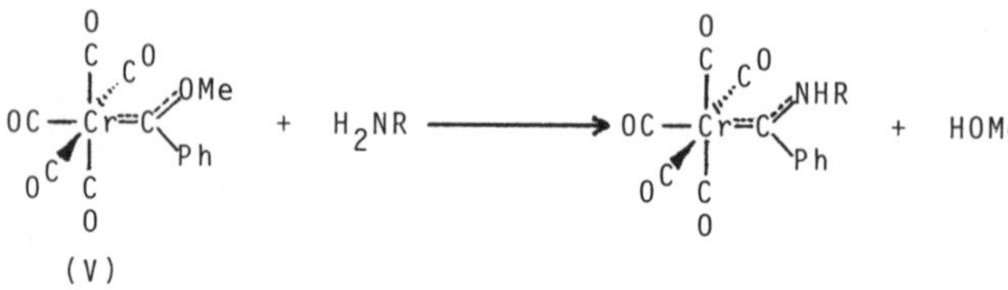

(V)

Scheme 7.

This reaction follows a fourth order rate law (R = Bu^n, C$_6$H$_{11}$ or CH$_2$Ph): $-d[V]/dt = k[V][H_2NR]^3$ (in n-decane). For a variety of solvents, this rate expression can be generalized to: $-d[V]/dt = k[V][H_2NR][HX][Y]$, where HX represents a proton donating and Y a proton accepting agent (33). There is no correlation between the strength of the amines used as bases and the reaction rate. The rate, however, decreases as the steric bulk of the amines increases. A consecutive step mechanism has been proposed. The reaction mechanism involves nucleophilic attack of an amine molecule (activated by the proton acceptor Y, *e.g.* H$_2$NR or dioxane) at the carbene carbon atom of the complex (activated by a hydrogen bond between a proton donor HX, *e.g.* H$_2$NR or MeOH, and the oxygen atom of the alkoxy group). A negative Arrhenius activation energy characterizes this aminolysis, *i.e.* the reaction rate increases with decreasing temperature.

Closely related to aminolysis are exchange reactions of OCD$_3$ (34), SR (35) or SeMe (36) for OCH$_3$ in methoxycarbene complexes or of OCH$_3$ for OEt (10) in ethoxycarbene complexes. The reactions are again most probably initiated by nucleophilic attack of an alcohol, thiol ("thiolysis") or methylselenol molecule ("selenolysis") at the carbene carbon to form an adduct. Elimination of MeOH and EtOH will yield new alkoxy-, organylthio- and methylselenocarbene complexes, respectively. In the case of methylselenol the methanol elimination is acid catalyzed.

In contrast to these reactions, isomerization is observed with the HPMe$_2$-carbene complex adduct (in which the P-H proton is acidic) to give a phosphine complex, (CO)$_5$Cr[Me$_2$PCH(Ph)OMe] (24). This reaction is paralleled by that of the adducts formed (in nonpolar solvents) from HSePh and I (37) or from HBr and $(CO)_5M[C(Me)SMe]$ (M = Cr or W) (38). Isomerization occurs to yield $(CO)_5Cr-Se(Ph)[CH(OMe)Me]$ and $(CO)_5M-S(Me)[CH-(Me)Br]$, respectively.

The reactions of other nucleophiles, *e.g.* hydrazines (39), benzophenoneoxime (40), aldoximes (40) or hydroxylamine (39), with alkoxycarbene complexes probably also proceed by initial attack of the nucleophile on the carbene carbon. Anionic nucleophiles such as $[C_6H_4R]^-$ (41, 42), H^- in $[HB(OPr^i)_3]^-$ (43), SR^- (44) and F^- (34) also add to the carbene carbon of arylcarbene complexes, in most cases giving isolatable anionic adducts (Scheme 8).

$$(CO)_5M{=}C{\overset{OR^1}{\underset{R^2}{\diagdown}}} \quad + \quad Nu^- \longrightarrow \left[(CO)_5M{-}C{\overset{Nu}{\underset{R^2}{\overset{\ldots}{\diagdown}}}}OR^1 \right]^-$$

R^1 = Me, Et; R^2 = aryl; Nu = H, aryl, SR, F.

Scheme 8.

The elimination of $[OR^1]^-$ from these anions can generally be achieved by the use of acids (41-44).

If alkylcarbene complexes containing an alpha hydrogen atom, *viz.* alpha with respect to the carbene carbon, are used in the reaction with LiBu or $NaOCH_3$, the abstraction of a proton is observed instead of nucleophilic addition thereby generating a carbene anion (10, 11).

The competition between a dissociative and an associative mechanism (paths a and b) in the reaction of carbene complexes with tertiary phosphines is also observed in the reaction of carbene complexes with double bond systems, *e.g.* olefins, $R_2C = PR'_3$ $O = SR_2$, $S = C = NR$ or $Se = C = NR$. Which path (a or b or a and b simultaneously) is adopted in a specific reaction depends on the carbene complex as well as on the type of double bond system.

On warming $(CO)_5M[C(OMe)Ph]$ (M = Cr, Mo or W) in simple alkenes such as cyclohexene or tetramethylethylene, the same organic products are obtained as from thermolysis in hydrocarbon solvents (8, 45). From the reaction of methyl *trans*-crotonate or other α,β-unsaturated esters with these carbene complexes at 90 – 140°C, a mixture of two isomeric cyclopropanes (derived from formal addition of the carbene ligand to the olefinic double bond) is isolated. The cyclopropanes are formed stereospecifically. The ratio of isomers depends on the metal, which excludes the intermediacy of any free carbenes (46). Two mechanisms have been proposed for the initial reaction step:

(a) Direct coordination of the activated, polarized double bond of the olefin to the equally polarized M–C (carbene) bond (46), and

(b) CO dissociation and subsequent coordination of the olefin to the penta-coordinated intermediate, since ^{13}CO exchange in these complexes is significantly faster under the conditions employed than cyclopropane formation (47). The mechanistic problem is still not completely solved.

In the case of carbene complex reactions with vinyl ethers, the type of product depends on the reaction conditions (48) (Scheme 9).

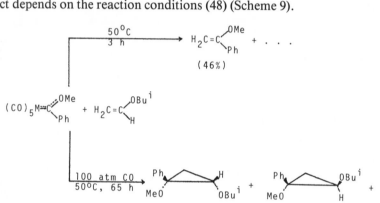

Scheme 9.

Again, the dissociative and the associative mechanisms compete. At atmospheric pressure, the initial M–CO dissociation is followed by fast addition of a vinyl ether molecule to the coordinatively unsaturated complex intermediate. Subsequent formation of a metallacyclobutane and ultimate expulsion of the alkene scission product gives rise to α-methoxystyrole. At 100 atm CO pressure re-addition of CO to the unsaturated intermediate is much faster than the addition of vinyl ether. Thus, the essentially slower nucleophilic attack (compared to CO dissociation) of vinyl ether on the electron-deficient carbene carbon atom of the coordinatively saturated carbene complex (path b) predominates. This in turn leads to the formation of cyclopropanes.

The same competition between the associative and the dissociative paths may explain the results obtained in the reaction of $(CO)_5W[CPh_2]$ (VI) with vinyl ethers and olefins. Ethyl vinyl ether with VI affords a mixture of diphenylethylene ($\approx 11\%$), *i.e.* the metathesis-like scission product, and 1-ethoxy-2,2-diphenylcyclopropane (60–70%) (49). This reaction is approximately ten times faster at 40°C in pure vinyl ethers than the ^{13}CO exchange with VI and gives as the initial complex product mainly $(CO)_5W[vinyl\ ether]$. The latter then slowly decomposes to form $(CO)_6W$ (34). When the reaction is carried out under 1 atm pressure of ^{13}CO, almost no ^{13}CO is incorporated into the vinyl ether complex (34). If simple olefins are employed, the reaction is significantly slower (even slower than ^{13}CO exchange with VI) and the metathesis-like alkene scission product predominates. The formation of cyclopropanes can again be explained as arising from nucleophilic attack of vinyl ether (olefin) at the carbene carbon and formation of diphenylethylene from the initial M-CO dissociation.

Compared to VI, the carbene carbon in $(CO)_5W[C(H)Ph]$ (VII) is more electrophilic and more easily accessible sterically (exchange of H for Ph). It is thus not surprising that, in the reaction of VII with olefins, only cyclopropane derivatives (resulting from initial nucleophilic attack at the carbene carbon) are obtained. No metathesis-like products have been detected. To explain the stereoselectivity of the reaction, interaction of the β-carbon of the olefin with the phenyl group has been proposed (50).

Similarly, the products and observations on the reactions of carbene complexes with other compounds containing polar double bonds or strongly polarized bonds can be rationalized. The reaction proceeds by inital nucleophilic attack at the carbene carbon, *e.g.* the reactions of carbene complexes with diazomethane (51), dimethylsulfoxide (6, 34), $OSC(Me)NH_2$ (52), iodosobenzene (53), pyridine *N*-oxide (53), $ONMe_3$ (34) and $H_2C = PPh_3$ (6). In most cases the carbene ligand is liberated in the form of $O = C(R^1)R^2$ or $H_2C = C(R^1)R^2$. When cyclic vinyl ethers or enamines are employed, a formal insertion of the $C = C$ double bond into the $M = C$(carbene) bond may be observed (54, 55).

In the reaction of $(CO)_5W[C(Ph)aryl]$ with organylisothiocyanates, insertion of the sulfur atom of $R-N = C = S$ into the $W = C$(carbene) bond occurs. As might be deduced from the kinetic investigation of the reaction (56), nucleophilic attack of $R-N = C = S$ at the carbene carbon is rate determining [as it probably also is in the analogous reactions with $R-N = C = Se$ and $[N = C = Te]^-$ giving, respectively, selenoketone complexes and a telluroketone complex (57)].

If molecules containing a polar triple bond, such as $R_2^1N-C \equiv C-R^2$, $RO-C \equiv C-H$, $R-C \equiv N$ ($R = NR_2$, OR, SR or $p-C_6H_4NMe_2$) or $R-N \equiv C$ are employed in the reaction with carbene complexes (preferably non-heteroatom stabilized) the products are new amino- (58), alkoxy- (50), alkylidenaminocarbene complexes (59) and ketenimine complexes (60, 61). The nucleophilic attack of the organic molecule on the carbene carbon is again rate limiting. These adducts subsequently rearrange in fast steps with insertion of the triple bond into the $M = C$ (carbene) bond and redistribution of the π-electrons (59, 62). These reactions are generally stereoselective.

However, in the reaction of R-C≡C-R′ (R, R′ = alkyl, aryl or H) with $(CO)_5Cr[C(OMe)Ph]$ to yield tricarbonyl(naphthol)chromium complexes (63), nucleophilic attack at the carbene carbon is no longer observed. This is probably due to a small or negligible polarity of the triple bond. The competition reaction – CO dissociation – now predominates and requires considerably higher temperatures. The coordinatively unsaturated intermediate resulting from loss of one (*cis*) CO ligand either recaptures carbon monoxide to reform the starting complex or adds an alkyne to yield an alkyne(carbene) complex which rearranges in fast subsequent steps to yield a naphthol compound (Scheme 10).

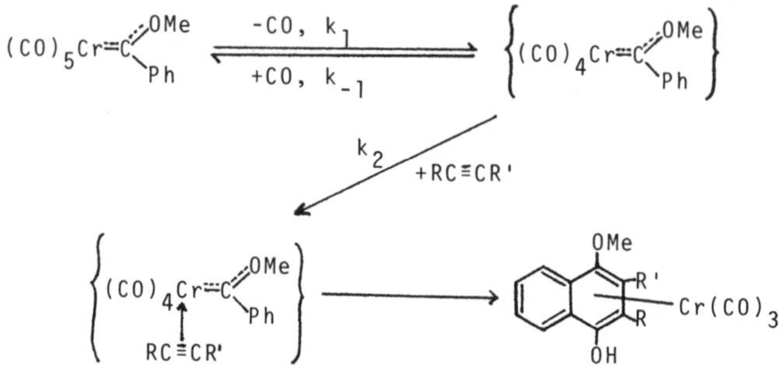

Scheme 10.

Whenever R and R′ are different substituents, the reaction is stereoselective. Kinetic measurements (64) reveal that the rate of alkyne addition is determined primarily by the electron density within the C≡C triple bond. The alkyne addition is slower than CO re-addition by a factor of $k_{-1}/k_2 = 230$ (R = Ph, R′ = p-$C_6H_4CF_3$) to $k_{-1}/k_2 = 39$ (R = Ph, R′ = Me). The stereoselectivity of alkyne incorporation on the contrary is dominated by steric factors (64).

From the large number of experimental results now available, the following generalizations can be made (though some exceptions are known):

(a) Strong nucleophiles and molecules containing a strongly polarized bond attack preferentially at the carbene carbon atom.

(b) Although a large number of carbene complex reactions have nucleophilic attack at the carbene carbon in common, the final products determined by the subsequent fast steps may be quite different.

(c) The reaction of carbene complexes with weak nucleophiles and molecules containing an nonpolar or weakly polar double or triple bond can be initiated by M-CO dissociation or by nucleophilic attack on the carbene carbon atom. Both paths may be observed simultaneously.

(d) In determining the rate and equilibrium constants for nucleophilic addition to the carbene carbon, steric factors seem to play a significant role.

(e) Similarly, the stereoselectivity of carbene complex reactions with double and triple bond systems appears to be profoundly influenced by steric factors.

2.3 Reactions with Electrophiles

Carbene complexes of the Fischer type, *i.e.* $(CO)_5M[C(R)YR']$, carry a nucleophilic centre alpha to the carbene carbon. They thus react with halides of Group IIIB elements (AX_3) even at low temperatures $(> -25°C)$ with formal elimination of $[YR']^-$ $(YR' = OR, SR$ or $NR_2)$ and one CO ligand and addition of X^- to yield carbyne complexes (65, 66) (Scheme 11).

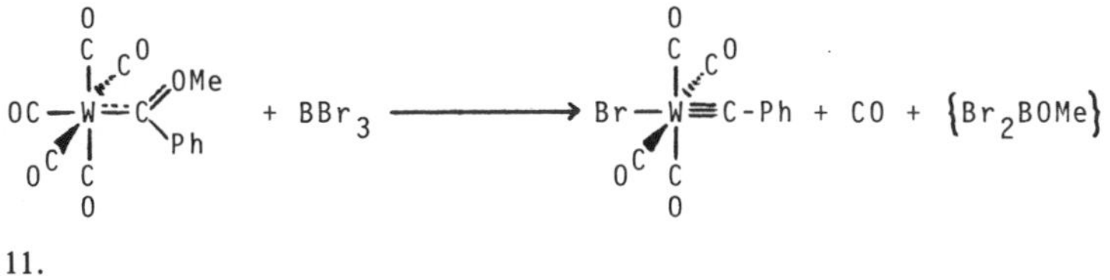

Scheme 11.

The reaction is very likely initiated by electrophilic attack of AX_3 on the free electron pair of Y. By reacting *cis*-$Br(CO)_4Mn[C(OH)Me]$ with BBr_3, a compound which derives from the resulting adduct via HBr elimination could be isolated (67). The next reaction step involves C(carbene)-Y bond dissociation leading to formation of cationic carbyne complexes. The latter can be isolated if (a) the position *trans* to the new carbyne ligand is occupied by a ligand with a σ-donor/π-acceptor ratio higher than that of CO [*e.g.* PR_3 (68), Cp (69)] or (b) the substituent at the carbyne ligand is a π-donor group [*e.g.* NR_2 (70)].

For cationic pentacarbonyl(carbyne)metal complexes there are now two paths for the next reaction steps:

(a) Dissociation of a CO ligand (very likely a *trans* CO group since the carbyne ligand has a lower σ-donor/π-acceptor ratio than that of CO) and subsequent addition of a nucleophile X^- (*e.g.* from $[X_3BOR]^-$) to the penta-coordinated intermediate to yield neutral carbyne complexes.

(b) Addition of X^- to the carbyne carbon atom to give neutral carbene complexes, *i.e.* $(CO)_5M[C(R)X]$, which either rearrange immediately with CO elimination and C/M migration of X (to form again neutral carbyne complexes) or are stable enough to be isolated. With some $(CO)_5Cr[C(NR_2)X]$ compounds the last step is slow, though it can be observed at elevated temperatures (see 2.1). Kinetic investigations (17-23) indicate that the rearrangement is intramolecular (Scheme 12).

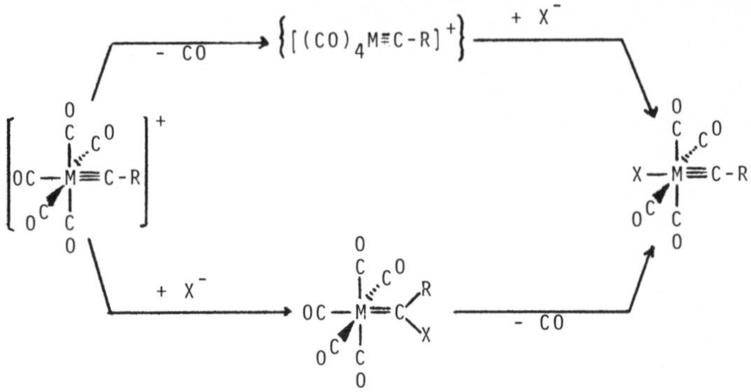

Scheme 12

To date, the formation of neutral carbyne complexes by means of sequence (b) has been established with certainty only for aminocarbyne complexes of chromium ($R = NR_2'$) whereas it is still unclear whether formation of aryl- and alkylcarbyne complexes follows sequence (a) or (b).

3 Reactions of Nucleophilic Carbene (Alkylidene) Complexes

The reactivity of nucleophilic carbene complexes can be understood by considering the reversed polarity of the metal-carbene bond (compared to that in Fischer-type complexes). Thus, electrophilic AlMe$_3$ adds to the methylene carbon in Cp$_2$(Me)Ta[=CH$_2$] (71), and nucleophiles such as PMe$_3$ or R-C≡N add to the metal in Cp(Cl)$_2$Ta[=C(H)CMe$_3$] (72). These R-C≡N adducts then react further to form a metallacycle; subsequent ring opening yields mixtures of *E* and *Z* isomers of the insertion products, Cp(Cl)$_2$Ta[=N-C(R)=C-(H)CMe$_3$] (72). (Note that electrophilic carbene complexes react with R-C≡N to give new carbene complexes.) However, only one isomer is obtained when the C≡C bond of tolane is inserted into the Ta=C(H)CMe$_3$ bond (72). The reaction of Np$_3$Ta[=C(H)CMe$_3$] with ketones, aldehydes, amides and esters was also thought to be initiated by nucleophilic attack of the carbonyl oxygen atom on the metal to form an adduct. "Metathesis" then affords an olefin and a complex containing Ta=O (73). Similarly, a metallicycle has also been proposed as an intermediate in the reaction of *e.g.* X$_3$(THF)$_2$Ta[=C(H)-CMe$_3$] (X = Cl or Br) with RN=CHPh to give X$_3$(THF)$_2$Ta[=NR] and Me$_3$C(H)C=C(H)Ph (74).

The type of product resulting from reaction of alkylidene complexes with terminal olefins depends on the complexes themselves and on the alkenes. Cp(Cl)$_2$Ta[=C(H)CMe$_3$] reacts with terminal olefins to give a tantalacyclobutane complex which rapidly rearranges to an unobservable olefin complex by migration of a β-proton to an α-carbon atom. Two equivalents of the smaller, more strongly coordinating olefin (present in excess) then displace this new olefin to give a tantalacyclopentane complex (75). Analogous organic products of β-hydride rearrangement of the intermediate metallacyclobutane complexes are also obtained by the reaction of terminal olefins with octahedrally coordinated X$_3$L$_2$Ta[=C(H)CMe] (X = Cl or Br; L = a tertiary phosphine). No metathesis products or cyclopropanes are observed here. When this complex is modified (L = THF, py), however, differing amounts of metathesis products are obtained in addition to the β-hydride rearrangement products. The former products depend on the type of olefin used. Cl(PMe$_3$) (Me$_3$CO)$_2$Ta[=C(H)CMe$_3$], finally, reacts with terminal as well as with internal olefins to yield only metathesis products. Tantalacyclobutanes are suggested as intermediates for all these reactions (76).

4 References

(1) T. F. Block, R. F. Fenske, C. P. Casey, J. Am. Chem. Soc. **98** (1976) 441 – 443.

(2) H. Werner, H. Rascher, Helv. Chim. Acta **51** (1968) 1765 – 1775.

(3) C. P. Casey, M. C. Cesa, Organometallics **1** (1982) 87 – 94.

(4) E. O. Fischer, K. H. Dötz, J. Organomet. Chem. **36** (1972) C4 – C6.

(5) E. O. Fischer, D. Plabst, Chem. Ber. **107** (1974) 3326 – 3331.

(6) C. P. Casey, T. J. Burkhardt, C. A. Bunnell, J. C. Calabrese, J. Am. Chem. Soc. **99** (1977) 2127 – 2134.

(7) C. P. Casey, R. L. Anderson, J. Chem. Soc., Chem. Commun. (1975) 895 – 896.

(8) E. O. Fischer, A. Maasböl, J. Organomet. Chem. **12** (1968) P15 – P17.

(9) E. O. Fischer, M. Leupold, Chem. Ber. **105** (1968) 599 – 608.

(10) C. G. Kreiter, Angew. Chem. **80** (1968) 402; Angew. Chem. Int. Ed. Engl. **7** (1968) 390 – 391.

(11) C. P. Casey, R. A. Boggs, R. L. Anderson, J. Am. Chem. Soc. **94** (1972) 8947 – 8949.

(12) E. Moser, E. O. Fischer, J. Organomet. Chem. **15** (1968) 147 – 155.

(13) E. O. Fischer, W. Held, J. Organomet. Chem. **112** (1976) C59 – C62.

(14) C. P. Casey, W. H. Miles, H. Tukada, J. M. O'Connor, J. Am. Chem. Soc. **104** (1982) 3761 – 3762.

(15) E. O. Fischer, A. Maasböl, Chem. Ber. **100** (1967) 2445 – 2456.

(16) E. O. Fischer, A. Riedel, Chem. Ber. **101** (1968) 156 – 161.

(17) H. Fischer, A. Motsch, W. Kleine, Angew. Chem. **90** (1978) 914 – 915; Angew. Chem. Int. Ed. Engl. **17** (1978) 842-843.

(18) E. O. Fischer, W. Kleine, F. R. Kreissl, H. Fischer, P. Friedrich, G. Huttner, J. Organomet. Chem. **128** (1977) C49 – C53.

(19) H. Fischer, E. O. Fischer, D. Himmelreich, R. Cai, U. Schubert, K. Ackermann, Chem. Ber. **114** (1981) 3220 – 3232.

(20) H. Fischer, E. O. Fischer, R. Cai, D. Himmelreich, Chem. Ber. **116** (1983) 1009 – 1016.

(21) E. O. Fischer, H. Fischer, U. Schubert, R. B. A. Pardy, Angew. Chem. **91** (1979) 929 – 930; Angew. Chem. Int. Ed. Engl. **18** (1979) 871.

(22) H. Fischer, J. Organomet. Chem. **195** (1980) 55 – 61.

(23) H. Fischer, E. O. Fischer, R. Cai, Chem. Ber. **115** (1982) 2707 – 2713.

(24) F. R. Kreissl, E. O. Fischer, C. G. Kreiter, H. Fischer, Chem. Ber. **106** (1973) 1262 – 1276.

(25) H. Fischer, E. O. Fischer, C. G. Kreiter, H. Werner, Chem. Ber. **107** (1974) 2459 – 2467.

(26) H. Fischer, J. Organomet. Chem. **170** (1979) 309 – 317.

(27) H. Werner, H. Rascher, Inorg. Chim. Acta **2** (1968) 181 – 185.

(28) E. O. Fischer, H. Fischer, Chem. Ber. **107** (1974) 657 – 672.

(29) H. Fischer, E. O. Fischer, Chem. Ber. **107** (1974) 673 – 679.

(30) H. Fischer, E. O. Fischer, H. Werner, J. Organomet. Chem. **73** (1974) 331 – 342.

(31) F. R. Kreissl, E. O. Fischer, Chem. Ber. **107** (1974) 183 – 188.

(32) U. Klabunde, E. O. Fischer, J. Am. Chem. Soc. **89** (1967) 7141 – 7142.

(33) H. Werner, E. O. Fischer, B. Heckl, C. G. Kreiter, J. Organomet. Chem. **28** (1971) 367 – 389.

(34) H. Fischer, unpublished results.

(35) E. O. Fischer, M. Leupold, C. G. Kreiter, J. Müller, Chem. Ber. **105** (1972) 150-161.

(36) E. O. Fischer, G. Kreis, F. R. Kreissl, C. G. Kreiter, J. Müller, Chem. Ber. **106** (1973) 3910 – 3919.

(37) E. O. Fischer, V. Kiener, Angew. Chem. **79** (1967) 982 – 983; Angew. Chem. Int. Ed. Engl. **6** (1967) 961.

(38) E. O. Fischer, G. Kreis, Chem. Ber. **106** (1973) 2310 – 2314.

(39) E. O. Fischer, R. Aumann, Chem. Ber. **101** (1968) 963 – 968.

(40) E. O. Fischer, L. Knauss, Chem. Ber. **103** (1970) 1262 – 1272.

(41) C. P. Casey, T. J. Burkhardt, J. Am. Chem. Soc. **95** (1973) 5833 – 5834.

(42) E. O. Fischer, W. Held, F. R. Kreissl, A. Frank, G. Huttner, Chem. Ber. **110** (1977) 656 – 666.

(43) C. P. Casey, S. W. Polichnowski, J. Am. Chem. Soc. **99** (1977) 6097 – 6099.

(44) C. T. Lam, C. V. Senoff, J. E. H. Ward, J. Organomet. Chem. **70** (1974) 273 – 281.

(45) E. O. Fischer, B. Heckl, K. H. Dötz, J. Müller, H. Werner, J. Organomet. Chem. **16** (1969) P29 – P32.

(46) K. H. Dötz, E. O. Fischer, Chem. Ber. **105** (1972) 1356 – 1367.

(47) C. P. Casey in H. Alper (Ed.): Transition Metal Organometallics in Organic Synthesis. Academic Press, New York 1976, Vol. 1, pp. 189 – 233.

(48) E. O. Fischer, K. H. Dötz, Chem. Ber. **105** (1972) 3966 – 3973.

(49) C. P. Casey, T. J. Burkhardt, J. Am. Chem. Soc. **96** (1974) 7808 – 7809.

(50) C. P. Casey, S. W. Polichnowski, A. J. Shusterman, C. R. Jones, J. Am. Chem. Soc. **101** (1979) 7282 – 7292.

(51) C. P. Casey, S. H. Bertz, T. J. Burkhardt, Tetrahedron Letters (1973) 1421 – 1424.

(52) C. M. Lukehart, J. V. Zeile, Inorg. Chim. Acta **17** (1976) L7 – L8.

(53) C. M. Lukehart, J. V. Zeile, J. Organomet. Chem. **97** (1975) 421 – 428.

(54) J. Levisalles, H. Rudler, D. Villemin, J. Organomet. Chem. **146** (1978) 259 – 265.

(55) K. H. Dötz, I. Pruskil, Chem. Ber. **114** (1981) 1980 – 1982.

(56) H. Fischer, J. Organomet. Chem. **222** (1981) 241 – 250.

(57) H. Fischer, S. Zeuner, unpublished results.

(58) K. H. Dötz, C. G. Kreiter, J. Organomet. Chem. **99** (1975) 309-314.

(59) H. Fischer, J. Organomet. Chem. **197** (1980) 303 – 313.

(60) R. Aumann, E. O. Fischer, Chem. Ber. **101** (1968) 954 – 962.

(61) C. G. Kreiter, R. Aumann, Chem. Ber. **111** (1978) 1223 – 1227.

(62) H. Fischer, K. H. Dötz, Chem. Ber. **113** (1980) 193 – 202.

(63) K. H. Dötz, Angew. Chem. **87** (1975) 672 – 673; Angew. Chem. Int. Ed. Engl. **14** (1975) 644 – 645.

(64) H. Fischer, J. Mühlemeier, R. Märkl, K. H. Dötz, Chem. Ber. **115** (1982) 1355 – 1362.

(65) E. O. Fischer, G. Kreis, C. G. Kreiter, J. Müller, G. Huttner, H. Lorenz, Angew. Chem. **85** (1973) 618 – 620; Angew. Chem. Int. Ed. Engl. **12** (1973) 564 – 565.

(66) E. O. Fischer, U. Schubert, J. Organomet. Chem. **100** (1975) 59 – 81.

(67) E. O. Fischer, Angew. Chem. **86** (1974) 651 – 663; Adv. Organomet. Chem. **14** (1976) 1 – 32.

(68) E. O. Fischer, K. Richter, Angew. Chem. **87** (1975) 359 – 360; Angew. Chem. Int. Ed. Engl. **14** (1975) 345 – 346.

(69) E. O. Fischer, E. W. Meineke, F. R. Kreissl, Chem. Ber. **110** (1977) 1140 – 1147.

(70) E. O. Fischer, W. Kleine, F. R. Kreissl, Angew. Chem. **88** (1976) 646 – 647; Angew. Chem. Int. Ed. Engl. **15** (1976) 616 – 617.

(71) R. R. Schrock, J. Am. Chem. Soc. **97** (1975) 6577 – 6578.

(72) C. D. Wood, S. J. McLain, R. R. Schrock, J. Am. Chem. Soc. **101** (1979) 3210 – 3222.

(73) R. R. Schrock, J. Am Chem. Soc. **98** (1976) 5399 – 5400.

(74) S. M. Rocklage, R. R. Schrock, J. Am. Chem. Soc. **104** (1982) 3077 – 3081.

(75) S. J. McLain, C. D. Wood, R. R. Schrock, J. Am. Chem. Soc. **101** (1979) 4558 – 4570.

(76) S. M. Rocklage, J. D. Fellmann, G. A. Rupprecht, L. W. Messerle, R. R. Schrock, J. Am. Chem. Soc. **103** (1981) 1440 – 1447.